SEX DIFFERENTIATION and CHROMOSOMAL ABNORMALITIES

BIRTH DEFECTS: ORIGINAL ARTICLE SERIES

Daniel Bergsma, *Series Editor*

Volumes in the series published by Alan R. Liss, Inc.

1975 – VOLUME XI

No. 7 **Morphogenesis and Malformation of Face and Brain,** Jan Langman, *Editor*

1976 – VOLUME XII

No. 1 **Cancer and Genetics,** R. Neil Schimke, Robert L. Summitt, and David J. Harris, *Editors*

No. 3 **The Eye and Inborn Errors of Metabolism,** Anthony J. Bron and Edward Cotlier, *Editors*

No. 4 **Developmental Disabilities: Psychologic and Social Implications,** Ann E. Pulver, *Editor*

No. 5 **Cytogenetics, Environment and Malformation Syndromes,** R. Neil Schimke, *Editor*

No. 6 **Growth Problems and Clinical Advances,** R. Neil Schimke, *Editor*

No. 8 **Iron Metabolism and Thalassemia,** Anthony Cerami, Charles M. Peterson, and Joseph H. Graziano, *Editors*

1977 – VOLUME XIII

No. 1 **Morphogenesis and Malformation of the Limb,** Widukind Lenz, *Editor*

No. 2 **Morphogenesis and Malformation of the Genital System,** Richard J. Blandau, *Editor*

No. 3 **Annual Review of Birth Defects, 1976,** R. Brian Lowry, *Editor*
Proceedings of the 1976 Vancouver Birth Defects Conference.
Published in 4 volumes:
- 3A Numerical Taxonomy of Birth Defects *and* Polygenic Disorders
- 3B New Syndromes
- 3C Natural History of Specific Birth Defects
- 3D Embryology and Pathogenesis *and* Prenatal Diagnosis

No. 5 **Urinary System Malformations in Children,** John W. Duckett, *Editor*

No. 6 **Trends and Teaching in Clinical Genetics,** Frederick Hecht, Gerald H. Prescott, and Joan H. Marks, *Editors*

1978 – VOLUME XIV

No. 1 **Genetic Effects on Aging,** David E. Harrison, *Editor*

No. 2 **The Molecular Basis of Cell-Cell Interaction,** Richard A. Lerner, *Editor*

No. 3 **The Genetics of Hand Malformations,** by Samia Temtamy and Victor A. McKusick

No. 5 **Neurochemical and Immunologic Components in Schizophrenia,** Allan L. Goldstein, *Editor*

No. 6 **Annual Review of Birth Defects, 1977,** Robert L. Summitt, *Editor*
Proceedings of the 1977 Memphis Birth Defects Conference.
Published in 3 volumes:
- 6A Cell Surface Factors, Immune Deficiencies, Twin Studies
- 6B Recent Advances *and* New Syndromes
- 6C Sex Differentiation *and* Chromosomal Abnormalities

No. 7 **Morphogenesis and Malformation of the Cardiovascular System,** Glenn C. Rosenquist, *Editor*

IN PRESS

Sex Chromosome Anomalies in Children, Arthur Robinson, *Editor*

Recent Advances in the Developmental Biology of Central Nervous System Malformations, Ntinos C. Myrianthopoulos, *Editor*

The National Foundation–March of Dimes
Birth Defects: Original Article Series, Volume XIV, Number 6C, 1978

SEX DIFFERENTIATION and CHROMOSOMAL ABNORMALITIES

Part C of Annual Review of Birth Defects, 1977

Sponsored by The National Foundation–March of Dimes at
The Hyatt-Regency Memphis Hotel
Memphis, Tennessee

Editors: **Robert L. Summitt, MD**, Professor of Pediatrics, Anatomy and Child Development, University of Tennessee, Memphis

Daniel Bergsma, MD, Series Editor
The National Foundation

Associate Editors: **Joe Leigh Simpson, MD**, Prentice Women's Hospital and Maternity Center, Chicago, Illinois

Natalie W. Paul,
The National Foundation

ALAN R. LISS, INC., NEW YORK

To enhance medical communication in the birth defects field, The National Foundation publishes the *Birth Defects Atlas and Compendium*, an *Original Article Series, Syndrome Identification,* a *Reprint Series,* and provides a series of films and related brochures.

Further information can be obtained from:
The National Foundation
1275 Mamaroneck Avenue
White Plains, New York 10605

Published by:
Alan R. Liss, Inc.
150 Fifth Avenue
New York, New York 10011

Views expressed in articles published are the authors', and are not to be attributed to The National Foundation or its editors unless expressly so stated.

Library of Congress Cataloging in Publication Data

Main entry under title:

Sex differentiation and chromosomal abnormalities.

(Birth defects original article series; v. 14, no. 6C) (Part C of Annual review of birth defects, 1977)

Includes bibliographical references and index.

1. Sex differentiation disorders--Congresses.
2. Human chromosomal abnormalities--Congresses.
I. Summitt, Robert L. II. Bergsma, Daniel.
III. National Foundation. IV. Series. V. Series:
Annual review of birth defects; 1977, pt. C.
RG626.B63 vol. 14, no. 6C [RC883.5.S47] 78-17057
ISBN 0-8451-1022-5 616'.043'08s [616.6]

Printed in the United States of America

THE NATIONAL FOUNDATION is dedicated to the long-range goal of preventing birth defects. Our interim goal is to search for ways to ameliorate those birth defects which cannot be prevented.

As part of our efforts to achieve these goals, we sponsor, or participate in, a variety of scientific meetings and symposia where all questions relating to birth defects are freely discussed. Through our professional educational program we speed the dissemination of information by publishing the proceedings of these meetings and symposia. From time to time, we also reprint pertinent journal articles to help achieve our goal. Now and then, in the course of these articles or discussions, individual viewpoints may be expressed which go beyond the purely scientific and into controversial matters. It should be noted, therefore, that personal viewpoints about such matters will not be censored but this does not constitute an endorsement of them by The National Foundation.

Contents

SECTION II: Chromosome Abnormalities

Contributors

Arthur S. Aylsworth [317, 349]
Department of Pediatrics
The University of North Carolina
Chapel Hill, NC 27514

Patricia I. Bader [383]
Caylor-Nickel Research Foundation
Bluffton, IN 46714

Martha S. Berberich [287]
Fellow in Genetics
University of California
San Francisco, CA 94143

Denys Beshara [393]
Laboratoire de Cytogénétique
Clinique Infantile
Hôpital Civil
67000 Strasbourg, France

W. Roy Breg [269]
Yale University School of Medicine
New Haven, CT 06510

Philip D. Buchanan [317, 349]
Department of Pediatrics
The University of North Carolina
Chapel Hill, NC 27514

John C. Carey [287]
Fellow in Genetics
University of California
San Francisco, CA 94143

Mary Esther Carlin [327, 399]
Department of Pediatrics
Mailman Center for Child Development
University of Miami School of Medicine
Miami, FL 33124

Dorothy Clow [269]
Yale University School of Medicine
New Haven, CT 06510

June Cole [377]
Department of Medical Genetics
Indiana University School of Medicine
Indianapolis, IN 46202

David M. Cook [109]
Department of Medicine
Division of Endocrinology
University of Oregon Health Sciences Center
Portland, OR 97207

Doris K. Cummings [97]
Department of Gynecology and Obstetrics
The Johns Hopkins University School of Medicine
Baltimore, MD 21205

The number in brackets following each contributor's name is the opening page number of that author's paper.

Robert I. DeMars [365]
Departments of Pediatrics, Medical Genetics, Preventive Medicine, and Pathology
University of Wisconsin Center for Health Services
Madison, WI 53706

Carolyn L. Doerr [317]
Department of Pediatrics
The University of North Carolina
Chapel Hill, NC 27514

Stephen Duck [123]
Department of Pediatrics
Medical College of Wisconsin
Milwaukee Childrens Hospital
Milwaukee, WI 53233

B. Rafael Elejalde [365]
Departments of Pediatrics, Medical Genetics, Preventive Medicine, and Pathology
University of Wisconsin Center for Health Services
Madison, WI 53706

Marvin Fishman [343]
Department of Pediatrics
Divisions of Medical Genetics and Neurology
St. Louis Children's Hospital
St. Louis, MO 63110

Elisabeth Flori [393]
Laboratoire de Cytogénétique
Clinique Infantile
Hôpital Civil
67000 Strasbourg, France

Uta Francke [191]
Department of Pediatrics
University of California, San Diego
School of Medicine
La Jolla, CA 92093

Jaime L. Frias [355]
Department of Pediatrics
University of Florida College of Medicine
Gainesville, FL 32610

R.J.M. Gardner [359]
Department of Genetics
Hospital for Sick Children
555 University Avenue
Toronto M5G IX8, Canada

Helen Gerber [383]
Caylor-Nickel Research Foundation
Bluffton, IN 46714

J. D. Gilbert [327]
Department of Cytogenetics
Mailman Center for Child Development
University of Miami School of Medicine
Miami, FL 33124

Zoë R. Graves [167]
Genetics Laboratory
Letchworth Village Developmental Center
Thiells, NY 10984

Joan Grzegocki [303]
Department of Pediatrics
Washington University School of Medicine
St. Louis, MO 63110

Bryan D. Hall [287]
Assistant Professor of Pediatrics
University of California
San Francisco, CA 94143

Judith G. Hall [297]
Division of Medical Genetics
Children's Orthopedic Hospital and Medical Center
Seattle, WA 98105

Frederick Hecht [269, 281]
University of Oregon Health Sciences Center
Portland, OR 97201

Marion E. Hodes [377]
Department of Medical Genetics
Indiana University School of Medicine
Indianapolis, IN 46202

Donald Hoffman [383]
Parkview Hospital
Randalia Drive
Fort Wayne, IN 46805

Ernest B. Hook [249]
Division of Laboratories and Research
Birth Defects Institute
New York State Department of Health
Albany, NY 12237

Patricia Hunter [309]
Queen's University
Kingston, Ontario, Canada

Herbert A. Huss [167]
Department of Obstetrics and Gynecology
Good Samaritan Hospital
Suffern, NY 10901

Stanley L. Inhorn [365]
Departments of Pediatrics, Medical Genetics, Preventive Medicine, and Pathology
University of Wisconsin Center for Health Services
Madison, WI 53706

Marjorie Jansch [383]
Saint Joseph Hospital
Fort Wayne, IN 46802

Howard W. Jones, Jr. [63, 97]
Department of Gynecology and Obstetrics
The Johns Hopkins University School of Medicine
Baltimore, MD 21205

Conrad G. Julian [97]
Department of Gynecology and Obstetrics
The Johns Hopkins University School of Medicine
Baltimore, MD 21205

Barbara Kaiser-McCaw [281]
Crippled Children's Division
Genetics Clinic
University of Oregon Health Sciences Center
P.O. Box 574
Portland, OR 97201

Paul Katayama [123]
Department of Gynecology and Obstetrics
Milwaukee County General Hospital
Milwaukee, WI 53226

William Kimberling [269]
University of Colorado School of Medicine
Denver, CO 80220

Charles R. King [109]
Department of Medical Genetics
Child Development Center
University of Oregon Health Sciences Center
Portland, OR 97207

Gloria C. Koo [1]
Memorial Sloan-Kettering Cancer Center
New York, NY 10021

Boris G. Kousseff [323]
Center for Developmentally Disabled
Queens Hospital
Jamaica, NY 11432

Helen J. Lawce [287]
Staff Research Associate in Cytogenetics
University of California
San Francisco, CA 94143

S. Leon [327]
Department of Psychology
Mailman Center for Child Development
University of Miami School of Medicine
Maimi, FL 33152

Herbert A. Lubs [269]
University of Colorado School of Medicine
Denver, CO 80220

David W. Macfarlane [269]
University of Oregon Health Sciences Center
Portland, OR 97207

R. Ellen Magenis [269]
University of Oregon Health Sciences Center
Portland, OR 97207

Paula R. Martens [219]
Department of Obstetrics and Gynecology
Naval Regional Medical Center
Portsmouth, VA 23708

James A. Monteleone [303]
Cardinal Glennon Memorial Hospital
1465 S. Grand Blvd.
St. Louis, MO 63104

Patricia L. Monteleone [303]
Cardinal Glennon Memorial Hospital
1465 S. Grand Blvd.
St. Louis, MO 63104

Robert M. Nelson [355]
Department of Pediatrics
University of Florida College of Medicine
Gainesville, FL 32610

Charles Norman [399]
Department of Pediatrics
Mailman Center for Child Development
University of Miami School of Medicine
Miami, FL 33152

John M. Opitz [133, 365]
Departments of Genetics and Pediatrics
Genetics Bldg.
University of Wisconsin
Madison, WI 53706

Susan O'Shaughnessy [309]
Queen's University
Kingston, Ontario, Canada

Roberta A. Pagon [297]
Departments of Pediatrics and Medicine
University of Washington School of Medicine
Seattle, WA 98195

Philip D. Pallister [133]
Shodair Regional Genetics and Birth Defects Center
Shodair Hospital
Helena, MT 59601

Catherine G. Palmer [269, 377, 383]
Indiana University School of Medicine
Indianapolis, IN 46202

Shivanand Patil [281]
Crippled Children's Division
University of Oregon Health Sciences Center
Genetics Clinic
P.O. Box 574
Portland, OR 97207

Richard M. Pauli [297]
Departments of Pediatrics and Medicine
University of Washington School of Medicine
Seattle, WA 98195

Jeffrey P. Phelan [219]
Departments of Pediatrics and Anatomy
University of Tennessee Center for The Health Sciences
Memphis, TN 38163

Leonard Pinsky [73]
Lady Davis Institute
Jewish General Hospital
Montreal, Quebec H3T 1E2
Canada

Gerald H. Prescott [269]
Division of Medical Genetics
University of Oregon Health Sciences Center
Box 574
Portland, OR 97201

Kathleen W. Rao [317, 349]
Department of Pediatrics
The University of North Carolina
Chapel Hill, NC 27514

Jack M. Rary [97]
The Johns Hopkins University School of Medicine
Biophysics Building, Room 203
Baltimore, MD 21205

Shari L. Ray [355]
Department of Pediatrics
University of Florida College of Medicine
Gainesville, FL 32610

Terry Reed [377]
Department of Medical Genetics
Indiana University School of Medicine
Indianapolis, IN 46202

Vincent M. Riccardi [123]
The Kleberg Genetics Center
Baylor College of Medicine
Texas Medical Center
Houston, TX 77030

Marian L. Rivas [269]
University of Oregon Health Sciences Center
Portland, OR 97201

John A. Rock [97]
Department of Gynecology and Obstetrics
The Johns Hopkins University School of Medicine
Baltimore, MD 21205

N. L. Rudd [359]
Department of Genetics
The Hospital for Sick Children
Toronto, M5G 1X8, Ontario
Canada

Gloria E. Sarto [37]
Prentice Women's Hospital and Maternity Center of Northwestern Memorial Hospital
333 East Superior Street
Chicago, IL 60611

Gurbax S. Sekhon [303, 343]
Department of Pediatrics
Washington University School of Medicine
St. Louis, MO 63110

Hari Sekhon [303]
Department of Pediatrics
Washington University School of Medicine
St. Louis, MO 63110

Linda Shanbeck [269]
University of Oregon Health Sciences Center
Portland, OR 97201

Lawrence R. Shapiro [167]
Departments of Pediatrics and Pathology
Westchester County Medical Center and Genetics Laboratory, Valhalla, NY
Letchworth Village Development Center
Thiells, NY 10984

M. H. K. Shokeir [147, 171]
Division of Medical Genetics
University of Saskatchewan
Saskatoon, Saskatchewan, Canada

Joe Leigh Simpson [9, 37]
Prentice Women's Hospital and Maternity Center of Northwestern Memorial Hospital
333 East Superior Street
Chicago, IL 60611

Nancy E. Simpson [309]
Queen's University
Kingston, Ontario, Canada

David W. Smith [57]
Department of Pediatrics
Dysmorphology Unit
University of Washington School of Medicine
Seattle, WA 98195

Dushan Soudek [309]
Queen's University
Kingston, Ontario, Canada

Vera Soudek [309]
Queen's University
Kingston, Ontario, Canada

L.J. Stevens [359]
Department of Genetics
Hospital for Sick Children
555 University Avenue
Toronto M5G IX8, Canada

Claude G. Stoll [393]
Laboratoire de Cytogénétique
Clinique Infantile
Hôpital Civil
67000 Strasbourg, France

Robert L. Summitt [219, 269]
Departments of Pediatrics and Anatomy
University of Tennessee Center for The Health Sciences
Memphis, TN 38163

Gail Taylor [383]
Caylor-Nickel Research Foundation
Bluffton, IN 46714

Kutay Taysi [343]
Department of Pediatrics
Washington University School of Medicine
St. Louis, MO 63110

Michael Tietjens [303]
Department of Pediatrics
Washington University School of Medicine
St. Louis, MO 63110

Robert E. Tipton [219]
Departments of Pediatrics and Anatomy
University of Tennessee Center for The Health Sciences
Memphis, TN 38163

Sister Leo Rita Volk [303]
St. Mary's Health Center
St. Louis, MO 63117

Stephen S. Wachtel [1]
Memorial Sloan-Kettering Cancer Center
New York, NY 10021

Dorothy Warburton [167]
Department of Obstetrics and Gynecology
College of Physicians and Surgeons
622 W. 168 Street
New York, NY 10032

R. S. Wilroy, Jr. [219]
Departments of Pediatrics and Anatomy
University of Tennessee Center for The Health Sciences
Memphis, TN 38163

R. G. Worton [359]
Department of Genetics
The Hospital for Sick Children
555 University Avenue
Toronto, M5G 1X8, Ontario, Canada

Herman E. Wyandt [269, 281]
University of Oregon Health Sciences Center
Portland, OR 97201

SECTION I:
SELECTED ABNORMALITIES OF SEX DIFFERENTIATION

H-Y Antigen and Abnormal Sex Differentiation*

Stephen S. Wachtel, PhD, and Gloria C. Koo

H-Y antigen is a cell surface component that is found in the tissues of all mammalian males but not in those of normal females. It is called H-Y (histocompatibility Y) antigen because it is responsible for the incompatibility of male-to-female skin grafts exchanged among members of the same highly inbred strains of laboratory mice; the only genetic difference between males and females in these strains is the Y chromosome (reviewed in [1]).

Serum from male-grafted female mice (H-Y antiserum) kills mouse sperm and male epidermal cells. Other male cells are not killed by H-Y antiserum, but it is nonetheless possible to demonstrate the presence of H-Y antigen on these cells by measuring their capacity to "absorb" H-Y antibodies, thereby decreasing the cytotoxicity of H-Y antiserum for sperm and male epidermal cells. The ability of H-Y antiserum to kill sperm or male epidermal cells is significantly reduced following its absorption with male but not female spleen cells, for example. Thus, serologic absorption of H-Y antiserum allows for the routine typing of most tissues [1]. Using the sperm cytotoxicity test and another serologic assay, the mixed hemadsorption-hybrid antibody (MHA.HA) test (concerned with the labeling rather than killing of sperm cells) [2] we have observed that H-Y antigen of the mouse cross-reacts with antigen found in males of all mammalian species so far tested, including the human [3].

*Supported in part by NIH grants CA 08748, AI 11982, HD 10065, and HD 00171. This paper is dedicated to Dr. C. Everett Koop, Professor of Pediatric Surgery, Children's Hospital, Philadelphia.

Birth Defects: Original Article Series, Volume XIV, Number 6C, pages 1–7

H-Y ANTIGEN AND SEX DETERMINATION

Based on a) the widespread phylogenetic conservation of H-Y antigen, b) its association with the *Sxr* gene, which "sex reverses" XX and XO mice causing them to develop as phenotypic (but sterile) males [4], and c) the putative role of plasma membrane components in organogenesis, we have proposed that the plasma membrane component that we recognize serologically as H-Y antigen is in fact specified by the mammalian testis-determining gene [5]. According to this proposal, H-Y antigen directs the initially indifferent embryonic gonad to differentiate as a testis. The testis secretes testosterone which governs differentiation of the wolffian derivatives (vasa deferentia, seminal vesicles, and epididymides) and müllerian inhibition factor which prevents differentiation of the müllerian derivatives (fallopian tubes, uterus, and cephalad portion of the vagina). Some of the testosterone is converted enzymatically to dihydrotestosterone which induces orderly differentiation of the penis and scrotum (from the same embryonic tissues that would otherwise give rise to the clitoris, labia minora, and labia majora) (Fig. 1).

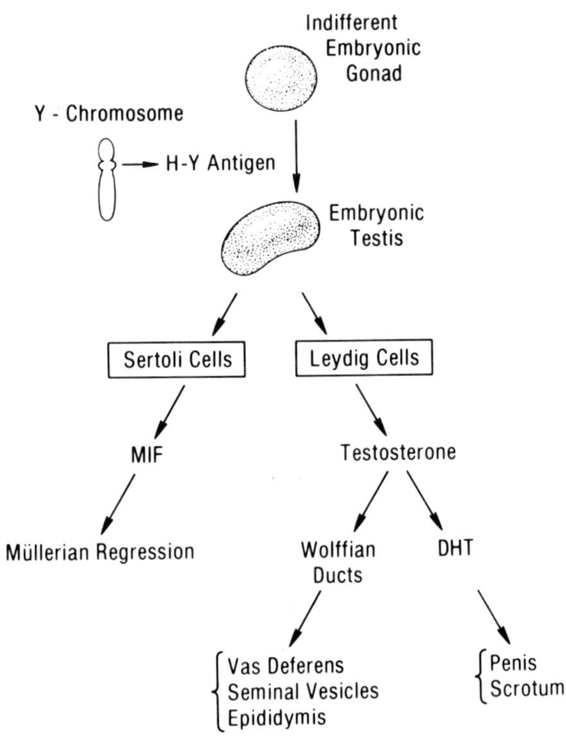

Fig 1. Mammalian sex determination and differentiation.

Thus H-Y antigen initiates only primary (gonadal) male differentiation; secondary male differentiation is imposed upon the embryo by the action of testicular hormones, against the inherent tendency towards female differentiation [6]. In the absence of H-Y (and in the presence of the appropriate ovarian inducer?) the indifferent gonad becomes an ovary; in the absence of testosterone the embryo becomes a female.

H-Y ANTIGEN IN TESTICULAR FEMINIZATION SYNDROME

Feminization also occurs in cases of testosterone insensitivity. Genotypic males (46,XY) with testicular feminization syndrome (TFS) develop testes under the influence of the Y chromosome and the testes secrete testosterone. But these patients cannot respond to testosterone due to mutation of the cellular testosterone receptor and so they become phenotypic females [7]. Müllerian derivatives are absent, however, presumably because of müllerian inhibition factor secreted by Sertoli cells of the fetal testes. H-Y antigen is prominent on the white blood cells of subjects with TFS [8], so it cannot be claimed that retention of the H-Y$^+$ cellular phenotype depends on the interaction of testosterone with its receptor. Presence of H-Y in TFS is precisely what we should expect if H-Y antigen were the determinant of male gonadal sex, and not a secondary manifestation of male sexuality.

Moreover, presence of H-Y antigen in the tissues of androgen-insensitive patients implies that a genetic determinant for H-Y expression is situated on the Y chromosome. Otherwise the determinant would be common to both males and females, and expression of H-Y would require some effect of testosterone, albeit one that is independent of the testosterone receptor. But H-Y antigen is found in the 8-cell male embryo [9] and this argues convincingly against *any* testosterone dependency of its expression.

MAPPING THE H-Y GENE

H-Y antiserum which has been absorbed with white blood cells (WBC) from human males with two Y chromosomes (47,XYY and 48,XXYY) is less cytotoxic for mouse sperm than H-Y antiserum which has been absorbed with WBC from normal 46,XY males [10]. This implies that the amount of H-Y antigen on the cell surface is related to the number of Y chromosomes in the nucleus. And this in turn strengthens our hypothesis that a Y-chromosomal gene determines expression of H-Y antigen.

More precise localization of the H-Y determinant has been accomplished by correlating presence or absence of H-Y antigen with presence or absence of particular portions of the Y chromosome as identified cytogenetically. In a survey of 17 patients exhibiting various deletions and other structural abnormalities of the Y chromosome, presence of H-Y antigen was most often associated with presence of the centromeric region of the short arm. However in a single individual the entire long arm of the Y appeared to be translocated to the X chromosome, and H-Y antigen was expressed despite absence of both the Y short arm (Yp-) and the Y centromere. These observations indicate two possible sites for the H-Y gene, one on the short arm and one on the long arm [11].

For the present our serologic assays cannot tell us whether the Y-chromosomal H-Y genes actually code for the structure of H-Y antigen or regulate its synthesis. It seems wise to assume that these are structural genes, however, because there is no reason to believe that supernumerary regulatory genes would elicit production of excess H-Y antigen.

H-Y ANTIGEN AND XX SEX REVERSAL

The problem with the assumption that H-Y (testis-determining) structural genes are on the Y chromosome is the occurrence of H-Y antigen (and testes) in subjects that apparently lack the Y chromosome, for example, XX male mice sex-reversed by the autosomal dominant *Sxr* gene [4]. If *Sxr* is a translocation of the Y-linked H-Y gene then why is it that this translocation remains undetected despite repeated karyotype analysis?

The discrepancy may be more apparent than real. In dogs, Selden has observed bilateral ovotestes in the *mother* of an XX male. Despite apparently normal female karyotypes (78,XX in the dog) both hermaphrodite mother and pup were H-Y+. However, WBC from the father of the true hermaphrodite showed significantly elevated levels of H-Y antigen, and subsequent preliminary chromosome analysis has revealed the possibility of a Y-to-autosome translocation in the cells of all three dogs [12].

This study is noteworthy for the following reasons: a) It confirms our earlier observation that excess H-Y antigen, as detected serologically, is associated with excess Y-chromosomal material, as determined karyotypically; b) it demonstrates that the H-Y$^+$ phenotype need not rule out development of functional ovaries; c) it indicates Y-to-autosome translocation as a possible cause of the XX male syndrome and XX true hermaphroditism in dogs and, by inference, in other mammals including mice and men; and d) it shows that both of these conditions are associated with abnormal inheritance of H-Y genes.

Given the same chromosome constitution why does one individual become an XX true hermaphrodite and the other an XX male, or more specifically, why does one gonad become an ovotestis and the other a testis? The concurrence of ovarian and testicular structure in the true hermaphrodite (whether in contralateral gonads or within the same gonad) indicates the presence of two disparate populations of cells. Indeed, cells cultured from the ovarian portion of a human ovotestis were H-Y$^-$ and those cultured from the testicular portion were H-Y+; the ovotestis was an H-Y$^-$/H-Y$^+$ mosaic (Shenin et al, manuscript in preparation). But the cause of this H-Y antigen mosaicism remains to be determined.

H-Y ANTIGEN IN THE WOOD LEMMING: EVIDENCE FOR A REGULATORY H-Y GENE

The Scandinavian wood lemming (Myopus schisticolor) is characterized by an excess of females. In this rodent species, many XY embryos develop ovaries and mature as fertile females, anatomically indistinguishable from their "normal" XX sisters [13]. How does one reconcile these observations with the putative male-determining role of the mammalian Y chromosome? The fact is that even though their Y chromosome appears to be intact, these females do not synthesize H-Y antigen [14]. But the XY female wood lemming condition cannot be due to mutation of H-Y structural genes on the Y chromosome because this chromosome is not transmitted to their progeny. Somatically XY female wood lemmings have an XX germ line and produce only X-bearing germ cells.

Indeed, the XY female condition is inherited as an X-linked trait, and this implies existence of an X-linked mutant gene that blocks the function of the Y-linked testis-determinant [15]. Ohno's hypothesis concerning the conservation of the mammalian X chromosome [16] implies that a similar regulatory gene occurs in man. Whereas XO rodents are fertile females with normal ovaries, XO humans are infertile females with undifferentiated streak gonads. This predicts an analogous mutant in humans with the phenotypically female proposita exhibiting a normal male karyotype (46,XY), H-Y$^-$ cellular phenotype, and undifferentiated streak gonads.

PLASMA MEMBRANE RECEPTORS FOR H-Y ANTIGEN

In cattle chorionic vascular anastomosis between male and female twin fetuses modifies development of the female gonad, often causing it to develop as a small testis [17]. Because testosterone cannot modify ovarian differentiation [6] it has been proposed that XY cells from the male twin are responsible for this "freemartin" condition. But how can a small minority of XY cells in the gonad (often less than 10%) entice a majority of XX cells to engage in testicular differentiation? Ohno [18] has proposed that a few specifically programmed XY cells disseminate H-Y antigen in hormone-like fashion. According to this proposal, disseminated H-Y antigen is bound by nonspecific anchorage sites and specific receptors in the plasma membrane of host XX cells. In confirmation of this proposal, H-Y antigen is prominent in the developing freemartin gonad [19].

If we assume that any normal morphogenetic function of H-Y antigen requires the existence of a plasma membrane receptor, then we must assume a gene that determines the structure of that receptor. And this leads us to the possibility of mutation of the receptor determinant. Given a mutated receptor in an otherwise normal 46,XY human embryo, we should expect development of an H-Y+ phenotypic female with undifferentiated gonads that may or may not virilize depending on any residual ability of the altered receptor to bind H-Y antigen.

CONCLUSION

We have reviewed evidence for three genes involved in the initial development of the mammalian testis: a) the Y-situated H-Y gene that codes for the antigenic plasma membrane component that induces the indifferent embryonic gonad to differentiate as a testis; b) the X-linked gene that activates or regulates the function of H-Y structural gene on the Y chromosome; c) an autosomal or X-linked gene that determines the structure of a plasma membrane receptor for H-Y antigen.

Mutation at any one of these genetic loci would affect the normal function of H-Y antigen, and by inference, the normal course of testicular differentiation. However, altered expression or absence of H-Y antigen need not lead to normal ovarian differentiation, a process which depends, so far as we understand, on the presence of two X chromosomes in man.

Just as absence of H-Y antigen does not necessarily indicate the presence of ovaries, its occurrence on the cell membrane does not necessarily indicate presence of testes. We have observed H-Y antigen expression in several 46,XY human females with streak gonads exhibiting little or no masculine architecture. Whereas cryptic mosaicism (45,X [H-Y−] cell line in the gonad) may account for some of these cases, and regression of testicular tissue for others, mutation of a plasma membrane receptor could certainly account for the remainder.

The importance of H-Y antigen in the virilization of the indifferent embryonic gonad is perhaps best illustrated by the fact that testicular differentiation does not occur in its absence. It follows that study of H-Y antigen expression in cases of abnormal gonadal differentiation may not only provide insights into the genetic nature of the disorders themselves, but may also clarify the role of this cell surface component in testicular differentiation generally.

REFERENCES

1. Silvers WK, Wachtel SS: H-Y antigen: Behavior and function. Science 195:956, 1977.

2. Koo GC, Boyse EA, Wachtel SS: In Edidin M, Johnson MH (eds): "Immunobiology of Gametes." Oxford: Alden Press, 1977, p 73.
3. Wachtel SS, Koo GC, Zuckerman EE, Hammerling U, Scheid MP, Boyse EA: Serological cross reactivity between H-Y (male) antigens of mouse and man. Proc Natl Acad Sci USA 71:1215, 1974.
4. Cattanach BM, Pollard CE, Hawkes SG: Sex-reversed mice: XX and XO males. Cytogenetics 10:318, 1971.
5. Wachtel SS, Ohno S, Koo GC, Boyse EA: Possible role for H-Y antigen in the primary determination of sex. Nature 257;235, 1975.
6. Jost A: Hormonal factors in the sex differentiation of the mammalian foetus. Philos Trans R Soc Lond Biol 259:119, 1970.
7. Keenan BS, Meyer WJ, Hadjian AJ, Jones HW, Migeon CJ: Syndrome of androgen insensitivity in man: Absence of 5-α-dihydrotestosterone binding protein in skin fibroblasts. J Clin Endocrinol Metab 38:1143, 1974.
8. Koo GC, Wachtel SS, Saenger P, New MI, Dosik H, Amarose AP, Dorus E, Ventruto V: H-Y antigen: Expression in human subjects with the testicular feminization syndrome. Science 196:655, 1977.
9. Krco CJ, Goldberg EH: H-Y (male) antigen: Detection on eight-cell mouse embryos. Science 193:1134, 1976.
10. Wachtel SS, Koo GC, Breg WR, Elias S, Boyse EA, Miller OJ: Expression of H-Y antigen in human males with two Y chromosomes. N Engl J Med 293:1070, 1975.
11. Koo GC, Wachtel SS, Krupen-Brown K, Mittl LR, Breg WR, Genel M, Rosenthal IM, Borgaonkar DS, Miller DA, Tantravahi R, Schreck RR, Erlanger BF, Miller OJ: Mapping the locus of the H-Y gene on the human Y chromosome. Science 198:940, 1976.
12. Selden JR, Wachtel SS, Koo GC, Haskins ME, Patterson DF: Genetic basis of XX male syndrome and XX true hermaphroditism evidence in the dog. (Submitted for publication.)
13. Fredga K, Gropp A, Winking H, Frank F: Fertile XX- and XY-type females in the wood lemming (Myopus schistocolor). Nature 261:225, 1976.
14. Wachtel SS, Koo GC, Ohno S, Gropp A, Dev VG, Tantravahi R, Miller DA, Miller OJ: H-Y antigen and origin of XY female wood lemmings (Myopus schistocolor). Nature (Lond) 264:638, 1976.
15. Fredga K, Gropp A, Winking H, Frank F: A hypothesis explaining the exceptional sex ratio in the wood lemming (Myopus schistocolor). Hereditas 85:101, 1977.
16. Ohno S: "Sex Chromosomes and Sex-Linked Genes." Berlin, Heidelberg, and New York: Springer-Verlag, 1967.
17. Short RV: The bovine freemartin: A new look at an old problem. Philos Trans R Soc Lond Biol 259:141, 1970.
18. Ohno S: Major regulatory genes for mammalian sexual development. Cell 7:315, 1976.
19. Ohno S, Christian LC, Wachtel SS, Koo GC: Hormone-like role of H-Y antigen in bovine freemartin gonad. Nature 261:597, 1976.

True Hermaphroditism: Etiology and Phenotypic Considerations

Joe Leigh Simpson, MD

True hermaphrodites have ovarian as well as testicular tissue. The diagnosis has traditionally been applied only if an individual has 1) histologically verified ovarian follicles or proof of their prior existence (eg corpora albicantia), and 2) seminiferous tubules or spermatozoa. Fibrous stroma does not suffice in lieu of follicles, nor do Leydig cells suffice in lieu of tubules. The diagnosis is applied irrespective of chromosome complement. True hermaphroditism has been reported in humans and in many animals. In fact, in some lower animals hermaphroditism is functional. This review will, however, be restricted to true hermaphroditism in humans, animal data having been summarized elsewhere [1].

Klebs (1876) [2] is usually credited with the first scientific consideration of true hermaphroditism in humans. Earlier cases were tabulated by Hinman [3], Young [4], Jones and Scott [5], Overzier [6], Neiman and Fonder [7], and Van Niekerk [8]. Other aspects of the disorder were reviewed by Inhorn and Optiz [9], Polani [10], Hamerton [11], Pfeiffer [12], Fournier et al [13], Van Niekerk [14], and Simpson [1].

By the beginning of the modern cytogenetic era (circa 1958) approximately 200 true hermaphrodites had been documented. Analysis of these earlier cases led to certain clinical generalizations that still retain some validity. For example, most true hermaphrodites were raised as males, despite frequently having hypospadias and incomplete labioscrotal fusion. Their gonads consisted of either 1) bilateral ovotestes (so-called *bilateral* gonadal hermaphroditism), 2) one ovary and one testis (*alternating* gonadal hermaphroditism), or 3) one ovotestis and a contralateral testis or ovary (*unilateral* gonadal hermaphorditism). Ovarian and testicular portions of ovotestes usually were juxtaposed end-to-end, thereby facilitating diagnosis by inspection or palpation. A uterus was usually present. A fallopian

Birth Defects: Original Article Series, Volume XIV, Number 6C, pages 9–35

tube was usually present ipsilateral to an ovary and often ipsilateral to an ovotestis. By contrast, a vas deferens and epididymis were usually present ipsilateral to a testis. At puberty breast development usually occurred (Fig. 1) and menstruation often occurred.

In 1959 Harnden and Armstrong [15] and Hungerford et al [16] reported the first true hermaphrodites subjected to cytogenetic analysis. Both groups surprisingly reported the karyotype to be 46,XX. Since then cytogenetic studies have been performed on at least 240 cases, a group that will serve as the basis for the present analysis.

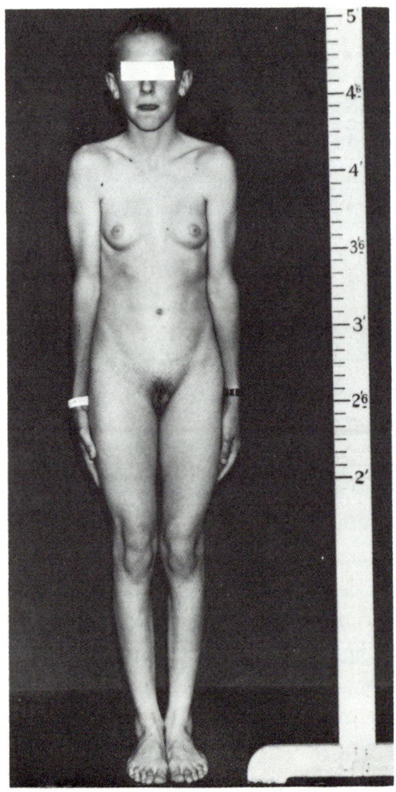

Fig. 1. Photograph taken at 13 years of age of the true hermaphrodite whose gonadal tissue is shown in Figure 3. Breast development occurs in many true hermaphrodites, including those reared as males.

DERIVATION OF SAMPLE

In at least 240 live-born cases traditional histologic criteria of hermaphroditism were fulfilled and cytogenetic studies were performed. Lee [17] believes that embryos show a relatively high frequency of hermaphroditism (1.08% of 1,525 embryos); however, Lee employed criteria different from those applied to live-born individuals. In this review a case was included only if live-born and only if it fulfilled traditional histologic criteria, ie oocytes and tubules or spermatozoa. Thus, several cases [18–29] considered by the reporting investigators to represent true hermaphroditism were excluded because either oocytes were not detected or tubules were not present. The diagnosis was especially likely to be applied erroneously to 45,X/46,XY individuals. On the other hand, restricting the appellation true hermaphroditism only to cases in which both oocytes and tubules were present is doubtless arbitrary because oocytes may have been present during embryogenesis or neonatal life, only to undergo attrition thereafter. This would be analogous to the ontogenesis of streak gonads in 45,X individuals [30]. If not ascertained until a later age, a true hermaphrodite may no longer have detectable oocytes, whereas at an earlier age oocytes could have been present. If so, the true nature of the disorder would not be appreciated and the appellation true hermaphroditism not applied. In addition, quality of descriptions varied greatly, especially descriptions of müllerian and wolffian derivatives. The prevalence of unicornuate or bicornuate uteri seemed especially likely to be underestimated.

Only cases subjected to chromosome studies were included. In most reports the authors' cytogenetic diagnosis was accepted, although often the quality of published karyotypes was poor and sometimes the number of cells analyzed was not stated. Often only a few cells were studied, leaving in doubt the exclusion of chimerism or mosaicism. In one case the diagnosis of 46,XX/46,XY mosaicism was based upon only a single 46,XY cell [31]; in a 2nd case the diagnosis was based upon only two 46,XY cells [32]. In one 46,XX/47,XXY case only two 47,XXY cells were detected [33]. These three cases were included in the analysis. However, cases studied only by analysis for X or Y chromatin were excluded, as was one case [34] in which lymphocytes were 46,XY and buccal epithelial cells X-chromatin positive.

APPROACH

The etiology of 46,XX and 46,XY true hermaphrodites is unknown. It thus seems appropriate to discuss initially those true hermaphrodites whose etiology is presumably known because of the presence of chimerism or mosaicism –

46,XX/46,XY, 46,XX/47,XXY, or 45,X/46,XY. Therefore, we shall discuss those true hermaphrodites whose etiology is unknown, namely those with 46,XY or 46,XX complements. Such an approach [1] not only allows consideration of genetic and cytogenetic data, but also permits a search for clinical and, hence, perhaps etiologic heterogeneity through comparison of the phenotypes of an etiologically known group (46,XX/46,XY) with the phenotypes of apparently etiologically unknown groups (46,XX or 46,XY). Similar phenotypes would suggest similar etiologies; dissimilar phenotypes might imply etiologic heterogeneity and, hence, suggest testable hypotheses.

46,XX/46,XY TRUE HERMAPHRODITISM

Twenty-eight cytogenetically documented 46,XX/46,XY true hermaphrodites have been reported [31, 32, 35–59]. Included are the two cases in which only one or two 46,XY cells were present [31, 32], but not included is the possible case in which lymphocytes were 46,XY and the buccal smear X-chromatin-positive [34]. Interestingly, the latter case [34] was a fertile male. True hermaphroditism of the type 46,XX/46,XY is rarer than 46,XX true hermaphroditism, but about as common as 46,XY true hermaphroditism. Not all 46,XX/46,XY individuals are true hermaphrodites, some being phenotypically normal or having gonadal dysgenesis or male psuedohermaphroditism [60–64]. In fact, fusion of XX and XY mouse zygotes usually produces not true hermaphorditism but rather clearly male or clearly female offspring [65].

Cytogenetic Data

Ascertainment of 46,XX/46,XY true hermaphrodites has been accomplished by studies restricted to lymphocytes as well as by studies of both lymphocytes and other tissues (eg skin or gonadal fibroblasts). Two cell lines were present in the lymphocytes of all except 3 [32, 38, 49] of the 28 known 46,XX/46,XY true hermaphrodites; however, among cases in which multiple tissues were analyzed the proportion of 46,XY cells in blood ranged from 0 to 94% (Fig. 2). In 3 cases [32, 38, 49] cytogenetic studies limited to lymphocytes would have failed to detect more than 1 line. If additional tissues are studied, the minority cell line is most likely to be detected in gonadal fibroblasts and relatively less likely to be detected in skin fibroblasts. In addition, the proportion of 46,XY cells in lymphocytes bears no ostensible relationship to gonadal status. That is, a relatively high proportion of 46,XY cells and lymphocytes does not necessarily indicate that the gondal tissue is predominantly testicular, nor conversely does a high proportion of 46,XY cells indicate that gonadal tissue is predominantly ovarian. In conclusion, analysis of multiple tissues is the only way to detect certain 46,XX/46,XY true hermaphrodites, although many can be detected by analyzing lymphocytes alone.

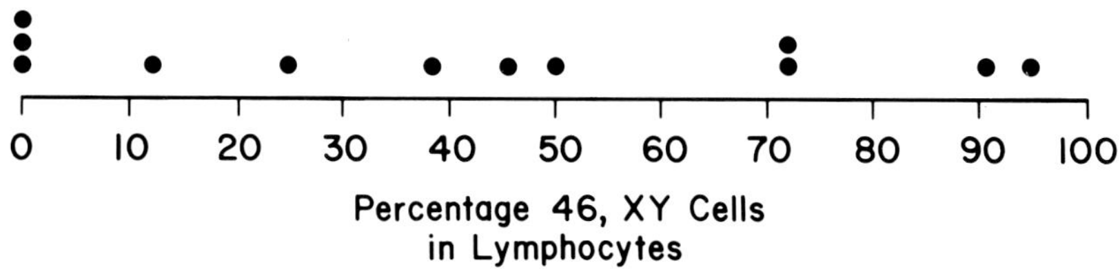

Fig. 2. Histographic representation showing the proportions of 46,XY cells in lymphocytes of those 46,XX/46,XY true hermaphrodites in which examinations of multiple tissue were performed.

Etiology

A 46,XX/46,XY chromosome complement could arise by 1) nondisjunction involving a 47,XXY or 46,XY zygote, with loss of certain cell lines and retention of others, or 2) chimerism.

In humans it is sometimes difficult to prove that nondisjunction, as opposed to chimerism, is the cause of a given 46,XX/46,XY true hermaphrodite. However, the existence of true hermaphrodites with certain karyotypes (eg 46,XX/47,XXY or 45,X/46,XY) suggests strongly that true hermaphroditism can result from mitotic nondisjunction. In fact, Whitten [66] showed that the relatively high frequency of true hermaphroditism in BALB/cWt mice results from nondisjunction.

In human 46,XX/46,XY true hermaphroditism, the etiology is usually suspected to be chimerism, rather than nondisjunction. Chimerism connotes the presence in a single individual of cells derived from different zygotes. The phenomenon is proved by detection of 2 populations of cells, usually erythrocytes with different blood types, in a single individual. At least 3 types of chimerism are known [67]: 1) blood chimerism, interchange of blood cells between co-twins through placental anastomoses; 2) transplacental chimerism, interchange of fetal and maternal blood cells; and 3) whole body chimerism. Whole body chimerism, presumably the type of chimerism responsible for some 46,XX/46,XY true hermaphrodites, could result from 1) fertilization of both an ovum and its polar body, 2) fertilization of each of 2 ova contained within a single binucleated follicle, 3) fertilization of ova derived from different follicles, followed by fusion, or 4) other related phenomena. Whole body chimerism can be assumed if 2 or more genotypes are 1) present in nonhematogenous tissues (skin or gonads) or 2) persist in hematogenous tissues. The frequency of whole body chimerism may be underestimated because chimeric individuals are usually ascertained because of abnormal sexual development (eg true hermaphroditism). Thus almost all recognized whole body chimeras are heterosexual, whereas one might expect an equal number of isosexual chimeras. That whole body chimerism has been detected in phenotypically normal individuals emphasizes that a 46,XX/46,XY karyotype is not associated invariably with true hermaphroditism.

Six 46,XX/46,XY true hermaphrodites are known chimeras [35–37, 39, 44, 52, 54, 58]. In 6 others, chimerism was sought yet not demonstrated; however, parents were heterozygous at only a few of the tested loci.

Phenotype

The external genitalia of 46,XX/46,XY true hermaphrodites are usually ambiguous or sufficiently masculinized to suggest to attending physicians that the sex of rearing should be male. A female sex of rearing was chosen in only 3 of 13 cases. Even if reared as males, affected individuals usually show perineal or penoscrotal hypospadias and incomplete labioscrotal fusion.

The distribution of gonadal tissue is shown in Table 1. No single type predominates, contrary to an opinion [1] stated previously after analysis of a smaller sample; however, unilateral or alternating distributions are more common than bilateral distributions. Bilateral ovotestes occurred in only 6 of 24 informative cases. One might expect the proportions of alternating or unilateral gonadal distributions to be higher in 46,XX/46,XY true hermaphrodites if the relative distribution of ovarian and testicular tissue reflected the frequencies of 46,XX and 46,XY cells, respectively. The chromosome constitution of gonadal tissue has been determined in only a few cases, but 46,XY cells indeed appear more likely to be present in a testis or ovotestis than in an ovary; however, additional studies are necessary to confirm this hypothesis.

In aggregate, the right gonad is much more likely to contain testicular tissue than the left, a characteristic that 46,XX/46,XY true hermaphrodites share with other true hermaphrodites. There is no obvious explanation for the predilection for testicular development on the right. Interestingly, however, in many species the right gonad is vestigial. Furthermore, following extirpation of the left ovary from newborn hens, the right gonad differentiates into a testis [68]. One possible explanation is that the embryonic right gonad contains a higher proportion of medullary tissue, thus being more likely to differentiate into a testis. However, in humans the corticomedullary dichotomy may or may not be a valid explanation for differentiation of the human indifferent gonads into either ovaries or testes.

The ovarian and testicular portions of an ovotestis usually exist in end-to-end juxtaposition. Both seminiferous tubules and ovarian follicles are present, by definition. Testicular tissue is characterized by relatively few normal germ cells, whether existing as a separate testis or as one component of an ovotestis. The tubules are usually hyalinized and composed only of Sertoli cells (Fig. 3); Leydig cells may be hyperplastic. Spermatozoa are rare; however, noted previously was one possible 46,XX/46,XY case [34] who was fertile and had a sperm count of $20 \cdot 10^6$/ml. By contrast, ovarian tissue often contains numerous primordial follicles in various developmental stages. That ovarian tissue is often normal is also evidenced by breast development, cyclic menses, and histologic evidence of

TABLE 1. Clinical Characteristics of True Hermaphrodites With Various Chromosomal Complements*

Chromosomal complement (total cases)[a]	Sex of rearing (female total)[b]	Presence of uterus[c]	Unicornuate or bicornuate uterus[d]	Gondal distribution (left:right)								Total informative cases
				OT:OT	O:T	T:O	O:OT	OT:O	T:OT	OT:T	Other	
46,XX/46,XY (28)	3/13	20/22	1/20	6	7	1	4	4	1	2	0	24
46,XX/47,XXY (13)	3/10	8/10	2/8	4	2	3	1	0	0	0	2	12
45,X/46,XY (11)	3/8	5/5	0/5	0	4	1	0	2	0	1	0	8
46,XY (37)	2/17	17/19	2/17	0	12	4	4	0	4	2	2	28
Familial 46,XX[e] (8)	0/8	2/8	0/2	5	0	0	2	0	1	0	0	8
Nonfamilial 46,XX (131)	34/89	90/108	18/90	23	24	8	36	9	2	10	5	117

*Not all cases proved informative; sometimes few details having been published. O, ovary; T, testis; OT, ovotestis.
[a]Complete descriptions not available for all cases.
[b]Includes only those cases in which attending physicians assigned a sex of rearing.
[c]Includes cases in which uterus described as "rudimentary."
[d]Probably an underestimate because descriptions of uteri often not complete.
[e]References [124–177] exclude cases of Kasdan et al [168] and Berger et al [178].

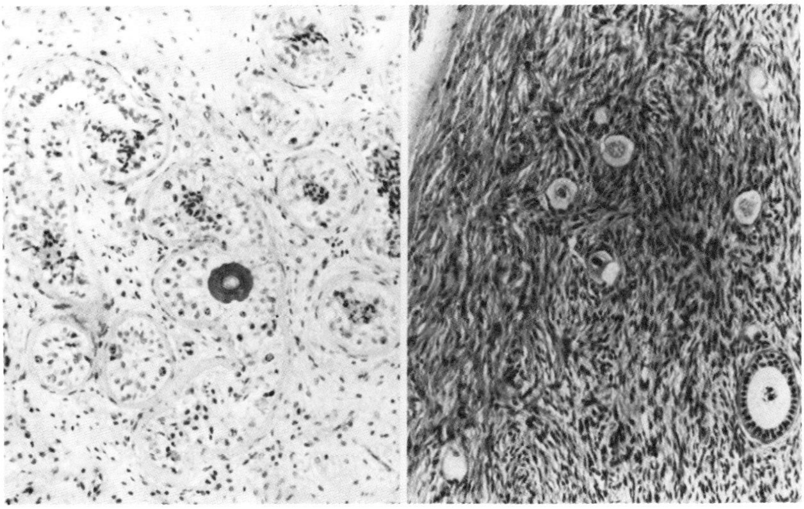

Fig. 3. Photomicrograph of testicular (left) and ovarian (right) portions of an ovotestis of 46,XX/47,XX,r(Y) true hermaphrodite. The ovarian portion contains not only stroma but also oocytes. (From Sarto GE, et al: Considerations of sex chromosome abnormalities in man. In Benirschke K (ed): "Comparative Mammalian Cytogenetics." New York: Springer-Verlag, 1969 with permission.)

ovulation. However, histologic criteria for diagnosis introduces some biases favoring presence of normal oocytes in true hermaphrodites.

True hermaphrodites 46,XX/46,XY with karyotypes usually have müllerian derivatives, namely a uterus and one or more fallopian tubes. In only 2 of 22 cases was the uterus absent, although a few authors commented that uterine development was rudimentary or "merely a remnant." Menstruation occurred in 3 of 7 pubertal individuals with a uterus. The presence or absence of fallopian tubes or wolffian derivatives reflects the ipsilateral gonad. A fallopian tube is invariably present ipsilateral to an ovary, whereas a vas deferens, epididymis, and often a seminal vesicle are usually present ipsilateral to a testis. Either fallopian tubes or wolffian derivatives may be present on the side of an ovotestis, although a fallopian tube is most often present. Both fallopian tubes and wolffian derivatives are usually not present on the same side, even on the side of an ovotestis; however, wolffian and müllerian derivatives have occasionally been observed on the same side [37, 39], contrary to opinions stated by some authors. Two cases developed gonadal tumors [41, 48].

46,XX/47,XXY TRUE HERMAPHRODITISM

True hermaphroditism associated with a 46,XX/47,XXY chromosome complement probably results from nondisjunction involving a 46,XY or 47,XXY zygote. The karyotype can be explained readily by postulating loss of certain cell lines and retention of others. However, chimerism has not been excluded nor usually even sought vigorously. Thirteen cases have been detected [33, 69–80], usually on the basis of cultured lymphocytes; in only 2 cases [73, 77] were multiple tissues studied. In one case [33] the mosaicism was diagnosed on the basis of only two 47,XXY cells. In another case [77] the Y chromosome was present in the form of a ring, but still sufficient to direct testicular differentiation.

The sex of rearing was chosen by the attending physicians in 10 cases; in 3 a female assignment was chosen [33, 72, 74]. The gonadal distribution, described in 12 cases, shows no obvious pattern (Table 1). A uterus was present in 8 of 10 cases, 2 being unicornuate [73, 77]. Prevalences of bicornuate and unicornuate uteri seem likely to be underestimated because the uterus was often not completely described. Moreover, the occurrence of unicornuate or bicornuate uteri would be predicted because androgens and the müllerian inhibitory factor (MIF) influence embryonic ductal differentiation through local diffusion from the fetal testes.

45,X/46,XY TRUE HERMAPHRODITISM

Individuals 45,X/46,XY with a karyotype display a spectrum of phenotypes, ranging from almost normal males to females indistinguishable from those with 45,X gonadal dysgenesis and the Turner stigmata [1]. The different phenotypes presumably reflect different tissue distributions of 45,X and 46,XY cells. However, this assumption has not been proved. In fact, it often cannot be demonstrated, although one should recall that in gonadal cultures not germ cells but fibroblasts are cultured.

The phenotype most often associated with 45,X/46,XY mosaicism is mixed or asymmetric gonadal dysgenesis, characterized by a unilateral streak gonad and a contralateral testis. In most 45,X/46,XY individuals no ovarian follicles are detectable and, hence, the diagnosis of true hermaphroditism is inappropriate. Nonetheless, the diagnosis is not infrequently applied inappropriately to 45,X/46,XY individuals who fail to fulfill traditional histologic criteria. On the other hand, we noted previously that 45,X embryos contain oocytes which undergo attrition more rapidly than 46,XX oocytes. Thus, the presence of an occasional 45,X/46,XY individual with oocytes (ie a true hermaphrodite) would not be totally unexpected. The above nosologic confusion emphasizes the arbitrary criteria for the diagnosis of true hermaphroditism, inasmuch as several cases illustrate that the 45,X/46,XY spectrum includes true hermaphroditism.

Eleven 45,X/46,XY true hermaphrodites have been reported [11, 73, 81–90], but only 8 were completely described. A male sex of rearing was chosen in 5 of the 8 cases. Five of 5 cases had a uterus. Five of 8 showed one ovary and testis, and in 4 of these 5 the right gonad was a testis. No case had bilateral ovotestes (Table 1).

TRUE HERMAPHRODITISM ASSOCIATED WITH OTHER FORMS OF MOSAICISM OR CHIMERISM

True hermaphroditism has been associated with various other chromosome abnormalities, including 45,X/46,XX [89]; 46,XX/47,XYY [91]; 46,XX/46,XY/47,XXY [92]; 46,XX/47,XXY/49,XXYYY [93]; 45,X/46,XY/47,XYY [94]; 46,XX/48,XXYY [95]. These karyotypes probably arise by mitotic nondisjunction, although chimerism has not been excluded. Because only a few cases have been associated with a given karyotype, generalizations would be unwise; however, alternating gonadal hermaphroditism occurs relatively frequently.

46,XY TRUE HERMAPHRODITISM

Phenotype

Thirty-seven cases of 46,XY true hermaphroditism have been reported [11, 47, 73, 75, 84, 96–114], although a complete description is available in fewer cases. Over half were Japanese. This investigator was unable to obtain detailed descriptions of 9 Japanese cases, attributed to Kadotani [89] by Makino [47]. Nonetheless, 46,XY true hermaphroditism appears to be the most common type of true hermaphroditism in Japan, in contrast to its relative rarity in Europe, North America, South America, Africa, and Australia. In only 3 cases [11, 73, 108] were multiple tissues studied; in 2 of the 3 culture ovarian fibroblasts were 46,XY [73, 108].

A sex of rearing was assigned in 19 cases, and only twice [107, 114] was the female role chosen. Thus, the external genitalia more nearly resemble the male than in certain other forms of true hermaphroditism, a suggestion consistent with published descriptions of external genitalia; however, quantitation of genital virilization is difficult. The gonadal distribution (Table 1) shows a high frequency of the alternating type, specifically a left ovary and a right testis. Twelve of 28 cases had a left ovary and right testis; 4 cases had a left testis and a right ovary. No 46,XY true hermaphrodites definitely had bilateral ovotestes, although a possible case was reported by Sandberg et al [96]. The uterus was present in 17 of 19 cases for whom complete descriptions were available; 2 uteri [110, 111] were unicornuate and a third [112] bicornuate. Development of wolffian and müllerian derivatives was similar to other true hermaphrodites.

Two 46,XY true hermaphrodites [112, 114] developed gonadoblastomas. Other gonadal tumors, both ovarian and testicular [115–120] have been reported in true hermaphrodites not subjected to cytogenetic studies, but none had gonadoblastomas. These 2 cases [112, 114] suggest that true hermaphrodites show the same predilection for neoplastic transformation that individuals with 45,X/46,XY mosaicism or XY gonadal dysgenesis do, in contrast to a previous opinion [121] stated prior to publication of those reports. Breast carcinoma has also been reported in 2 true hermaphrodites; one was 46,XX [8] and the karyotype of the other [122] is unknown.

Etiology

The presence of oocytes in 46,XY true hermaphrodites could result from 1) undetected chimerism or mosaicism, 46,XX cells not being readily detectable; 2) translocation of ovarian determinants from the X chromosome to either the Y chromosome or to an autosome; or 3) a mutant gene(s).

Certain phenotypic features suggest to this investigator that undetected chimerism or mosaicism is often, although not necessarily invariably, the cause of 46,XY true hermaphrodites. For example, the frequency of alternating gonads is relatively high – 16 of 28 (58%) cases. Moreover, the gonadal tissue appears more likely to be testicular than ovarian, and the external genitalia are relatively more virilized than in true hermaphrodites with other karyotypes. True hermaphrodites of type 46,XY are especially likely to differ from 46,XX true hermaphrodites (see below) with respect to gonadal distribution, extent of virilization, and sex of rearing. The features that characterize 46,XY true hermaphrodites would be expected if such cases had mosaicism or chimerism in which the proportion of 46,XX cells was relatively small and, hence, difficult to detect. That is, if some cells contained the factor responsible for the true hermaphroditism and others did not, dissimilar gonads might be expected. Indeed, cytogenetic studies have rarely been extensive enough to exclude mosaicism or chimerism in 46,XY true hermaphrodites.

By contrast, if 1) X-Y interchange, 2) a Y-autosome translocation, or 3) a mutant allele were present, both gonads would theoretically seem likely to be morphologically similar (ie bilateral ovotestes) because every cell would presumably possess the factor responsible for abnormal gonadal development. X-Y or Y-autosome translocations remain theoretic explanations for 46,XY true hermaphroditism; however, no cytogenetic data support their existence. The relatively high frequency of 46,XY true hermaphroditism among Japanese suggests presence of a mutant gene because differences in racial prevalences are often the first clue to an underlying genetic etiology. However, the lack of familial aggregates and the absence of parental consanguinity fail to support the hypothesis of recessive inheritance. On the other hand, Milner et al [123] reported X-chromatin-negative sibs with true hermaphroditism; additional cytogenetic studies were unavailable. Furthermore, Lowry et al [28] believed that 46,XY true hermaphroditism was present in 2 paternal first cousins; however, neither cousin had definite oocytes.

46,XX TRUE HERMAPHRODITES

The karyotype most often detected among true hermaphrodites is 46,XX. About 140 cases [7, 8, 11, 13, 15, 16, 33, 40, 43, 75, 78, 79, 82, 84, 86, 89, 104, 113, 124–177] have been reported, including 40 in the Bantu and other African blacks. Any explanation for 46,XX true hermaphroditism must account for the presence of testicular tissue in individuals who apparently lack a Y chromosome, a phenomenon apparently contrary to the axiom that a Y chromosome is required for testicular differentiation. Possible explanations include 1) undetected chimerism or mosaicism, or, alternatively, loss of a Y chromosome after initiation of testicular differentiation; 2) translocation of testicular determinants from the Y to the X chromosome; 3) translocation of testicular determinants from the Y chromosome to an autosome; or 4) a mutant gene.

Familial Cases

In contrast to the apparent lack of heritability in other forms of true hermaphroditism, several familial aggregates of 46,XX true hermaphroditism have been reported [124–128]. In 3 families multiple sibs had 46,XX true hermaphroditism (Fig. 4). Five of 8 individuals in the 3 families had bilateral ovotestes, 2 had a left ovary and a right ovotestis, and one had a left testis and a right ovotestis (Table 2). In contrast to 46,XX/46,XY or 46,XY true hermaphrodites, a uterus was present in only 2 of 8 patients, both in the same family [126]. A male sex of rearing was chosen in all 8 cases; pubertal development was similar to other true hermaphrodites. Irrespective of whether testicular or ovarian tissue was incubated, endocrine studies in one family [128] showed that estrogens, especially estrone, were the predominant sex steroid formed in vitro. Plasma hormone fluctuations were said to resemble the female "pattern" more than the male. Possibly a 4th aggregate of affected sibs was reported by Milner et al [123]; however, chromosome studies were not performed. In yet another family [149], an affected patient was said to have a sib with clitoral hypertrophy.

These families indicate that at least one form of 46,XX true hermaphroditism results from an autosomal recessive gene. Indeed, both dominant and recessive sex-reversal genes exist in animals (Table 3). However, an X-Y or Y-autosome

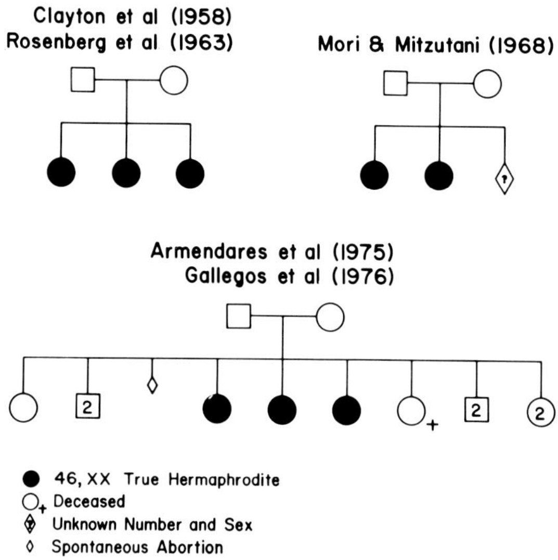

Fig. 4. Pedigrees of 3 families in which 2 or more members had 46,XX true hermaphroditism.

translocation in a paternal gonad could also explain sibship aggregates. No H-Y antigen data are available in the human familial aggregates.

If true hermaphroditism results from a mutant gene, especially a recessive gene, both gonads might be expected to be similar (ie ovotestes) because the factor responsible for the gonadal abnormality would presumably be present in every cell, rather than being present in some yet not all cells and, hence, organs. By contrast, if different organs showed different frequencies of, say, 46,XX or 46,XY cells, one would predict a higher frequency of alternating hermaphroditism (eg left ovary:right testis). Finally in one [168] and possibly two [178] families, 46,XX males and 46,XX true hermaphrodites occurred in the same sibship, observations that we shall return to below. In animals, both dominant and recessive genes cause sex reversal [179–186].

Nonfamilial Cases

Phenotype. One hundred thirty-one 46,XX true hermaphrodites had no apparent affected relative [7, 8, 11, 13, 15, 16, 33, 40, 43, 75, 78, 79, 82, 84, 86, 89, 104, 113, 129–177]. In 20 cases multiple tissues were analyzed. Interestingly, although doubtless coincidentally, the gonadal distribution was bilateral (ovotestes) in only 2 [174, 175] of the 20. Thus, the group investigated by multiple

TABLE 2. Clinical Characteristics of Affected Sibs in Four Families in Which Multiple Sibs Had True Hermaphroditism*

Investigators	Sex of rearing	Pubertal breast development	Left gonad	Right gonad	Uterus	Cytogenetic studies
Clayton et al [124]						
Rosenberg et al [125]	Male	+	T	OT	–	46,XX
	Male	+	OT	OT	–	46,XX
	Male	+	OT	OT	–	46,XX
Mori and Mitzutani [126]	Male		O	OT	+	46,XX
	Male		O	OT	+	46,XX
Armendares et al [127]						
Gallegos et al [128]	Male	+	OT	OT	–	46,XX
	Male	+	OT	OT	–	46,XX
	Male	–	OT	OT	–	46,XX
Milner et al [123]	Male	+	OT	OT	–	X-chromatin negative
	?	–	OT	OT	–	X-chromatin negative

*+, Present; –, absent; blank, no data; O, ovary; T, testis; OT, ovotestis.

TABLE 3. Possible Sex-Reversal Genes in Animals*

Species	Sex-reversal genes		
	Gene	Inheritance	Phenotype
Mouse [179, 180]	Sxr (sex-reversal)	Sex-limited autosomal dominant	Heterozygous XX = phenotypic males
Saanen Goat [181, 182]	Polled	Autosomal recessive	Homozygous XX = phenotypic males or true hermaphrodites Homozygous XY = epididymal occulusion
Pig [83]	Sex-reversal	Autosomal recessive	Homozygous XX = phenotypic males or true hermaphrodites
Dog [186]	Sex-reversal	?	Homozygous XX = phenotypic males
Drosophila [184, 185]			
D. melanogaster	tra	Autosomal recessive	Homozygous XX =
D. virillis	lse	Autosomal recessive	phenotypic males

*Sxr mice are H-Y-antigen positive [180], thus, they may result not from an autosomal dominant gene but from X-Y or Y-autosome translocation. Preliminary results on canine hermaphroditism are discussed elsewhere in this volume by Wachtel [186].

tissue analysis exemplifies the group theoretically most likely to have chimerism or mosaicism.

A female sex of rearing was chosen in 34 of 89 nonfamilial cases in which a sex assignment was made, more often than in the 46,XY or 46,XX/46,XY groups. In fact, one true hermaphrodite [174] was delivered of a male infant. Somatic anomalies were occasionally present [145, 149].

The gonadal distribution most commonly observed was a left ovary and a right ovatestis. The next most common was bilateral ovotestes (Table I). A left ovary and a right testis, the most common distribution in 46,XY true hermaphrodites, was present in only 23 of 117 cases; 8 cases had a left testis and right ovary. Thirty-six cases had a left ovary and a right ovotestis. An alternating gonadal distribution thus occurs less frequently in 46,XX true hermaphrodites than in 46,XY true hermaphrodites. Histologic and gross appearance of the gonads is similar to that of other true hermaphrodites, and the predilection of the right gonad to be testicular rather than ovarian is also observed. One affected individual developed a dysgerminoma [8], and a 2nd developed a gonadoblastoma [175].

A uterus was present in 90 of 108 cases. Although still relatively high, this frequency is lower than the frequency among 46,XY or 46,XX/46,XY true hermaphrodites. The uterus was described as either unicornuate or bicornuate in 18 of 90 cases. Moreover, the frequency of uterine anomalies is probably underestimated because müllerian and wolffian derivatives were often described only perfunctorily. As in true hermaphrodites with other karyotypes, a vas deferens and an epididymis were usually present ipsilateral to a testis. A fallopian tube was usually present ipsilateral to an ovotestis, although occasionally only wolffian derivatives or both müllerian and wolffian derivatives were present. In addition, Van Niekerk [8, 14] observed that 1) the fimbriated end of the fallopian tube was often occluded, and 2) the cervical canal was often obliterated or characterized by squamous metaplasia. These observations have apparently not been observed often by others [187], leading one to wonder whether Van Niekerk's sample (all Bantu) indicates genetic herterogeneity among 46,XX true hermaphrodites.

In summary, nonfamilial 46,XX true hermaphrodites are more likely than other true hermaphrodites to have either a left ovary and a right ovotestis, or bilateral ovotestes. In particular, alternating hermaphroditism is less common than in 46,XY true hermaphrodites and a uterus is less likely to be present.

Etiology. The presence of testicular tissue in individuals who apparently lack a Y chromosome could be explained by 1) undetected mosaicism or chimerism, 2) translocation of the Y testicular determinants to the X chromosome, 3) translocation of Y testicular determinants to an autosome, or 4) a mutant allele(s). Let us consider each hypothesis.

First, in some cases undetected 46,XY cells are doubtless present. However, cultures of testicular fibroblasts sometimes show no 46,XY cells, although the complement in germ cells is unknown. In addition, cases with alternating hermaphroditism would theoretically appear most likely to have chimerism or mosaicism; thus, the relatively low frequency of alternating gonads in 46,XX true hermaphrodites suggests that it would be unwise to assume in all cases that mosaicism was present yet undetected.

Second, the presence of testicular tissue in 46,XX true hermaphrodites could result from translocation of Y-linked testicular determinants to an X chromosome or to an autosome (X-Y or X-autosome translocation). Supporting this hypothesis are 1) anomalous familial distributions of Xg alleles [151] and 2) presence of H-Y antigen in certain cases [177]. On the other hand, H-Y antigen has been sought in only a few cases, and anomalous distribution of Xg alleles has likewise been observed infrequently. Moreover, translocation of testicular determinants does not alone explain the true hermaphroditism phenotype, for consideration of the following shows that other phenomena must simultaneously be postulated. If 1) H-Y antigen is the gene product of the testicular determinant, and 2) the H-Y locus is translocated to an X chromosome or to an autosome, why is testicular differentiation abnormal in individuals who carry the translocation? That is, why are H-Y-antigen positive 46,XX true hermaphrodites not normal males? (A similar statement applies to 46,XX males who are H-Y-antigen positive.) One possible explanation is coexisting chimerism or mosaicism. A second is a position effect that might alter the manner in which the Y-testicular determinant(s) exert their action. A third possibility is a qualitative or quantitative abnormality involving the product of the H-Y gene.

Finally, the occurrence of familial aggregates suggests that some sporadic cases of 46,XX true hermaphroditism result from a mutant gene. The postulated mutant could interfere with H-Y antigen, or it could repress a male-determining locus controlled in regulatory fashion by the Y-testicular determinant(s). Unfortunately, few authors publish adequate genealogic data. Morever, the disparate nature of ascertainment precludes meaningful segregation analysis. Nonetheless, data are available concerning 29 of the 129 probands with nonfamilial XX true hermaphrodites. Reporting investigators described 10 unaffected male sibs, 22 unaffected female sibs, and 67 unaffected sibs of unknown sex. Parental consanguinity is also apparently rare, although reported [167]. Thus, it seems unlikely that all sporadic 46,XX true hermaphrodites could be explained on the basis of a mutant autosomal recessive allele; however, some sporadic cases, especially those with bilateral ovotestes and no uterus, might result from such factors.

In addition to a recessive gene, an autosomal dominant sex-reversal gene might exist. Kasdan et al [168] reported a kindred in which the proband was a 46,XX true hermaphrodite; both a "male" sib and a paternal "uncle" showed complete sex reversal (46,XX males). In this family (Fig. 5) there thus appeared to be segregating a dominant allele capable of causing complete sex reversal (46,XX male), or, perhaps if less completely expressed, 46,XX true hermaphroditism. A 46,XX male proband and probably a 46,XX true hermaphrodite sib were also present in the family reported by Berger et al [178]; (Fig. 5); the true hermaphrodite sib, said to be 46,XX/46,XY, had only two 46,XY cells, studied prior to availability of banding techniques. Thus, it is uncertain whether the complement was 46,XX or 46,XX/46,XY. In addition, De La Chapelle et al reported a kindred (Fig. 5) in which 2 paternal second cousins were 46,XX males [188]. In animals, both dominant and recessive sex-reversal genes [178–186] have been reported, lending credence to hypotheses that such genes may exist in humans. In conclusion, additional data are clearly necessary in order to explain the presence of testicular tissue in individuals who apparently lack a Y chromosome. Several possibilities can be cited: 1) an autosomal recessive or autosomal dominant gene; 2) mosaicism or chimerism characterized by only a small percentage of 46,XY cells, which are thus difficult to detect; and 3) inapparent translocation

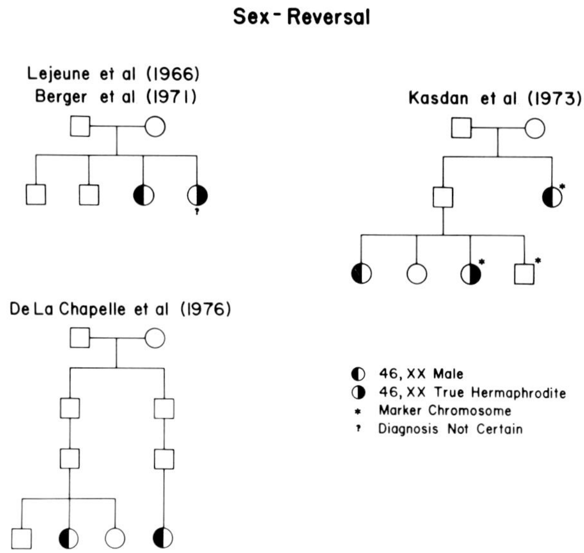

Fig. 5. Pedigrees of 3 families in which multiple members were either 46,XX males or 46,XX true hermaphrodites.

of testicular determinant(s) to the X chromosome or to an autosome. Each possibly could be important, indicating heterogeneity among 46,XX true hermaphrodites.

CONCLUSIONS

1) Approximately 240 true hermaphrodites have been subjected to chromosome studies. One hundred forty are 46,XX; 37 are 46,XY; 28 are 46,XX/46,XY; 13 are 46,XX/47,XXY; 11 are 45,X/46,XY; and at least 10 have other karyotypes. Adherence to traditional histologic criteria was followed in this report, although such criteria are probably arbitrary.

2) In aggregate, true hermaphrodites are usually reared as males and usually develop breasts at puberty. Their external genitalia are most likely to resemble an incompletely virilized male. A separate ovary and a separate testis may be present, or one or more ovotestes may be present. A uterus is usually present. The presence or absence of fallopian tubes or vasa deferentia reflects ipsilateral gonadal composition. The diagnosis is usually not made prior to laparotomy; however, the following features suggest the diagnosis of true hermaphroditism in a child with genital ambiguity: a) a gonad characterized by two consistencies (ovotestis), b) a unicornuate or bicornuate uterus, c) occlusion of fallopian tube fimbria, and d) atresia or squamous metaplasia of the uterine cervix.

3) Neoplastic transformation of gonads can occur in true hermaphrodites. At least 7 ovarian and testicular neoplasma were described prior to the availability of cytogenetic methods. More recently, gonadoblastomas or dysgerminomas have been reported in two 46,XY, two 46,XX/46,XY, and two 46,XX true hermaphrodites. Two true hermaphrodites developed carcinoma of the breast; the karyotype of one was 46,XX, and the karyotype of the other was not determined.

4) Six of the 28 46,XX/46,XY true hermaphrodites are proven chimeras. The other 22 cases could have resulted from either chimerism or mosaicism. In 3 of the 28 cases analysis of lymphocytes alone would have failed to detect the 46,XY cell line. Almost all 46,XX/46,XY cases have a uterus, and no particular pattern of gonadal distribution predominated.

5) True hermaphroditism of the types 46,XX/47,XXY and 45,X/46,XY probably results from nondisjunction. Neither type shows a distinct phenotype.

6) 46,XY true hermaphrodites, the most common form of true hermaphroditism in Japan, are usually reared as males and usually have a uterus. Sixteen of 28 informative cases had alternating gonadal hermaphroditism, whereas none or possibly one had bilateral ovotestes. These clinical features, considered in view of the frequent lack of thorough chromosome studies, suggests that some 46,XY true hermaphrodites result from chimerism or mosaicism in which 46,XX cells are not readily detectable.

7) The occurrence of a least 3 (sibship) aggregates of 46,XX true hermaphroditism indicates that an autosomal recessive allele is responsible for at least one form of 46,XX true hermaphroditism. The familial cases are more likely to have bilateral ovotestes and more likely to lack a uterus that the nonfamilial cases.

8) Nonfamilial 46,XX true hermaphrodites are probably of heterogenous etiology. Some probably have mosaicism or chimerism in which 46,XY cells are not readily detectable; however, analysis of testicular fibroblasts may fail to detect 46,XY cells. In other cases anomalous familial distributions of Xg alleles suggest Y-X or Y-autosome translocations, as does the presence of the H-Y antigen in 46,XX true hermaphrodites. Finally, some sporadic cases might result from a mutant recessive allele, particularly cases with bilateral ovotestes and no uterus. An autosomal dominant sex-reversal gene might also exist in humans.

REFERENCES

1. Simpson JL: "Disorders of Sexual Differentiation: Etiology and Clinical Delineation." New York: Academic Press, 1976.
2. Klebs E: "Handbuch der pathologischen Anatomie." Berlin: Hirschwald, 1876.
3. Hinman F: "The Principles and Practice of Urology." Philadelphia: WB Saunders, 1935.
4. Young HH: "Genital Abnormalities, Hermaphroditism and Related Adrenal Diseases." Baltimore: Williams & Wilkins, 1937.
5. Jones HW Jr, Scott WM: "Hermaphroditism, Genital Anomalies, and Related Endocrine Disorders," 2nd Ed. Baltimore: Williams & Wilkins, 1971.

6. Overzier C: True hermaphroditism. In Overzier C (ed): "Intersexuality." New York: Academic Press, 1963.
7. Neiman N, Fonder A: L'hermaphrodisme. Pédiatrie 12:499, 1967.
8. Van Niekerk WA: "True Hermaphroditism." New York: Harper & Row, 1974.
9. Inhorn SL, Opitz JM: Abnormalities of sex development. In Bloodworth JMB Jr (ed): "Endocrine Pathology." Baltimore: Williams & Wilkins, 1968, p 529.
10. Polani PE: Hormonal and clinical aspects of hermaphroditism and the testicular feminizing syndrome in man. Philos Trans R Soc Lond (Biol) 259:187, 1970.
11. Hamerton JL: "Human Cytogenetics." New York: Academic Press, 1971, vol 2.
12. Pfeiffer RA: True hermaphroditism. Helv Paediatr Acta (Suppl) 12/13:99, 1974.
13. Fournier JL, Saint-Aubert P, Ponte C, Gaudier B, Waldbaun R, Farriaux JP, Fontaine G: Étude nosologique de l'hermaphrodisme vrai et de la dysgénésie gondéque mixte et asymétrique. Ann Pédiatr 23:763, 1976.
14. Van Niekerk WA: True hermaphroditism. Am J Obstet Gynecol 126:890, 1976.
15. Harnden DG, Armstrong CN: The chromosomes of a true hermaphrodite. Br Med J 2:1287, 1959.
16. Hungerford DA, Donnelly AJ, Nowell PC, Beck S: The chromosome constitution of a human phenotypic intersex. Am J Hum Genet 11:215, 1959.
17. Lee S: High incidence of true hermaphroditism in the early human embryos. Biol Neonate 18:418, 1971.
18. Hirschhorn K, Decker WH, Cooper HL: Human intersex with chromosome mosaicism of type XY/XO. Report of a case. N Engl J Med 263:1044, 1960.
19. Miller OJ: Sex determination. The sex chromosomes and the sex chromatin pattern. Fertil Steril 13:93, 1962.
20. Ober WB: Embryonal carcinoma of testicular type arising in the gonad of a true hermaphrodite. Jew Mem Hosp Bull (NY) 67:95, 1962.
21. Overzier C: Ein XX/XY-Hermaphrodit mit einen "intratubualren Ei" und einem Gonadoblastom (Gonocytome III). Klin Wochenschr 42:1056, 1964.
22. Schmidt BJ, Queiroz AS, Becak W, Merzel J, Barbante PJ, Pinus J, Saad FA, Yaroslavsky W, Carvalho AA: Hermaph-Matern Infancia (Sao Paulo) 25:473, 1966.
23. Tsurgi K, Sakuma M, Wakie M, Sekiguchi T, Kumamoto Y, Takayasu H: True hermaphrodite presenting with scrotal ovotestes, hypospadias and gynecomastia. J Urol 104:350, 1970.
24. Barta L, Mittner I, Regoly-Merei A: Pseudohermaphroditismus masculinus. Arch Kinderheilkd 183:353, 1971.
25. Olsson CA, Tessier PA, Brown ML, Austen G Jr: True hermaphroditism. J Urol 105: 856, 1971.
26. Eberle P, Gallasch E, Truss F: XX/XY Hermaphroditismus verus mit Blutchimare. Blut 25:255, 1972.
27. Librik L, Clayton GW Jr: Diagnosis of true hermaphroditism. J Pediatr 80:1060, 1972.
28. Lowry RB, Honore LM, Arnold WJD, Johnson HW, Kliman MR, Marshall RH: Familial true hermaphroditism. In Bergsma D (ed): "Genetic Forms of Hypogonadism." Miami: Symposia Specialists for The National Foundation–March of Dimes, BD:OAS XI(4):105, 1975.
29. Yanagisawa S: Hermaphroditism associated with unusual gonadal finding in a mentally retarded child. J Ment Defic Res 19:37, 1975.
30. Boué A, Boué J: Chromosome abnormalities and abortion. In Coutinho EM, Fuchs F (eds): "Physiology and Genetics of Reproduction." New York: Plenum Press, 1975, Part B, p 317.

31. Michaud P, Maggiolo C, Beas F, Pierret T, Roa A, Chuagni B: Hermaphroditismo verdadero con mosaico XX-XY, Estudio clinico e psicologico. Rev Med Chile 100:665, 1972.
32. Lejeune J, Berger R, Rethore MO, Vialatte J, Salmon CH: Sur un cas d'hermaphrodisme XX/XY. Ann Genet (Paris) 9:171, 1966.
33. Zmerli S, Petrucci J-M, Lles S-E, Oucherif A: De l'hermaphrodisme vrai. J Urol Nephrol 1:57, 1972.
34. Shannon R, Nicolaides MJ: True hermaphroditism with oogenesis and spermatogenesis. Aust NZ J Obstet Gynecol 13:84, 1973.
35. Waxman SH, Gartler SM, Kelley VC: Apparent masculinization of the female fetus diagnosed as true hermaphroditism by chromosomal studies. J Pediatr 60:540, 1962.
36. Gartler SM, Waxman SH, Giblett E: An XX/XY human hermaphrodite resulting from double fertilization. Proc Natl Acad Sci USA 48:332, 1962.
37. de Grouchy J, Moullec J, Salmon C, Josso N, Frezal J, Lamy M: Hermaphroditisme avec caryotype XX/XY. Etude genetique d'un cas. Ann Genet (Paris) 7:25, 1964.
38. Brøgger A, Aagneaes O: The human Y chromosome and the etiology of true hermaphroditism, with the report of a case with XX/XY sex chromosome mosaicism. Hereditas 53:231, 1965.
39. Josso N, de Grouchy J, Auvert J, Nezelof C, Jayle MF, Moullec J, de Casaubon A, Lamy M: True hermaphroditism with XX/XY mosaicism, probably due to double fertilization of the ovum. J Clin Endocrinol Metab 25:114, 1965.
40. Massimo L, Vianelli MG: Personal comminication cited in Park IJ, Jones HW Jr, Bias WB: True hermaphroditism with 46,XX/46,XY chromosome complement. Obstet Gynecol 36:377, 1970.
41. Montero E, Salvatierra V, Guirao M, Sastre M: Analisis citogenetico de un hermafrodita verdadero alternante. Ann Desarrollo 13:277, 1965.
42. Segni G, Grossi-Bianchi ML: Un raro caso di ermafroditismo vero di tipo alterno can cromosomi XX/XY. Minerva Pediatr 17:630, 1965.
43. Josso N, Frezal J, Pellerin D, Cukier J, Nezelof C, de Grouchy J, Lamy M: Hypospadias et cryptorchidie: Expression possible d'une intersexualite. Presse Med 74:1999, 1966.
44. Corey MJ, Miller JR, MacLean JR, Chown B: A case of XX/XY mosaicism. Am J Hum Genet 19:378, 1967.
45. Deminatti M, Maillard E: Etudé d'un cas d'hermaphrodisme humain a caryotype 46,XY/46,XX. CR Acad Sci 265:365, 1967.
46. Okada M: Studies on sex-chromatin and chromosome abnormalities. (In Japanese). Presented at 16th Annual Meeting North Japan Obstetrical Gynecological Society, 1968. Cited in Makimo S:"Human Chromosomes." Amsterdam: North-Holland, 1975.
47. Makimo S: "Human Chromosomes." Amsterdam: North-Holland, 1975.
48. Saint-Aubert P, Millard E, Wallbaun R, Dellmar Marsalet Y, Deminatti M, Fontaine G: L'etude chromosomique de l'hermaphrodisme human. A propos d'un cas a caryotype 46,XY-46,XX. Ann Pédiatr 44:9, 1978.
49. Fitzgerald PH, Brehaut LA, Shannon FT, Angus HB: Evidence of XX/XY sex chromosome mosaicism in a child with true hermaphroditism. J Med Genet 7:383, 1970.
50. Grace HJ: An unusual case of haematuria in an XX/XY hermaphrodite. S Afr Med J 44:40, 1970.
51. Jackson WPU: Variability of the human XX/XY mosaic. S Afr Med J 44:208, 1970.
52. Park IJ, Jones HW Jr, Bias WB: True hermaphroditism with 46,XX/46,XY chromosome complement. Obstet Gynecol 36:377, 1970.
53. Kakati S, Sharma T, Udupa KN, Chaudhuri SPR: A true hermaphrodite with XX/XY mosacisim. Indian J Med Res 59:104, 1971.

54. Bernirschke K, Naftolin F, Gittes R, Khudr G, Yen SSC, Allen FH Jr: True hermaphroditism and chimerism. Am J Obstet Gynecol 113:449, 1972.
55. Carbonara AO, Mattiuz PL, Migiano VC, Molinatti G, Richiardi P, Ceppellini R: Clinical-endocrinological and immunogenetical study of a case of human XX/XY Atti Assoc Genet Ital 17:46, 1972.
56. Grace HJ, Edge WEB: A white hermaphrodite in South Africa. S Afr Med J 47:1553, 1973.
57. Shanfield I, Young RB, Hume DM: True hermaphroditism with XX/XY mosaicism: Report of a case. J Pediatr 83:471, 1973.
58. de la Chapelle A, Schröder J, Rantanen P, Thomasson B, Neimi M, Tiilikainen A, Sanger R, Robson EB: Early fusion of two human embryos?Am J Hum Genet 38:63, 1974.
59. De Marchi M, Carbonara AO, Carozzi F, Massara F, Belforte L, Molinatti GM, Bisbocci D, Passarino MP, Palestro G: True hermaphroditism with XX/XY sex chromosome mosaicism: Report of a case. Clin Genet 10:265, 1976.
60. Zuelzer WW, Beattie KM, Reisman LE: Generalized unbalanced mosaicism attributable to dispermy and probable fertilization of a polar body. Am J Hum Genet 16:38, 1964.
61. Forteza G, Bonilla F, Baguerra R, Monmeneu S, Galbis M, Zaragoza V: Un caso de mosaicismo XY/XX chromatin negativo, con disgenesia gonadal y sexo fenotipico femenino. Rev Clin Esp 24:394, 1963.
62. Bain AD, Scott JS: Mixed gonadal dysgenesis with XX/XY mosaicism. The evidence for the occurrence of fertilization by two spermatozoa in man. Lancet i:1035, 1965.
63. Myhre BA, Meyer T, Opitz JM, Race RR, Sanger R, Greenwalt TJ: Two populations of erythrocytes associated with XX/XY mosaicism. Transfusion 5:501, 1965.
64. Klinger R, Carbonara AO, Miggiano VC, Sanger R: Studio dei gruppi sanguini su di un caso di chimerismo XX/XY umano. Atti Assoc Genet Ital 15:100, 1970.
65. Mystkowska ET, Tarkowski AK: Observations on CBA-p/CBA-T_6T_6 mouse chimeras. J Embryol Exp Morphol 20:33, 1968.
66. Whitten WK: Chromosomal basis for hermaphroditism in mice. Dev Biol Reprod 5:189, 1975.
67. Bernirschke K: Chimerism, mosaicism, and hybrids. In de Grouchy J, Ebling FJG, Henderson IW (eds): "Human Genetics." ICS 250. Amsterdam: Excerpta Medica, 1972, p 212.
68. Jost A: Gonadal hormones in the sex differentiation of the mammalian fetus. In De Hann RL, Urspring H (eds): "Organogenesis." New York: Holt, Rinehart, and Winston, 1965, p 611.
69. Turpin R, Lejeune J, Greton A: Hermaphroditism. C R Acad Sci (Paris) 225:3088, 1962.
70. Botélla-Llusia J: Sobre un neuvo el caso de hermafroditismo verdade ro. An R Acad Nacl Med 127:261, 1960.
71. Botélla-Llusia J: Discussion generale des rapports de cytogenetique. J Pathol (Paris) 11:1185, 1963.
72. Clavero J-A, Nogales F, Moncada E, Sopena A: Sur un cas d'hermaphrodisme vrai avec constitution mosaique XX/XXY. Ann Endocrinol 26:77, 1965.
73. McMahon RA: Hermaphroditism in infancy and childhood. Surgery 59:290, 1966.
74. Knorr D, Gey W, Fendel H, Devens K, Meister P: Echter Hermaphroditismus mit chromosomen Mosaik 46,XX/47,XXY. Munch Med Wochenschr 110:522, 1968.
75. Butler LJ, Snodgrass GJAI, France NE, Russell A, Swain VAJ: True hermaphroditism or gonadal intersexuality. Arch Dis Child 44:666, 1969.
76. Denes J, Gorácz G, Léb J, Csabay L, Lehoczky I: Hermaphroditismus verus. Zbl Gynäk 94:585, 1969.
77. Sarto GE, Opitz JM, Inhorn SL: Considerations of sex chromosome abnormalities in man. In Benirschke K (ed): "Comparative Mammalian Cytogenetics." New York: Springer-Verlag, 1969, p 390.

78. Thorburn MJ, Carpenter RA, Williams SI: Clinical and cytogenetic observations on four true hermaphrodites. West Indian Med J 18:5, 1969.
79. Sandrini Neto R, Rivarola MA, Coco R, Bergada C: The testes in patients with abnormalities of sex differentiation. Histology and endocrine function. Acta Endocrinol (Kbh) 73:179, 1973.
80. Pflueger OH Jr, Schmidt DM, Harrod CS: True hermaphroditism with unambiguous male phenotype, pathologic and cytogenetic findings: Report of a case. J Urol 112:406, 1974.
81. Duhamel B: Hermaphrodisme vrai alternant. Arch Obstet Ginecol 69:746, 1962.
82. Court Brown WM, Harnden DG, Jacobs PA, MacLean N, Mantle DJ: "Abnormalities of the Sex Chromosome Complement in Man." Med Res Counc No 305. London: Her Majesty's Stationery Office, 1964.
83. Sato S, Takano S, Anezake M, Saito R, Sakurai S, Takayasu: A case of true hermaphroditism. Jpn J Urol 55:407, 1964.
84. Tazaki H, Ikeda N, Omori S: True hermaphrodites in Japan. Report of a case and review of the literature. Keio J Med 13:143, 1964.
85. Capanna E, Civitelli MV, De Martino C: Early data in chromosomal constitution in human sexual abnormalities. Mamm Chrom Newsl 13:4, 1966.
86. Michon J, Fonder A, Carolus JM: A propos de deux cas d'hermaphrodisme vrai: Considerations generales et techniques. Ann Chir Plast 10:266, 1965.
87. Takata M, Tanaka H, Niijima T: Five cases of various hermaphroditism. Jpn J Fertil Steril 15:50, 1970.
88. Ponte G, DuPont A, Nuyts JP, Saint-Aubert P, Debruxelles P, Bombard E: Hermaphrodisme vrai alterne mosaique XO/XY. Pediatrie 22:831, 1967.
89. Kadotoni T: Cytogenetic studies on human pregnancy wastage and malformations. (Japanese) In 22nd Ann Cong Jpn Obstet Gynecol, 1970, p 1. Cited in Makimo S: "Human Chromosomes." Amsterdam: North-Holland, 1975.
90. Gerli M, Biagioni M, Bruschelli GM, Ferrares R, Rosi G: A case of true hermaphroditism with 45,X/46,XY mosaicism. Hum Genet 35:43, 1976.
91. Buyse M: 46,XX/47,XYY mosaicism in a true hermaphrodite. Lancet i:1300, 1975.
92. Chaudhuri A, Chatterjee SK, Sarkar S: Presumptive 46,XX/46,XY/47,XXY mosaicism in a hermaphrodite. J Med Genet 8:117, 1971.
93. Fraccaro M, Lindsten J, Klinger HP, Tiepolo L, Bergstrand CG, Herrlin KM, Livadites H, Pehrson M, Tillinger KG: Cytogenetical and clinical investigations in four subjects with anomalies of sexual development. Ann Hum Genet 29:281, 1966.
94. Thompson IM: Results of vaginal flap urethroplasty in hermaphrodites. J Urol 105:456, 1971.
95. Blank CE, Zachary RB, Bishop AM, Emery JL, Dewhurst CJ, Bond JH: Chromosome mosaicism in a hermaphrodite. Br Med J 2:90, 1964.
96. Sandberg AA, Koepf GF, Crosswhite LH, Hauschka TS: The chromosome constitution of human marrow in various developmental and blood disorders. Am J Hum Genet 12:231, 1960.
97. Domenci A, Romeo P: I chromosomi in un casodi ermafroditismo vero. Arch Ostet Ginecol 69:746, 1962.
98. Eliachar E, Duhaniel B, Tassy R, Benzaquen I: Un cas exceptionnel d'hermaphrodisme vrai. Ann Pédiatr 9:510, 1962.
99. Lonsdale D, Mercer RD, McCullogh EP: True hermaphroditism: Report of a case with chromosomal analysis. Cleve Clin Q 30:207, 1962.
100. Tonomura A, Honda T: Sex-chromosome conditions in four cases of true hermaphroditism. Ann Rep Inst Genet Jpn 13:1051, 1962.

101. Aisters-Bauer VL, Kleinhenz RJ: Chromosomes in 15 cases of congenital anomalies. Hum Chrom Newsl 10:5, 1963.
102. Komase M: Five cases of true hermaphroditism. (In Japanese.) Jpn J Urol 54:456, 1963. Cited by Tazaki et al [84].
103. Kumamoto E: A case of true hermaphroditism with an XY chromosome. (In Japanese.) Jpn J Urol 54:569, 1963. Cited by Tazaki et al [84].
104. Sakatoku J, Hongo H, Ebisuta K: Two cases of true hermaphroditism. (In Japanese.) Acta Urol Jpn 10:33, 1964. Cited by Tazaki et al [84].
105. Manchanda DP, Gard SC, Khare SB: True hermaphrodite with positive sex. J Indian Med Assoc 43:239, 1964.
106. Matsunaga T: A case of true hermaphroditism. (In Japanese.) Jpn J Urol 54:406, 1964. Cited by Tazaki et al [84].
107. Shearman RP, Singh S, Lee CWG, Hudson B, Ilberg PLT: Clinical, hormonal and cytogenetic findings in a true hermaphrodite. J Obstet Gynecol BR Cwlth 71:627, 1964.
108. Hung W, Jacobson CB, Wigger HJ, Randolph JG: Chromosome studies in a chromatin-negative, XY true hermaphrodite. J Urol 96:565, 1966.
109. Barriero E, Peralto A, Monero J, Usandizaga JA, Coutreras F, Swartz A: Two new cases of true hermaphroditism. J Med Genet 6:426, 1969.
110. Duhamel B: Hermaphrodisme vrai alternant (A propos de la communication de M. Nardi). Chirurgie 96:868, 1970.
111. Barinka L: The management of a case of true hermaphroditism with negative sex chromatin and an idiogram 46/XY. Br J Plast Surg 25:427, 1972.
112. Park IJ, Pyeatte JC, Jones HW Jr, Woodruff JD: Gonadoblastoma in a true hermaphrodite with 46,XY genotype. Obstet Gynecol 40:466, 1972.
113. Singh DN, Hara S, McArudor CE, Ashurst JT, Davis K, Crump EP: Cytogenetic studies of three true hermaphrodites. J Natl Med Assoc 68:496, 1976.
114. Szokol M, Kondrai G, Papp Z: Gonadal malignancy and 46,XY karyotype in a true hermaphrodite. Obstet Gynecol 49:358, 1977.
115. Polano O: Ueber wahre Zwitterbildung beim Menschen. Z Geburtsh Gynaekol 83:114, 1921.
116. Essenberg JM, Feinberg IM: A case of true hermaphroditism in man. West J Surg 45:474, 1937.
117. Weed JC, Segaloff A, Wiener WB, Doublas JW: True hermaphroditism. Endocrine studies in a case of ovotestis. J Clin Endocrinol Metab 7:741, 1947.
118. Bani U: Addome acuto de crisiti ovarica torta sul peduncolo in ermafrodite vero. Policlinico (Prat) 63:1238, 1956.
119. Stirling WC: Abnormal sex development. Report of a unique case of true ambisexuality. Med Ann DC 28:559, 1959.
120. Vaughn J, Gonzalez-Angulo A: True hermaphrodite with ovarian tumor. J Urol 86:776, 1961.
121. Simpson JL, Photopulos G: The relationship of neoplasia to disorders of abnormal sexual differentiation. In Bergsma D, Schimke RN, Summitt RL, Harris DJ (eds): "Cancer and Genetics." New York: Alan R Liss for The National Foundation–March of Dimes, BD:OAS XII(1):15, 1976.
122. Moriarty JD: True hermaphroditism. Report of a case with mammary carcinoma. Am J Pathol 20:799, 1961.
123. Milner WA, Garlick WB, Fink AJ, Stein AA: True hermaphrodite siblings. J Urol 79:1003, 1958.

124. Clayton GW, Smith JD, Rosenberg HS: Familial true hermaphroditism in pre- and postpubertal genetic females. Hormonal and morphologic studies. J Clin Endocrinol Metab 18:1349, 1958.
125. Rosenberg HS, Clayton GW, Hsu TC: Familial true hermaphroditism. J Clin Endocrinol Metab 23:203, 1963.
126. Mori Y, Mitzutani S: Familial true hermaphroditism in genetic females. Jpn J Urol 59:857, 1968.
127. Armendares S, Salamanca F, Cantu JM, del Castillo V, Nava S, Dominquez de la Piedra E, Cortes-Gallegos V, Gallegos A, Cervantes C, Parra A: Familial true hermaphroditism in three siblings. Humangenetik 29:99, 1975.
128. Gallegos AJ, Guizar E, Armendares S, Cortes-Gallegos V, Cervantes C, Bedolla N, Parra S: Familial true hermaphroditism in three siblings: Plasma hormonal profiles and in vitro steroid biosynthesis in gonadal structures. J Clin Endocrinol Metab 42:653, 1976.
129. Marshall VF Jr, and staff: True hermaphroditism: A conference on the management of one patient. J Urol 72:77, 1954.
130. Armstrong CN: Diversities of sex. Br Med J 1:1173, 1955.
131. Ferguson-Smith MA, Johnston AW, Weinberg AN: The chromosome complement in true hermaphroditism. Lancet ii:126, 1960.
132. De Assis LM, Epps DR, Battura C: Chromosomal constitution and nuclear sex of a true hermaphrodite. Lancet ii:129, 1960.
133. Gordon RR, O'Gorman FJP, Dewhurst CJ, Blank CE: Chromosome count in a hermaphrodite with some features of Klinefelter's syndrome. Lancet ii:736, 1960.
134. Sasaki M, Makimo S: The chromosomal constitution of a human hermaphrodite. Tex Rep Biol Med 18:493, 1960.
135. Civantos F: Human chromosomal abnormalities. Bull Tulane Univ Fac 20:241, 1961.
136. German JL, Bearn AG, McGovern JH: Chromosomal studies of three hermaphrodites. Am J Med 33:83, 1962.
137. McGovern JH, Marshall VF: Three cases of true hermaphroditism. J Urol 88:680, 1962.
138. Bregman RU, Gregman ET, Cushner GB, Woods FM: Chromosome analysis, nuclear sex, clinical and endocrine studies of a true hermaphrodite. J Urol 89:475, 1963.
139. Crossfield JH: True hermaphroditism: Including a case with chromosome analysis. J Urol 88:674, 1962.
140. Klempman S, Wilton E: True hermaphroditism: Chromosomal analysis. S Afr Med J 37:1096, 1963.
141. Merill JA, Ramsey JE: True hermaphroditism. Report of a case and review of the literature. Obstet Gynecol 22:505, 1963.
142. Solomon IL, Green OC: The chromosome count in a true hermaphrodite. J Pediatr 63:333, 1963.
143. Di Lorenzo M, Ciampalini L, Lotti P: L'ermafroditismo vero. Gazz Int Med Chir 69:3118, 1964.
144. Pena J, Rivas D: Un caso de hermafroditismo verdadero unilateral con cariotipo XX. El papel de los cromosomas sexuales en la determinacion del sexo. Rev Esp Pediatr 20:325, 1964.
145. Root AW, Eberlein WR, Briebart S, Moorhead PS, Mellman WJ: Chromosomal analysis of multiple tissues from a true hermaphrodite. J Clin Endocrinol Metab 24:467, 1964.
146. Baldachin BJ, White A: Mpilo Hospital Staff Clinical Meeting. Cent Afr J Med 9:276, 1965.
147. Friedberg SH, Rosenberg EE: 46,XX chromosome constitution of a true hermaphrodite. S Afr Med J 39:327, 1965.

148. Hungerford DA, Donnelly AJ, Nowell PC: The chromosome constitution of a human phenotypic intersex: Reconfirmation of a 46-chromosome, XX, apparently non-mosaic "true hermaphrodite." Hereditas 52:379, 1965.
149. Jones HW Jr, Ferguson-Smith MA, Heller RH: Pathologic and cytogenetic findings in true hermaphroditism: Report of six cases and review of 23 cases from the literature. Obstet Gynecol 25:435, 1965.
150. Clavero Nunez JA, Llusia JB, Ortiz FN, Bonachera EM, Ruiz LM, Tornero MC, Sopena A: Un caso de hermaphroditismo verdadero con caritipo XX. Acta Ginecol (Madr) 17:331, 1966.
151. Ferguson-Smith MA: X-Y chromosomal interchange in the aetialogy of true hermaphroditism and of XX Klinefelter's syndrome. Lancet ii:475, 1966.
152. Fraser K: Hermaphroditus verus, with report of a case. Med J Aust 1:1003, 1966.
153. O'Mahoney MP: A true hermaphrodite with a sex chromatin anomaly. J Ir Med 59:151, 1966.
154. Mills AE, Tunbridge WMG: Intersex in an African child. Cent Afr J Med 13:51, 1967.
155. Salvatierra O Jr, Skaist LB, Morrow JW: True hermaphroditism discovered 10 years after hypospadias repair: Report of 2 cases. J Urol 98:111, 1967.
156. Kondo K, Kanagawa S, Akazawa Y: A case of an XX hermaphrodite. Tokushima J Exp Med 15:103, 1968.
157. McDaniel EC, Nadel M, Woolverton WC: True hermaphrodite with bilaterally descended ovotestes. J Urol 100:77, 1968.
158. Hefelfinger DC, Tyson KRT, Assor D, Bryan GT: True hermaphroditism unassociated with a common urogenital sinus. Pediatrics 43:896, 1969.
159. Koontz WW Jr, Young RB, Tucker HSt G, Prout GR Jr: True hermaphroditism: A report of 3 cases. J Urol 101:102, 1969.
160. Sasaki K, Takeuchi M, Ichijo S, Matsushita S, Shirai M: A case of true hermaphrodite with an XX sex chromosome constitution. Jpn J Clin Urol 23:311, 1969.
161. Tourneur R, Boudon P, Benhamou, Dairou F: Hermaphrodisme vrai alternant avec caryotype XX. A propos d'une nouvelle observation. Ann Med Interne 121:1073, 1970.
162. Dewhurst CJ, Haines RM, Lucas M, Delhanty J: 46,XX karyotype in four tissues of a true hermaphrodite. J Obstet Gynecol Br Cwlth 78:1084, 1971.
163. Levine LS, New MI: Preoperative detection of hidden testes. Am J Dis Child 121: 176, 1971.
164. Dougherty CM, Spencer R: "Female Sex Anomalies." New York: Harper & Row, 1972.
165. Tegenkamp TR, Tegenkamp I: True hermaphroditism: Cytogenetic analysis, surgical repair and social implications. J Med Genet 9:115, 1972.
166. Cheah JS: A case of true hermaphroditism with impaired growth and normal growth hormone levels in a Chinese. Med J Aust 2:28, 1973.
167. Diop A, Niang I, Fall M, Dioug B, Mensah A: A propos d'un nouveau cas d'hermaphrodisme atternant. Bull Soc Med Afr Noire Lang Fr 15:172, 1973.
168. Kasdan R, Nankin HR, Troen P, Wald N, Pan S, Yanaihara T: Paternal transmission of maleness in XX human beings. N Engl J Med 288:539, 1973.
169. Cremin BJ: Intersex states in young children: The importance of radiology in making a correct diagnosis. Clin Radiol 25:63, 1974.
170. Johnson RH: A case of true hermaphroditism presenting as an undescended testicle. S Afr Med J 48:1540, 1974.
171. Roux JP, Kowen D, Retief PJM: The management of pediatric hermaphroditism. S Afr Med J 48:2088, 1974.

172. Roth LM, Cleary RE, Hokum WL: Ultrastructure of an ovotestis in a case of true hermaphroditism. Obstet Gynecol 48:619, 1975.
173. Severi F, Bianchi E, Colombo A, Magrini U, Pasquall F, Scappaticci E: Ermafroditismo vero varieta laterale. Minerva Pediatr 27:1738, 1975.
174. Narita O, Manba S, Nakanishi T, Ishizuka N: Pregnancy and childbirth in a true hermaphrodite. Obstet Gynecol 45:593, 1975.
175. Byrd JR, Tho PT, Otken L: Gonadoblastoma in a true hermaphrodite with a 46,XX karyotype. Obstet Gynecol 47:355, 1976.
176. Saenger P, Levine LS, Wachtel SS, Korth-Schutz S, Doberne Y, Koo GC, Lavengood RW Jr, German JL III, New MI: Presence of H-Y antigen and testis in 46,XX true hermaphroditism, evidence for Y-chromosomal function. J Clin Endocrinol Metab 43:1234, 1976.
177. Wachtel SS, Koo GC, Breg WR, Thaler HT, Dillard CM, Rosenthal IM, Dosik H, Gerald PS, Saenger P, New M, Lieber E, Miller OJ: Serologic detection of a Y-linked gene in XX males and XX true hermaphrodites. N Engl J Med 295:750, 1976.
178. Berger R, Abonyi D, Nodot A, Vialatte J, Lejeune J: Hermaphrodisme vrai et "Garcon XX" dans une fratrie. Rev Eur Etud Clin Biol 15:330, 1970.
179. Cattanach BM, Pollard CE, Hawkes SG: Sex-reversed mice: XX and XO males. Cytogenet Cell genet 10:318, 1971.
180. Bennett D, Matheson BJ, Scheid M, Yanigisawo K, Boyse EA, Wachtel S, Cattanach BM: Serological evidence for H-Y antigen in Sxr, XX sex-reversed phenotypic males. Nature 265:255, 1977.
181. Soller M, Padek B, Wysoki M, Ayalon N: Cytogenetics of Saanen goats showing abnormal development of the reproductive tract associated with the dominant gene for polledness. Cytogenet Cell Genet 8:51, 1969.
182. Hamerton JL, Dickson JM, Pollard CE, Grieves SA, Short RV: Genetic intersexuality in goats. J Reprod Fertil 7:Suppl 25, 1969.
183. Gerneke WH: Cytogenetic investigations on normal and malformed animals, with special reference to intersexes. Onderstepoort J Vet Res 34:219, 1967.
184. Sturtevant AH: A gene in Drosophila melanogaster that transforms females into males. Genetics 30:297, 1945.
185. Lebedeff GA: A study of intersexuality in Drosphila virilis. Genetics 30:297, 1945.
186. Wachtel SS: H-Y antigen. This volume pp 1–8.
187. Jones HW Jr: Discussion following presentation by Van Niekerk [14]. Am J Obstet Gynecol 126:905, 1976.
188. de la Chapelle A, Schröder J, Murros J, Tallquist G: Two XX males in one family and additional bearing on the etiology of XX males. Clin Genet 11:91, 1977.

Abnormalities of the Müllerian and Wolffian Duct Systems

Gloria E. Sarto, MD, PhD, and Joe Leigh Simpson, MD

Although encountered not infrequently by gynecologists and urologists, anomalies limited to müllerian or wolffian derivatives have received relatively little attention from geneticists. Despite a paucity of formal investigations, heritable tendencies seem apparent in several abnormalities. In this review we shall delineate selected disorders and comment upon certain clinical and genetic aspects. Not considered are those disorders of sex differentiation in which there are abnormalities of either gonads or external genitalia, in addition to abnormalities of the internal ductal system [1]. Prior to discussion of these disorders, a review of a few salient features of reproductive embryology is in order.

EMBRYOLOGY OF MÜLLERIAN AND WOLFFIAN DUCTS

Like other organ systems in the genital tract, the genital ducts pass through an indifferent stage during which both male and female ductal structures coexist. The wolffian (mesonephric) ducts originate as pronephric ducts from ectodermal buds dorsolateral to the pronephric blastema in 2 to 3 mm embryos [25–30 days gestational age (GA)]. The wolffian ducts initially grow caudally as solid cellular structures, but later gradually develop lumina. Luminization occurs at about the same time that primordial germ cells migrate from the hindgut to the genital ridges. By contrast, the müllerian (paramesonephric) ducts are first detectable in older embryos, 13–15 mm long (36–42 days GA). The subsequent development of the müllerian ducts depends upon that of the wolffian ducts, as verified by observations that experimental interruption of the wolffian ducts

Birth Defects: Original Article Series, Volume XIV, Number 6C, pages 37–55

[2] results in failure of müllerian development. After the anterior lateral columnar epithelium of the wolffian duct thickens and grows laterally from the wolffian ducts to form müllerian ducts, lumina are gradually formed in müllerian ducts.

At the 10 mm stage (30–36 days GA) in the male embryo, the rete cords of the developing testes fuse with the mesonephric tubules to form ductuli efferentia. The cranial portion of the wolffian duct and the ductuli efferentia form the epididymis. The distal portion of the wolffian duct, which will become the ductus deferens, terminates at the müllerian tubercle in the posterior wall of the urogenital sinus and becomes the prostatic urethra. In male embryos, müllerian duct development ends at the 10 mm stage, evidenced only by vestiges at either end as the appendix testis or prostatic utricle, respectively. The terminal wolffian ducts become the adult ejaculatory ducts. Seminal vesicles appear as lateral outpouchings of the wolffian ducts at about the 13th week. Finally, the prostate gland is thought to have both an endodermal and mesodermal origin. The prostate develops during the 12th week. Its lateral and posterior lobes are formed as epithelial buds from the wall of the urethra; thus , these lobes are endodermal in origin. The median lobes are formed from the epithelium of the colliculus, which contains müllerian and wolffian epithelium and is thus of mesodermal origin.

The female genital ducts develop later than the male genital ducts. The paired müllerian ducts, having become luminized at 13–15 mm (36–42 GA), lie lateral to the wolffian ducts until they reach the caudal end of the mesonephros, at which site they cross the wolffian ducts ventrally to contact each other. By 30 mm (56–63 days GA) the müllerian ducts reach the urogenital sinus and invaginate into its wall to form the müllerian tubercle. By 48 mm (120 days GA) the paired ducts have fused completely into a single tube. As the fused tips of the müllerian ducts reach the urogenital sinus, concomitant elongation of the pelvic regions directs the urogenital sinus away from the point of the fused ducts. This results in a solid epithelial cord between the end of the tubular portion of the müllerian duct and the wall of the urogenital sinus. The solid vaginal cord enlarges as result of 1) proliferation and stratification of epithelial cells and 2) invasion of surrounding mesenchyme. By the 150 mm stage (19th–20th week), lacunae appear in the center of the cord; by 212 mm (28th week) the vagina is hollow. In the female, the wolffian ducts regress by 56 mm, persisting only as epoophoron, ductuli efferntia, and paroophoron.

The origin of the vaginal epithelium is uncertain. Postulated theories include 1) müllerian origin, 2) wolffian duct origin, 3) urogenital sinus origin, or 4) combinations thereof. The most popular theory is that the upper portion of the vagina is of müllerian origin, whereas the caudal portion is of urogenital sinus origin. However, experiments [3] employing tritiated thymidine in developing mouse vaginas showed that only the caudal portion is derived from the urogenital sinus. In humans, the presence of a normal caudal portion juxtaposed with complete atresia of the cranial portion of the vagina supports the theory that the vaginal epithelium may be derived from two sources: the müllerian tubercle and the urogenital sinus.

For more comprehensive reviews of the embryology of the genital ducts, see Gray and Skandalakis [4] and Jirásek [5, 6].

ANOMALIES OF THE FEMALE REPRODUCTIVE TRACT

Müllerian Aplasia

Congenital absence of the vagina. Congenital absence of the vagina is a term often applied without qualification to patients with total or partial absence of the vagina, with or without a functioning uterus [7–10]. In some instances, reported series have included persons with a 46,XY complement, ie the complete androgen insensitivity syndrome [10, 11]. In this review, congenital absence of the vagina will refer to 46,XX individuals whose external genitalia are otherwise normal. Two distinct entities should be distinguished from one another: 1) individuals with a blindly ending vagina and an absent or rudimentary nonfunctioning uterus (müllerian aplasia), and 2) individuals with a blindly ending vagina and a normal (functioning) uterus (vaginal atresia) (Fig. 1). Most individuals with absence of the vagina have müllerian aplasia, but perhaps 10% have vaginal atresia [7, 9–15]. The eponym Rokitansky-Kuster-Hauser (RKH) syndrome is

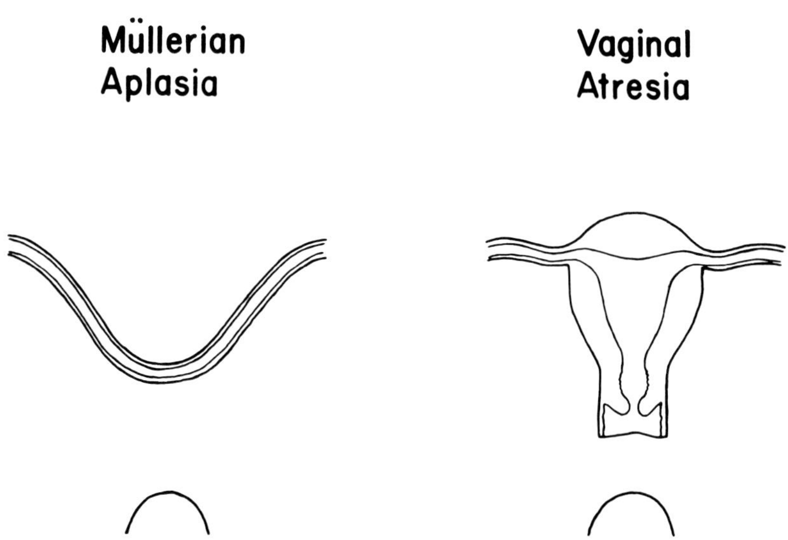

Fig. 1. Diagram of 2 distinct entities associated with congenital absence of the vagina.

commonly employed in the European literature to describe absence of most of the vagina in association with a rudimentary uterus characterized by bilateral noncanalized muscular buds. Müllerian aplasia and RKH syndrome are probably the same entity.

Embryology. Complete absence of the uterus is rare and is always associated with absence of the vagina. However, the uterus may be hypoplastic and represented by a small, solid mass; fallopian tubes are frequently well developed. Vaginal development occurs only if one or both müllerian ducts reach the endodermal epithelium of the urogenital sinus. Development of the uterus and the vagina do not depend on the ovary.

Clinical aspects. Individuals with müllerian aplasia often remain undiagnosed until puberty, at which time they consult a physician because of primary amenorrhea. For example, 11 of 50 in one series of women with primary amenorrhea had a developmental abnormality of the vagina [16]. In all series between one-third and one-fourth of the women ascertained because of primary amenorrhea have müllerian aplasia [1, 16]. Secondary sex characterics develop normally because the ovaries are normal and, hence, sex steroid levels are normal. The external genitalia are normal for females, but the vagina ends blindly at various depths. A definitive uterus is not palpable; however, a rudimentary uterus may persist as a muscle bundle. The fallopian tubes are usually but not invariably normal.

Urologic anomalies are associated with müllerian aplasia more frequently than one would expect by chance. Renal anomalies are present in about 40% of cases [1, 17]. The most common anomalies are unilateral renal agenesis and ectopia of the kidney and pelvic kidneys.

Vertebral anomalies occur in approximately 10% of the cases [1, 17], and the prevalence of cardiac anomalies may also be slightly increased [17]. Many vertebral anomalies occur at the cervical level. For example, 4 of 96 patients examined by Bryan et al [7] had skeletal anomalies: hemivertebra, cervical spina bifida, spina bifida occulta, or congenitally dislocated hip and malformed ribs. Other reported anomalies include scoliosis [10], pectus excavatum, and Klippel-Feil anomaly (fusion of the cervical vertebra). Skeletal anomalies involving the limbs, usually syndactyly, occur only rarely. Pinsky [18] noted that axial skeletal anomalies are characteristic of nonfamilial müllerian aplasia, but not of familial syndromes.

The differential diagnosis includes vaginal atresia and the complete androgen insensitivity syndrome (AIS). Individuals with AIS and müllerian aplasia have a female phenotype, a blindly ending vagina and no uterus; however, in AIS there is a lack of pubic and axillary hair growth. The diagnosis can be confirmed by chromosome studies or gonadal biopsy.

Genetic considerations. Cytogenetic studies of individuals with müllerian aplasia usually show a normal 46,XX karyotype [11, 19]; however, chromosome abnormalities of unknown significance have been detected [16]. Affected sibs have been reported. For example, Anger et al [20] reported 3 affected sibs. Each had normal external genitalia, except the urethral meatus was displaced posteriorly toward the posterior fourchette. In 2 of the 3 absence of the uterus was confirmed by laparoscopy. Ovarian biopsies, leukocyte cultures, and hormonal evaluations were normal in the 2 sisters studied. Jones and Mermut [21] likewise described two 46,XX sibs, aged 20 and 21 years old. Both had normal secondary sex characteristics, but neither had a uterus. One had a 15-cm, soft, ovarian cyst; otherwise, both had normal ovarian tissue. There was no evidence of the disorder in any other family members. Griffin et al [17] also reported 2 affected sibs; one had ptosis of the right kidney and scoliosis, whereas the other had hydronephrosis. The detection of müllerian aplasia in multiple affected sibs [17, 20, 21] suggests that müllerian aplasia may result from a rare autosomal recessive gene. On the other hand, Lischke et al [22] reported 3 patients, each of whom had a normal twin. The patients had the following features: 1) normal external genitalia, including the location of the urethra; 2) absence of the vagina; and 3) anomalies limited to the renal and skeletal systems. The report of these 3 patients, each of whom had a normal monozygotic twin [22] suggests that an exclusively genetic etiology is unlikely. Finally, Shokeir's [23] observations suggesting sex-limited autosomal dominant factors are described elsewhere in this volume.

Vaginal Atresia

Absence of the vagina with concomitant normal uterine development is rare. Vaginal atresia accounts for only 5–10% of such cases of "congenital absence of the vagina." However, the etiology is different from müllerian aplasia.

Embryology. After the müllerian ducts reach the urogenital sinus, elongation of the pelvic region displaces the urogenital sinus caudally, hence connecting a solid epithelial cord between the end of the tubular portion of the müllerian ducts and the wall of the sinus. This is the "solid stage" of the vagina. Initially, lacunae appear in the center of the cord, and by the seventh month the vagina is a hollow organ. Although the source of the vaginal epithelium is still not entirely clear, many believe that the caudal portion of the vaginal epithelium is of urogenital sinus origin, whereas the cranial portion is of müllerian origin. Failure of development of the urogenital sinus may thus result in failure of development of the lower one-fifth to one-third of the vagina, which is replaced by fibrous tissue. The uterus, cervix and upper vagina are normal.

Occasionally one may detect a small, 2 to 3 cm vaginal pouch, superior to which lies fibrous tissue. Above this is a cervix with a very minimal amount of vagina around it. The etiology may be failure of canalization of the solid vaginal cord, believed to be composed of squamous epithelium.

Clinical aspects. Individuals with vaginal atresia of either type have a uterus and consequently usually seek medical attention at puberty because of cyclic cramping abdominal pain. Menstrual blood and fluid accumulates in the small, upper vagina, producing hematocolpos and possibly hematometra. Affected individuals have normal secondary sex development, axillary and pubic hair. Their ovaries are likewise normally developed, as are the fallopian tubes. The size of the endometrial cavity, and consequently the functional capacity of the uterus, vary from normal to nonfunctioning. Urologic anomalies appear less common than in müllerian aplasia.

Genetic considerations. Vaginal atresia is most commonly reported as part of a large series of patients with "congenital absence of the vagina." Frequently no mention has been made of family history, consanguinity, or affected sibs. Familial aggregates have not definitely been described.

Atresia of the Uterine Cervix

Congenital absence of the cervical canal associated with a normal uterine fundus and a normal vagina is rare. Only about 20 cases have been reported [24–26]. The anomaly can occur without any aberration in development of the cranial or caudal pelvic organs. The vagina is usually normal, despite no demonstrable cervix; however, in some cases a rudimentary cervix without a patent os is present. In one case, uterus didelphys was present [26]. Atresia of the cervix probably results from aplasia or failure of canalization of the distal segments of the müllerian ducts.

Individuals with this anomaly have primary amenorrhea and frequently cyclic lower abdominal pain. Otherwise, they usually develop as normal females, although endometriosis is not uncommon.

Transverse Vaginal Septa

Transverse vaginal septa, one of the most common vaginal anomalies, may be complete or may have a small opening located either centrally [27] or eccentrically [28]. Septa may occur at several locations in the vagina. Lodi [29] reported 42 patients with transverse vaginal septa: 46% involved the upper third of the vagina, 40% the middle third, and 14% the lower third.

Embryology. The embryonic origin of this abnormality is not easily explained, a fact which is not surprising inasmuch as the embryology of the vagina has not been completely elucidated. Some believe that the müllerian and urogenital sinus components of the vagina fail to fuse, whereas others suggest incomplete degeneration of the previously solid vaginal cord. The theory that the vaginal epithelium arises from 2 sources, ie the upper portion derived from the fused müllerian ducts and the lower portion from the urogenital sinus, is believed by some to be supported by the anatomic defects noted with this anomaly. That is, septa with squamous epithelium on one surface and columnar epithelium on the other have been detected. However, septa with gland-like structures or smooth muscle have also been described [28, 30]. Thus, although it was once thought that studying the histology of transverse vaginal septa might help explain the etiology of the anomaly, it no longer seems possible to ascribe a single etiology for transverse vaginal septa.

Clinical considerations. Affected individuals appear normal externally and undergo normal secondary sex development except for amenorrhea. Most cases are detected soon after puberty [27, 28, 30, 31] because of amenorrhea associated with lower abdominal cramping or a pelvic mass. However, the disorder has been diagnosed in the newborn because hydrocolpos or hydrometrocolpos may produce a large abdominal mass that causes respiratory embarrassment and urinary, intestinal, and circulatory obstruction [32]. Hydrocolpos or hydrometrocolpos in the newborn may have another etiologic component. It is difficult to conceive that cervical mucous and uterine secretions secondary to maternal estrogens, as has been postulated by Spence [33], can account for the large amounts of fluid noted in the vaginas of these newborns. Instead, we would like to propose that the fluid is secondary to a functioning primitive mesonephros and patent mesonephric duct which empties into the vagina. In 1939, Tompkins [34] reported 113 cases of hematocolpos detected after puberty, whereas by 1940 only 20 cases of hydrometrocolpos in the newborn period has been reported [33]. Rectovaginal fistuli and imperforate anus may be present [35].

Genetics. A rare autosomal recessive gene causes at least some cases of transverse vaginal septum, specifically in the Amish. McKusick et al reported affected sibs in two Amish families with hydrometrocolpos in 1964 [35], and later in 1968 reported a similarly affected girl in another branch of the same kindred [36]. All parents traced their ancestry to the same couple born in the 18th century. McIlroy and Ward [37] reported 3 postpuberal sisters with imperforate hymen and hematocolpos. The cases of McIlroy and Ward could represent either transverse vaginal septa or imperforate hymen; however, the authors compared their cases to individuals with "imperforate vagina due to defective canalization of the urogenital sinus."

Longitudinal Vaginal Septa

Longitudinal vaginal septa, not associated with incomplete müllerian fusion, are rare. When they occur, they rarely produce clinical problems [38]. Heritable tendencies are not obvious in cases unassociated with somatic anomalies. However, Edwards and Gale [39] reported an autosomal dominant syndrome characterized by hand anomalies, longitudinal vaginal septum, and bladder neck anomalies causing urinary incontinence (Table 1).

TABLE 1. Multiple Malformation Syndromes With Anomalies of the Uterus or Vagina

Condition	Genital anomalies	Extragenital anomalies	Inheritance
Müllerian aplasia, middle ear anomalies, Klippel-Feil anomaly [6]	Müllerian aplasia	Malformation of malleus, incus, stapes; Klippel-Feil anomaly	Unknown
Winter syndrome [56]	Vaginal atresia	Middle ear anomalies; renal agenesis	Autosomal recessive
Fraser syndrome [59]	Vaginal atresia; bicornuate uterus	Cryptophthalmos; external ear anomalies	Possible autosomal recessive
Hand-foot-uterus syndrome [54]	Bicornuate uterus	Metacarpal and metatarsal anomalies; malformed thumbs	Autosomal dominant
Meckle syndrome [58]	Bicornuate uterus; genital ambiguity in males	Polydactyly; eye anomalies; cleft palate; polycystic kidneys; occipital exencephalocele	Autosomal recessive
Rüdiger syndrome [57]	Bicornuate uterus	Distal limb hypoplasia; agenesis of ear cartilage; brachydactyly; ureterovesical junction obstruction	Possible autosomal recessive
Edwards and Gale [39]	Longitudinal vaginal septum	Broad, short hands and feet; flexion contractures; interphalangeal joints; syndactyly	Autosomal dominant

Incomplete Müllerian Fusion

The incidence of incomplete müllerian fusion has been reported to be between 0.1% [40–42] and 2–3% of females [42, 44]. The reported differences in incidences depend upon the extent of prospective diagnoses and upon the prevalence of uterus arcuatus in a given series.

Embryology. Failure of müllerian primordia to fuse completely during the 9th to 12th weeks accounts for uterus didelphys, uterus duplex, and uterus septus (see below for terminology). Because uterus didelphys is not a normal embryonic stage of uterine development, that anomaly is not considered to be due solely to an arrested stage of development; instead, it is believed due to failure of fusion before the uterus and vagina.

If the müllerian duct on one side fails to develop or to reach the urogenital sinus, an ipsilateral single uterine horn results. If absence of a hemiuterus is associated with absence of the ipsilateral kidney, the underlying defect is failure of wolffian duct development, inasmuch as müllerian duct development is dependent upon wolffian development.

Terminology. In most reported series, the classification of anomalies has been based solely upon the anatomic defect [40, 43, 45]; however, in some instances, the classification has been based on physiologic capability [44, 46]. The latter method of classification, based upon obstetric performance, is probably of more use for prognostic value. Although most institutions use a modification of the Jarcho anatomic classification, confusion or variation in classification makes attempts to compare results of therapy almost impossible.

Uterine fusion anomalies are depicted in Figure 2. In uterus unicornis (absence of 1 uterine horn) there may or may not be a rudimentary hemiuterus. There is a single uterine tube, and the uterus opens into a normal vagina. Absence of the kidney ipsilateral to the defective uterine horn is not unusual. There is no septum, as such, in uterus arcuatus (arcuate uterus); however, the fundus is flatter than usual and may display a midline notch. In uterus septus, the uterus appears normal externally; however, a septum persisting the entire length of the uterus, ie to the cervix, thus divides the cavity into 2 lateral halves. Uterus subseptus is also associated with persistence of the septum, but not persistence to the most caudal portion of the uterus. In uterus bicornis unicolis, the uterine cavities are separate, although each cavity leads to the same cervix. In uterus bicornis unicolis, the fundus is deeply notched. In uterus bicornis bicolis, 2 cervices are present. The dividing septum may be perforated; the vagina is sometimes septate, although usually normal. In uterus didelphys, there are 2 completely separate cavities and fundi, but the 2 cervices are externally united. In about 75% of cases of uterus didelphys, the vagina is septate [4]. Finally, separate hemiuteri may occasionally be associated with separate vaginas; the vaginal openings are widely separated and the entire vulva may be duplicated, as may the colon and the anus.

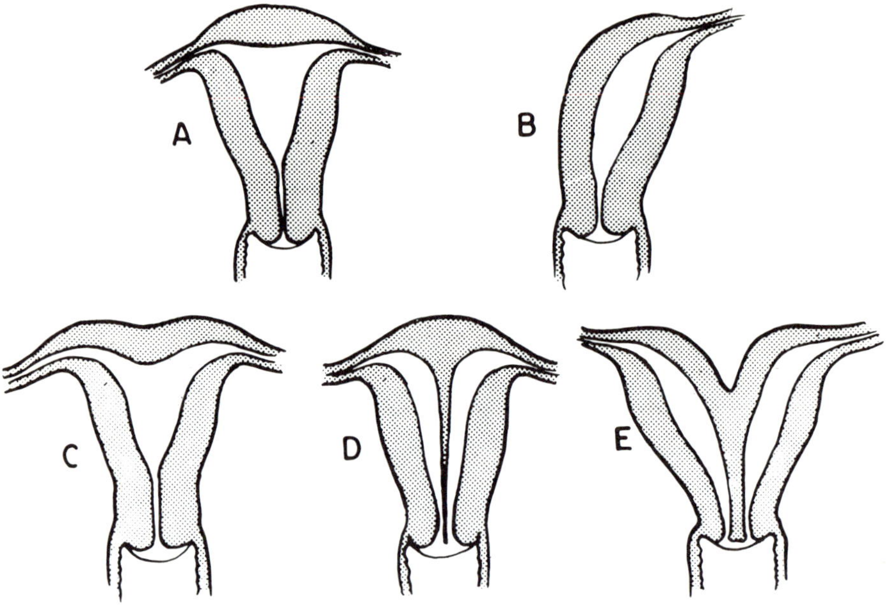

Fig. 2. Diagram of uterine fusion anomalies: A) normal uterus, B) uterus unicolis, C) uterus arcuatus, D) uterus septus, E) uterus bicornis unicolis.

Clinical considerations. Incomplete fusion of the müllerian ducts with unilateral obstruction is associated with dysmenorrhea beginning soon after menarche. Incomplete fusion with or without obliteration is associated with increased frequencies of abortions, premature labor, malpresentations, and postpartum hemorrhage [41–43]. The reproductive capacity of individuals with uterine anomalies is difficult to predict. Many affected patients have no apparent difficulty, yet others benefit from reconstructive surgery [13, 14].

Genetics. Affected sibs were reported by Nykiforuk [47], Tyler [48], Drescher [49], Jones and Wheeless [13], and Polishuk and Ron [50] (Fig. 3). Often the specific type of anomaly differed among various family members. For example, 1 of the 2 affected sibs reported by Nykiforuk had uterus didelphys; the other had at least a longitudinal vaginal septum, and possibly uterus didelphys as well. The cases reported by Tyler [48] both had uterus didelphys. The 2 familial aggregates reported by Jones and Wheeles [13] had either a bicornuate or septate uterus, either partial or complete. The case reported by Drescher [49] likewise had uterus bicornis unicolis. The anomaly in family No. 1 of Polishuk and Ron [50] had uterus didelphys, whereas in the other 2 families the affected members had bicornuate uteri.

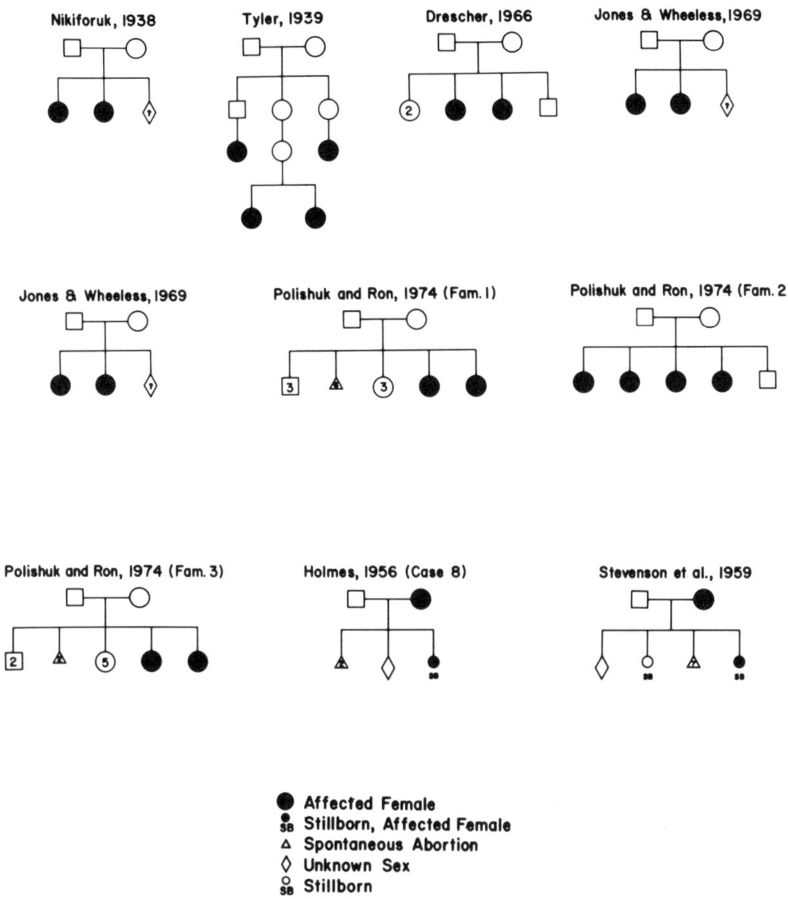

Fig. 3. Pedigrees of 10 families with incomplete müllerian fusion anomalies.

In 3 families both a mother and her daughter were affected [51–53]. In 2, the affected offspring had other associated anomalies and were stillborn. In the third [53], both mother and daughter had a bicornuate uterus and ipsilateral renal agenesis.

In view of these various types of familial aggregates, if no other anomalies are present a reasonable assumption is that incomplete müllerian fusion is multifactorial in etiology.

True Duplication of the Müllerian Ducts

True duplication of the uterus is rare and probably results from division of one or both müllerian ducts early in embryogenesis. In this anomaly, affected individuals have 2 separate uteri, to each of which may be attached 2 fallopian tubes. True duplication should be distinguished from incomplete müllerian fusion, in which each of 2 hemiuteri has a single fallopian tube.

Multiple Malformation Syndromes With Anomalies of the Uterus or Vagina

In several malformation syndromes both nongenital anomalies and abnormalities of the vagina or uterus are present. In some the mode of inheritance is known, but in others it is not (Table 1).

Hand-Foot-Uterus Syndrome. The hand-foot-uterus syndrome was first reported in 13 persons (both sexes) in 4 generations of a single kindred by Stern et al [54]. The nongenital malformations include short 1st metacarpal, pointed distal phalanx of the thumb, clinodactyly of the 5th finger and unusual carpal fusions, shortened 1st metatarsal with a pointed distal phalanx, short calcaneus, and unusual tarsal fusions. Anomalies of the genital tract include varying degrees of incomplete müllerian fusion, ranging from uterus duplex unicolis to uterus didelphys. This syndrome is inherited as an autosomal dominant. Later observations were reported by Poznanski et al [55].

Camptobrachydactyly-longitudinal vaginal septum. Edwards and Gale [39] described 18 individuals in families with this autosomal dominant syndrome. Affected persons had broad, short feet due to metatarsal shortening; webbing between either the 2nd and 3rd toes or the 4th and 5th toes; and abnormalities of the middle phalanges of the toes, consisting of either absence or shortening. Other anomalies included short, broad hands; flexion contractures of the distal interphalangeal joints; and short 5th fingers with an absent skin crease. Urinary incontinence was present in 9 persons, 2 males and 7 females. The 2 males had urinary incontinence with onset at the age of 12 years, whereas the 7 females had urinary incontinence since infancy. Urologic investigation failed to reveal the exact cause of the urinary incontinence; however, the affected individuals were believed to have an abnormality of the bladder neck. Four of the affected females were born with a longitudinal vaginal septum.

Syndrome of Winter et al. Winter et al [56] described a family in which 4 female sibs had varying degrees of renal dysgenesis, genital malformations, and anomalies of the middle ear. Two of the 4 females had bilateral renal agenesis and died neonatally; 1 had only a right kidney; the 4th had a hypoplastic right kidney. One of the individuals had vaginal atresia, normal fallopian tubes, and a rudimentary nodule of tissue representing the uterus; 2 had vaginal atresia but no other genital duct anomalies. The authors concluded that the distribution of affected individuals in the family suggested autosomal recessive inheritance; however, teratogenic factors could not be excluded.

Syndrome of Rüdiger et al. This syndrome, described by Rüdiger et al [57] in 2 sibs of unlike sex, is characterized by prominent forehead, absent pinnal cartilage, flat nasal bridge, stubby nose, protuberant upper lip, short fingers and toes, palmar flexion contractures, and bilateral ureterovesical stenosis. The affected female had a bicornuate uterus and cystic ovaries. A bicornuate uterus may also be present in the Meckel syndrome [58].

Cryptophthalmos syndrome. First described by Fraser [59], this syndrome is characterized by malformations of the globe and eyelids. The ocular involvement may be variable and may be unilateral. In addition to ocular anomalies, oral, renal, and genital duct anomalies similar to those of the Winter syndrome were reported. These anomalies include stenosis of the external auditory canals, anomalies of the middle ear ossicles, renal dysgenesis, and vaginal atresia. The trait is inherited in an autosomal recessive fashion.

Müllerian aplasia – Klippel-Feil deformity – deafness. In 1971 Park and Jones [60] described 2 unrelated patients with müllerian aplasia, Klippel-Feil deformity, and conductive deafness. Interestingly, Pinsky [18] observed that familial syndromes with abnormalities of the vagina or uterus show anomalies of the distal limbs, but not malformations of the trunk. By contrast, sporadic occurrences of müllerian duct anomalies are more commonly associated with anomalies of the trunk than anomalies of the distal limbs.

ANOMALIES OF THE WOLFFIAN DUCTS IN MALES

Anomalies of the wolffian ducts, particularly aplasia or absence, have been reported only rarely, probably because affected individuals are usually asymptomatic unless infertile.

Absence of Wolffian Derivatives

Absence of wolffian derivatives may or may not be associated with absence of upper urinary tract. Complete agenesis of both the wolffian duct structures and the upper urinary tract implies failure of the mesonephric development. Agenesis of wolffian derivatives associated with the presence of a normal kidney and ureter implies resorption of wolffian elements after the wolffian duct has reached the cloaca.

Only a few cases of the former, more extreme form of aplasia, have been reported. Absent are the epididymis, ductus deferens, seminal vesicle, ureter, and kidney [61–64]. The gonad is only rarely involved, although reported [61].

More frequently, only the epididymis, ductus deferens, and seminal vesicles are absent, or only the ductus epididymis and proximal portion of the deferens are missing. In such patients, infertility results if the anomaly is bilateral. If the anomaly is unilateral, an affected patient is asymptomatic. No heritable tendencies have been reported.

Failure of Fusion of Epididymis and Testis

Failure of fusion of the rete chords of the testis and the mesonephric tubules that form the ductuli efferentes can occur. The testes are usually undescended. In normal males the ductuli efferentia connecting the epididymis to the testis are not externally visible. A relative separation, ie elongation of ductuli efferenti or complete separation may occur; in the latter instance, spermatozoa cannot exit from the testis and, if bilateral, infertility results. One or both testes may fail to descend.

These fusion anomalies are not so rare as one might expect from the relative paucity of reports in the literature. Surveys of cryptorchid males and azoospermic males indicate that failure of fusion anomalies occurs in at least 1% of cryptorchid males [65, 66] and in about 1% of azoospermic men [67] (Table 2). No genetic data are available.

Persistence of Müllerian Derivatives in Males

Clinical considerations and embryology. Males with persistence of müllerian derivatives have relatively well-differentiated müllerian duct structures (uterus and fallopian tubes) but otherwise are often normal. Approximately 100 cases have been reported. Infertility probably occurs more than in the general population [68, 69] and testicular neoplasia occurs in 6% of cases [70]. The syndrome may not be diagnosed until a uterus and fallopian tubes are found in an inguinal hernial sac. The etiology is presumably failure of regression of the müllerian duct structures in the 50 to 60 day (30 mm) embryo, due either to failure of the fetal testis to elaborate müllerian inhibiting factor or to refractoriness of the müllerian duct anlagen to normal amounts of müllerian inhibiting factor.

Genetics. This trait probably results from a single mutant allele. In 6 families multiple sibs were affected [68, 71–73]. Parental consanguinity was not observed. In 1 additional family, maternal half sibs were affected [69] suggesting X-linked recessive inheritance.

TABLE 2. Prevalence of Epididymal-Testicular Fusion Anomalies in Several Surveys

Investigators	Sample	No. with failure of fusion
Badenoch [65]	42 cryptorchid males	3/42
Dean et al [66]	245 cryptorchid males	3/245
Hanley and Hodges [67]	400 azoospermic males	4/300+
	Total	10/587+

CONCLUSION

Anomalies of the müllerian and wolffian ducts may occur in otherwise normal individuals or as a component of a malformation syndrome. In both sexes the external genitalia are usually normal but these anomalies frequently cause infertility. In some disorders a specific mode of inheritance is suggested; however, conclusions regarding the mode of inheritance involving the genital ducts of most disorders awaits further delineation and more complete genetic investigations.

REFERENCES

1. Simpson JL: "Disorders of Sexual Differentiation, Etiology and Clinical Delineation." New York: Academic Press, 1976.
2. Gruenwald P: Zur Entwicklungsmechanik des Urogenital systems beim Huhn. Arch Entwick-Mech Org 136:786, 1937.
3. Forsberg J-G: Mitotic rate and autoradiographic studies on the derivation and differentiation of the epithelium in the mouse vaginal anlage. Acta Anat 62:266, 1965.
4. Gray SW, Skandalakis JE: "Embryology for Surgeons." Philadelphia: WB Saunders, 1972.
5. Jirásek JE: In Cohen MM (ed): "Development of the Genital System and Male Pseudohermaphroditism." Baltimore: Johns Hopkins University Press, 1971.
6. Jirásek JE: Principles of reproductive biology. In Simpson JL (Ed): "Disorders of Sexual Differentiation, Etiology and Clinical Delineation." New York: Academic Press, 1976.
7. Bryan AL, Nigro JA, Counseller VS: One hundred cases of congenital absence of vagina. Surg Gynecol Obstet 88:79, 1949.
8. Thompson JD, Wharton LR, Te Linde RW: Congenital absence of the vagina. An analysis of thirty-two cases corrected by the McIndoe operation. Am J Obstet Gynecol 74:397, 1957.
9. Cali RW, Pratt JH: Congenital absence of the vagina. Am J Obstet Gynecol 100:752, 1968.
10. Leduc B, Van Campenhout J, Simard R: Congenital absence of the vagina. Am J Obstet Gynecol 100:512, 1968.
11. Grover S, Solanki BR, Banerjee M: A clinopathologic study of Müllerian duct aplasia with specific reference to cytogenetic studies. Am J Obstet Gynecol 107:133, 1970.
12. Turunen A, Unnérus C-E: Spinal changes in patients with congenital aplasia of the vagina. Acta Obstet Gynecol 46:99, 1967.
13. Jones HW Jr, Wheeless CR: Salvage of the reproductive potential of women with anomalous development of the Müllerian ducts: 1868-1968-2068. Am J Obstet Gynecol 104: 348, 1969.
14. Jones HW Jr, Scott WM: "Hermaphroditism, Genital Anomalies, and Related Endocrine Disorders." 2nd Ed. Baltimore: Williams & Wilkins, 1971.
15. Garcia J, Jones HW Jr: The split thickness graft technic for vaginal agenesis. Obstet Gynecol 49:328, 1977.
16. Sarto GE: Cytogenetics of fifty patients with primary amenorrhea. Am J Obstet Gynecol 119:14, 1974.

17. Griffin JE, Edwards C, Madden JD, Harrod MJ, Wilson JD: Congenital absence of the vagina, the Mayer-Rokitansky-Kuster-Hauser Syndrome. Ann Intern Med 85:224, 1976.
18. Pinsky L: A community of human malformation syndromes involving the Müllerian ducts, distal extremities, urinary tract, and ears. Teratology 9:65, 1974.
19. Azoury RS, Jones HW Jr: Cytogenetic findings in patients with congenital absence of the vagina. Am J Obstet Gynecol 94:178, 1966.
20. Anger D, Hemet J, Ensel J: Forme familiale du syndrome de Rokitansky-Kuster-Hauser. Bull Fed Soc Gynecol Obstet Lang Fr 18:229, 1966.
21. Jones HW Jr, Mermut S: Familial occurrence of congenital absence of the vagina. Am J Obstet Gynecol 114:1100, 1972.
22. Lischke JH, Curtis CH, Lamb EJ: Discordance of vaginal agenesis in monozygotic twins. Obstet Gynecol 41:920, 1973.
23. Shokeir MN: This volume, pp 171–184.
24. Zarou GS, Acken HS, Brevetti RC: Surgical management of congenital atresia of the cervix. Am J Obstet Gynecol 82:923, 1961.
25. Scott JR, Galask R, Yannone ME: Congenital atresia of the uterine cervix. Int J Gynaecol Obstet 9:249, 1971.
26. Geary WL, Weed JC: Congenital atresia of the uterine cervix. Obstet Gynecol 42:213, 1973.
27. Bowman JA Jr, Scott RB: Transverse vaginal septum. Report of four cases. Obstet Gynecol 3:441, 1954.
28. Kanagasuntheram R, Dassanayake AGS: Nature of the obstructing membrane in primary cryptomenorrhea. J Obstet Gynaecol BR Cwlth 65:487, 1958.
29. Lodi A: Contributo clinco stastistico sulle malformazioni della vagina osservate nella Clinical Ostetrico e Gineocologica de Milano del 1906 al 1950. Ann Ostet Ginecol Med Perinat 73:1246, 1951.
30. Brenner P, Sedlis A, Cooperman H: Complete imperforate vaginal septum. Obstet Gynecol 25:135, 1965.
31. Deppisch LM: Transverse vaginal septum. Obstet Gynecol 39:193, 1972.
32. Dennison WM, Bacsich P: Imperforate vagina in the newborn. Neonatal hydrocolpos. Arch Dis Child 36:156, 1961.
33. Spence HM: Congenital hydrocolpos. A review with emphasis on urologic aspects and a report of four additional cases. JAMA 180:1100, 1962.
34. Tompkins P: The treatment of imperforate hymen with hematocolpos. A review of 113 cases in the literature and a report of five additional cases. JAMA 113;913, 1939.
35. McKusick VA, Bauer RL, Koop CE, Scott RB: Hydrometrocolpos as a simply inherited malformation. JAMA 189:813, 1964.
36. McKusick VA, Weilbaecher RG, Gragg GW: Recessive inheritance of a congenital malformation syndrome. JAMA 204:113, 1968.
37. McIlroy DM, Ward IV: Three cases of imperforate hymen occurring in one family. Proc R Soc Med 23:633, 1930.
38. Monie IW, Sigardson LA: A proposed classification for uterine and vaginal anomalies. Am J Obstet Gynecol 59:696, 1950.
39. Edwards JA, Gale RP: Camptobrachydactyly: A new autosomal dominant trait with two probable homozygotes. Am J Hum Genet 24:464, 1972.
40. Fenton AN, Singh BP: Pregnancy associated with abnormalities of the female reproductive tract. Am J Obstet Gynecol 63:744, 1952.
41. Blair RG: Pregnancy associated with congenital malformation of the reproductive tract. J Obstet Gynaecol Br Emp 67:36, 1960.

42. Semmens JP: Congenital anomalies of female genital tract. Functional classification based on review of 56 personal cases and 500 reported cases. Obstet Gynecol 19:328, 1962.
43. Hay D: The diagnosis and significance of minor degrees of uterine abnormality in relation to pregnancy. J Obstet Gynaecol Br Cwlth 65:557, 1958.
44. Greiss FC Jr, Mauzy CH: Genital anomalies in woman. An evaluation of diagnosis, incidence and obstetric performance. Am J Obstet Gynecol 82:330, 1961.
45. Jarcho J: Malformations of the uterus. Review of the subject, including embryology, comparative anatomy, diagnosis and report of cases. Am J Surg 71:106, 1946.
46. Jones WS: Obstetric significance of female genital anomalies. Obstet Gynecol 10:113, 1957.
47. Nykiforuk NE: Uterus didelphys. Can Med Assoc J 38:175, 1938.
48. Tyler GT: Didelphys in sisters. Am J Surg 45:337, 1939.
49. Drescher H: Sur Frage gehäuften familiären Vorkommens von Uterus-missbildungen. Zentralbl Gynaekol 88:1673, 1966.
50. Polishuk WZ, Ron MA: Familial bicornuate and double uterus. Am J Obstet Gynecol 119:982, 1974.
51. Holmes JA: Congenital abnormalities of the uterus and pregnancy. Br Med J 1:1144, 1956.
52. Stevenson AC, Dudgeon MY, McClure HI: Observations on the results of pregnancies in women resident in Belfast. II. Abortions, hydatidiform moles and ectopic pregnancies. Ann Hum Genet 23:395, 1959.
53. Wiersma AF, Peterson LF, Justema EJ: Uterine anomalies associated with unilateral renal agenesis. Obstet Gynecol 47:654, 1976.
54. Stern AM, Gall JC Jr, Perry BL, Stimson CW, Weitkamp LR, Poznanski AK: The hand-foot-uterus syndrome. J Pediatr 77:109, 1970.
55. Poznanski AK, Kuhns LR, Lapides J, Stern AM: A new family with the hand-foot-genital syndrome – a wider spectrum of the hand-foot-uterus syndrome. In Bergsma D (ed): "Genetic Forms of Hypogonadism." Miami: Symposia Specialists for The National Foundation–March of Dimes, BD:OAS XI(4): 127, 1975.
56. Winter JSD, Kohn G, Mellman WJ, Wagner S: A familial syndrome of renal, genital and middle ear anomalies. J Pediatr 71:88, 1968.
57. Rüdiger RA, Schmidt W, Loose DA, Passarge E: Severe developmental failure with coarse facial features, distal limb hypoplasia, thickened palmar creases, bifid uvula, and ureteral stenosis: A previously unidentified familial disorder with lethal outcome. J Pediatr 79:977, 1971.
58. Opitz JM, Howe JJ: The Meckel syndrome (dysencephalia splanchnocystica, the Gruber syndrome). In Bergsma D (ed): Part II. "Malformation Syndromes." New York: The National Foundation–March of Dimes, BD:OAS V(2):167, 1969.
59. Fraser GR: Our genetical "load." A review of some aspects of genetic malformations. Ann Hum Genet 25:387, 1962.
60. Park IJ, Jones HW Jr: A new syndrome in two unrelated females: Klippel-Feil deformity, conductive deafness, and absent vagina. In Bergsma D (ed): Part X. "The Endocrine System." Baltimore: Williams & Wilkins for The National Foundation–March of Dimes, BD:OAS VII(6):311, 1971.
61. Michelson L: Congenital anomalies of the ductus deferens and epididymis. J Urol 61: 384, 1949.
62. Nelson RE: Congenital absence of the vas deferens: A review of the literature and report of three cases. J Urol 63:176, 1950.

63. Whitehorn CA: Complete unilateral Wolffian duct agenesis with homolateral cryptorchism; a case report, its explanation and treatment; and the mechanism of "testicular descent." J Urol 72:685, 1954.
64. Emery CB, Goldstein AMB, Morrow JW: Congenital absence of vas deferens with ipsilateral urinary anomalies. Urology 4:201, 1974.
65. Badenoch AW: Failure of the urogenital union. Surg Gynecol Obstet 82:471, 1946.
66. Dean AL Jr, Major JW, Ottenheimer EJ: Failure of fusion of the testis and epididymis. J Urol 68:754, 1952.
67. Hanley HG, Hodges RD: The epididymis in male sterility: A preliminary report of microdissection studies. J Urol 82:508, 1959.
68. Brook CGD, Wagner H, Zachmann M, Prader A, Armendares S, Frenk S, Alemán P, Najjar SS, Slim MS, Genton N, Bozic C: Familial occurrence of persistent Müllerian structures in otherwise normal males. Br Med J 1:771, 1973.
69. Sloan WR, Walsh PC: Familial persistent Müllerian duct syndrome. J Urol 115:459, 1976.
70. Simpson JL, Photopulos G: The relationship of neoplasia to disorders of abnormal sexual differentiation. In Bergsma D, Schimke RN, Summitt RL, Harris DJ (eds): "Cancer and Genetics." New York: Alan R. Liss for The National Foundation–March of Dimes, BD:OAS XII(1):15, 1976.
71. Guell-Gonzales JR, Paramio-Ruibal A, Delgado-Morales B: Pseudohermafroditismo masculino con genitales internos bisexuales. Report de 2 hermanos. Rev Cubana Pediatr 43:579, 1971.
72. David L, Saez JM, Francois R: Male pseudohermaphroditism in two brothers. Acta Paediatr Scand 61:249, 1972.
73. Armendares S, Buentello L, Frenk S: Two male sibs with uterus and Fallopian tubes. A rare, probably inherited disorder. Clin Genet 4:291, 1973.

EDITORIAL COMMENTS

It seems worthwhile to mention one more anomaly of the wolffian duct system in males. In 1968, Kaplan et al [1] described absence of the vasa deferentia in autopsies on 10 and surgical exploration of 7 males with cystic fibrosis of the pancreas. They reported only 1 offspring of 15 married adult males with cystic fibrosis. In addition, semen analysis revealed aspermia in all samples from 25 affected males. The authors mentioned a personal communication which reported "deficient epididymides and absent vasa deferentia" in 10 males in a study by Landing and associates. In the cases adequately studied by Kaplan et al seminal vesicles were absent or hypoplastic and, in most, epididymides were rudimentary or absent. Other abnormalities included unilateral or bilateral small testes, and reduced numbers of spermatozoa on testicular sections along with maturation arrest. However, active mitosis was found in all sections examined.

In 1969, Valman and France [2] found absent, or hypoplasia with lack of patency of, the vasa deferentia in each of 10 males with cystic fibrosis between 56 hours and 15 years of age.

Cystic fibrosis of the pancreas is inherited in an autosomal recessive manner. The fact that a serum factor in affected patients inhibits ciliary movement of cells of isolated tracheae, and the observation that ciliary movement is associated with pinocytotic activity led Ohno et al [3] to suggest that this defect is responsible for aplasia or hypoplasia of the wolffian duct structures in males with cystic fibrosis. The differentiation of the mesonephric or wolffian duct into vas deferens, epididymis, and seminal vesicle is dependent upon locally elaborated androgen. Ohno suggested that testosterone elaborated by the embryonic testicle is normally secreted directly into the lumen of the wolffian duct and then taken up by the cells lining the duct by pinocytosis. In cystic fibrosis, this pinocytosis is defective, resulting in aplasia or hypoplasia of those structures derived from the wolffian duct.

While Ohno's hypothesis lacks definitive proof, the hypoplasia or aplasia of the vas in males with cystic fibrosis is extremely common, producing sterility in affected males.

REFERENCES

1. Kaplan E, Schwachman H, Perlmutter AD, Rule A, Khaw K-T, Holsclaw DS: Reproductive failure in males with cystic fibrosis of the pancreas. N Engl J Med 279:65, 1968.
2. Valman HB, France NE: The vas deferens in cystic fibrosis. Lancet 2:566, 1969.
3. Ohno S, Tettenborn U, Dofuku R: Molecular biology of sex differentiation. Hereditas 69:107, 1971.

Robert L. Summitt

Male Genital Defects in Patterns of Malformation

David W. Smith, MD

External genital malformations are a feature in a number of patterns of malformation [1]. These may be categorized into anomalads, syndromes, and nonrandom associations for which the distinction has not yet been made between anomalad versus syndrome.

ANOMALADS

An anomalad represents the multiple defects consequent to a single earlier defect in morphogenesis. Some of the examples which include external genital defects are set forth in Table 1. The genital tubercles normally arise bilaterally in the infraumbilical mesenchyme. Hence, it is understandable how there can be epispadias with deficient infraumbilical mesenchyme and extrophy of the bladder, a bifid rudimentary phallus in the male with extrophy of the bladder, and the absence of external genital development in the severe defect of early caudal blastema resulting in fused lower limbs, sirenomelia. Another presumed anomalad, the abdominal muscle deficiency anomalad (also termed prune belly or triad syndrome) may have incomplete morphogenesis of the phallus with a hypoplastic, flaccid phallus secondary to deficiency of the corpora cavernosa.

SYNDROMES

The most common syndrome in which genital anomalies may be a feature is the Down syndrome. Affected males appear to have hypogonadism (CA Paulson, personal communication) with a small penis and sometimes incomplete descent

Birth Defects: Original Article Series, Volume XIV, Number 6C, pages 57–61

TABLE 1. Anomalads With Genital Defect as One Feature

Anomalad	Genital defect
Extrophy of bladder	Epispadias
Extrophy of cloaca	Rudimentary bifid phallus
Sirenomelia	Rudimentary to absent phallus
Abdominal muscle deficiency	Occasionally a flaccid phallus with absent corpora cavernosa

of the labioscrotal folds yielding a "shawl" scrotum. As with the XXY Klinefelter syndrome [2], the Down syndrome boys tend to have a small penis and to have testicular insufficiency of testosterone production at adolescence (CA Paulson, personal communication). Though micropenis is a common feature in the XXY Klinefelter syndrome, they occasionally present with hypospadias and cryptorchidism as well. Table 2 summarizes some of the syndromes which have aberrant genitalia as one feature. The degree of incomplete development of the male genitalia for some of them can be so severe that they are occasionally mistaken for females or must by necessity be reared as females at birth. This applies to some instances of Smith-Lemli-Opitz syndrome, Robinow syndrome, and cryptophthalmos syndrome.

In some of these disorders the primary problem appears to be testicular deficiency as in the Down syndrome (CA Paulson, personal communication) and the XXY Klinefelter syndrome [2]. For the Prader-Willi syndrome [3], and possibly the Meckel-Gruber syndrome as well, LH insufficiency is the presumed reason for the micropenis. For the other syndromes listed in Table 2 the cause for the incomplete morphogenesis of the genitalia is undetermined. However, for many there is probably no endocrine problem, and the primary problem is most likely in the morphogenesis of the genitalia themselves, as in the cryptophthalmos syndrome and presumably the Aarskog syndrome.

Though there have initially been some indications of specificity of a particular type of genital anomaly in a particular syndrome, the individual defects have been relatively nonspecific, and the syndrome diagnosis is dependent on the total pattern of malformation. For example, shawl scrotum is not specific to the Aarskog syndrome.

ASSOCIATIONS

There are nonrandom associations of defects which include genital abnormality. The most common of these appears to be the VATER association [4] which in its expanded context [5] includes not only *v*ertebral anomalies, *a*nal

TABLE 2. Syndromes With Genital Defect as One Feature[a]

Genetic imbalance:	Down syndrome[a]	Micropenis
	XXY Klinefelter syndrome[a]	Micropenis, occ. hypospadias
	13 Trisomy syndrome	Aberrant scrotum
	Triploidy	Micropenis, hypospadias
	4p– syndrome	Hypospadias
Autosomal recessive	Smith-Lemli-Opitz syndrome	Micropenis, hypospadias
	Cryptophthalmos syndrome	Hypospadias (vaginal atresia)
	Meckel-Gruber syndrome	Micropenis
Autosomal dominant	Robinow syndrome[a]	Micropenis, (microclitoris)
? Autosomal dominant versus	Opitz syndrome	Hypospadias
X-linked	Opitz-Frias syndrome	Hypospadias
X-linked	Aarskog syndrome	Shawl scrotum, hypospadias
Environmental	Trimethadione syndrome	Hypospadias
Unknown	Prader-Willi syndrome[a]	Micropenis

[a] Early testosterone therapy has brought the penis to normal size for age.

atresia, *t*racheo-*e*sophageal atresia and fistula, *r*adial limb dysplasia, renal anomalies, cardiac anomalies, and single umbilical artery, but may also include aberrant genitalia, as shown in Figure 1. This includes a variety of aberrations, especially hypospadias and bifid scrotum in the male, plus bicornuate uterus and/or vaginal atresia in the female. This complex association probably accounts for the previously recognized nonrandom association of hypospadias with imperforate anus, vertebral anomalies, and renal anomalies. For the female, it encompasses many of the patients hitherto classed as Rokitansky syndrome, in which the association of missing uterus with blind proximal vagina was often found in the same patients who had vertebral defects, renal anomalies, and/or imperforate anus, for example [6].

COMMENTS RELATIVE TO MANAGEMENT

Occasionally the genital defect in some of the males with the foregoing disorders may be so severe that a female sex of rearing is indicated. For the XY individual with a micropenis +/– hypospadias a short course of systemic testosterone therapy may be indicated in an effort to bring the penis to normal size for age [7]. There is a high likelihood of success when the reason for the micropenis is considered to be primary testicular insufficiency as in the Down

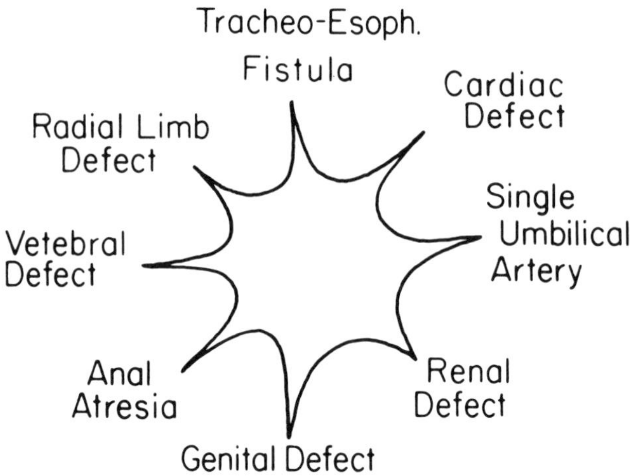

Fig. 1. The complex of anomalies which occur nonrandomly in the expanded VATER association.

syndrome and the XXY Klinefelter syndrome or secondary testosterone deficiency as in Prader-Willi syndrome. The disorders marked with superscript [a] in Table 2 are those in which some or all patients have been testosterone responsive. However, if the problem is considered to be secondary to an inborn error in the responsiveness of the phallic tissues to testosterone or to a basic nonendocrine defect in the morphogenesis of the penis itself, then the administration of testosterone would not be expected to be beneficial. A short course of testosterone therapy may not only be helpful in enlarging the penis to normal size for age in the first category, but it may assist in arriving at a decision toward sex-of-rearing in the latter category. The female sex-of-rearing would merit serious consideration in the XY patient with incomplete penile development in whom the penis was unresponsive to testosterone therapy.

The testosterone therapy is based on the premise that the growth deficiency of the penis was most likely secondary to fetal testosterone insufficiency. Theoretically the ideal therapy would have been fetal testosterone replacement therapy. Lacking this, the next best management would be testosterone therapy soon after birth. The therapy which I have utilized in the infant has been 25 mg of intramuscular depo-testosterone every 3 weeks for 3 months (4 doses). For the child the dosage utilized has been 35 mg every 3 weeks for 3 months. Personal experience with more than 20 testosterone responsive patients has shown an increase in both length and caliber of the penis during the 3 months of treatment. The enlargement is usually close to twofold, and occasionally the increase in size may be as great as fivefold in extent. Following the therapy there may be a slight reduction in size, about 10%, but the greater part of the enlargement is permanent [7]. Hence the testosterone-induced enlargement presumably represents an increase in penile cell number, since an increase in cell size alone would probably be transient.

Even though the patient may have mild to moderate mental deficiency, such as those with the XXY Klinefelter syndrome, Down syndrome, and Prader-Willi syndrome, the enlargement of the penis to normal size for age during infancy often assists in the acceptance of the child by the family. Continuous testosterone replacement therapy at the age of adolescence has been found to be generally beneficial for XXY Klinefelter syndrome individuals [2] with primary testicular insufficiency and for Prader-Willi syndrome males with secondary testicular insufficiency [3]. Such therapy merits trial in males of adolescent age with the Down syndrome, the majority of whom have primary testicular insufficiency.

REFERENCES

1. Smith DW: "Recognizable Patterns of Human Malformation." 2nd Ed. Philadelphia: W.B. Saunders, 1976.
2. Caldwell P, Smith DW: The XXY (Klinefelter's) syndrome in childhood: Detection and treatment. J Pediatr 80:250, 1972.
3. Hall BD, Smith DW: Prader-Willi syndrome. J Pediatr 81:286, 1972.
4. Quan L, Smith DW: The VATER association. A spectrum of associated defects. J Pediatr 82:104, 1973.
5. Temtamy SA, Miller JD: Extending the scope of the VATER association. J Pediatr 85: 345, 1974.
6. Bryan AL, Nigro JA, Counsellor VS: One hundred cases of congenital absence of the vagina. Surg Gynecol Obstet 88:79, 1949.
7. Guthrie RD, Smith DW, Graham CB: Testosterone treatment for micropenis during early childhood. J Pediatr 83:247, 1973.

Problems of Sex Differentiation—Surgical Correction *

Howard W. Jones, Jr., MD

The aims of surgery in patients with disorders of sex differentiation are to remove structures, especially gonads, which function in a manner contradictory to the sex of rearing, and to provide genitalia which are cosmetically acceptable and functionally useful to the assigned sex.

This discussion will be limited to patients seen by the gynecologist who, in the normal course of responsibilities, accepts for care those who are reared as females. If consideration be given only to those who by classic definition can be considered as examples of female or male pseudohermaphroditism, or true hermaphroditism, it follows that the gynecologist is concerned with essentially all patients with female pseudohermaphroditism, about 80% of all patients with male pseudohermaphroditism, as only about 20% of male pseudohermaphrodites have genitalia which lend themselves to functional male surgical augmentation, and at least 50% of the uncommon true hermaphrodites, this being the approximate percentage of those who have a uterus and vagina, and therefore the potential for reproduction as a female. The remarks to be made stem from 20-odd years of experience with well over 400 patients with problems of sex differentiation, of which 240 comprise the female and male pseudohermaphrodites, and the true hermaphrodites being discussed today (Table 1).

As each of the three main categories of intersexuality has special features, it is useful to consider them separately.

FEMALE PSEUDOHERMAPHRODITISM

Although it is possible to distinguish 5 separate groups of female pseudohermaphrodites (Table 1), the ovarian development of all groups is quite normal.

*An update of material appearing in Blandau RJ, Bergsma D (eds): "Morphogenesis and Malformation of the Genital System." New York: Alan R. Liss for The National Foundation–March of Dimes, BD:OAS XIII(2): 1977.

Birth Defects: Original Article Series, Volume XIV, Number 6C, pages 63–71

TABLE 1. Female and Male Pseudohermaphrodites and True Hermaphrodites*

Type of hermaphroditism	Number of patients
Female pseudohermaphrodites	
Adrenogenital syndrome	112
Maternal androgen	35
Special	2
Idiopathic	10
Familial	2
Male pseudohermaphrodites	
Ambiguous external genitalia with vagina	45
Ambiguous external genitalia without vagina	15
Female external genitalia	18
True hermaphrodites	9
Total	248

*1954–1975. All patients reared as females.

Furthermore, these patients have in common normal müllerian duct development. Also, the vaginal development is basically normal, except for the location of the external genitalia. This consists of abnormal fusion of the scrotolabial folds, which process in the normal male forms the urethra, but if abnormally fused in the female, forms a urogenital sinus, during the formation of which the vaginal orifice is enclosed within the sinus. The phallus is also generally abnormally enlarged. Thus, all types of female hermaphrodites have normal genitalia, except for masculinization of the external structures, due to abnormally high concentrations of androgen during intrauterine life. The types differ from each other only in the source and etiology of the excess androgen.

As the anatomic deformity is essentially the same in all female pseudohermaphrodites, except in degree of masculinization, it follows that the operative treatment is basically the same in all cases.

Laparotomy or laparoscopy is not necessary, except to distinguish true hermaphroditism from the idiopathic variety, which may very well be the result of transient maternal androgen excess due to a luteoma of pregnancy.

The surgical construction of the ambiguous external genitalia of the female pseudohermaphrodite is best accomplished prior to 18 months of age, so that the child has no recollection of the procedure. The operation was described a number of years ago [1, 2] and has been used in well over 200 individuals, but has been more recently modified to improve the cosmetic appearance by attempting once again to accomplish what was once abandoned because of sloughing, ie preservation of the glans. Recently, this has seemed to be successful by widening the base of the flap and attempting to preserve only an external shell of the glans (Figs. 1 and 2). With this process, the cosmetic results are better.

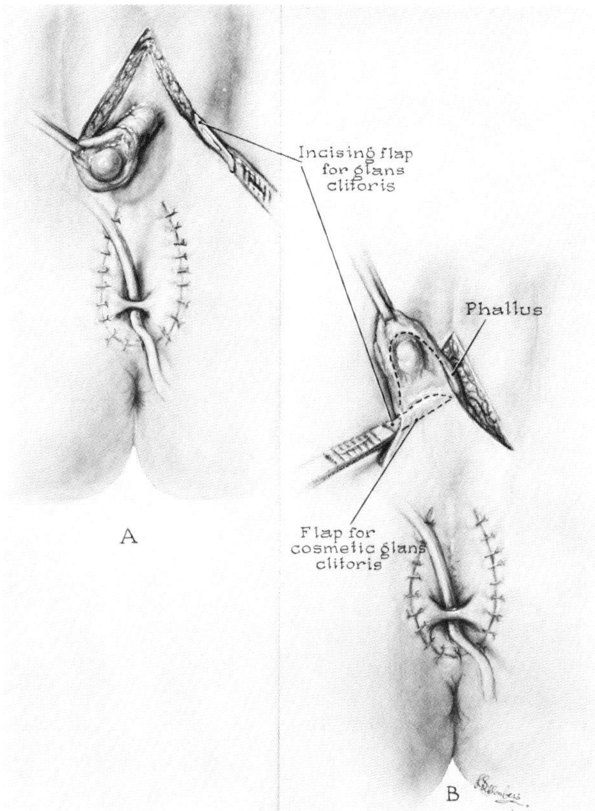

Fig. 1. A revised method for preserving the glans clitoris. It is necessary to have a very broad-based flap, and to include in the flap only a shell of glans.

As of June 30, 1977, 84 patients with the adrenogenital syndrome, treated prior to the end of 1968, were followed in an attempt to secure a long-range evaluation of the results of such an operative procedure. The follow-up of these patients as of 1974 was previously published [3].

Through June 30, 1977, 28 of the 84 patients had required secondary procedures and 5 of the 28 patients requiring secondary procedures had required a third procedure (Table 2).

TABLE 2. Patients According to Number of Surgical Procedures (84 patients)

	No.	%
One procedure	56	67
Two procedures	23	27
Three procedures	5	6

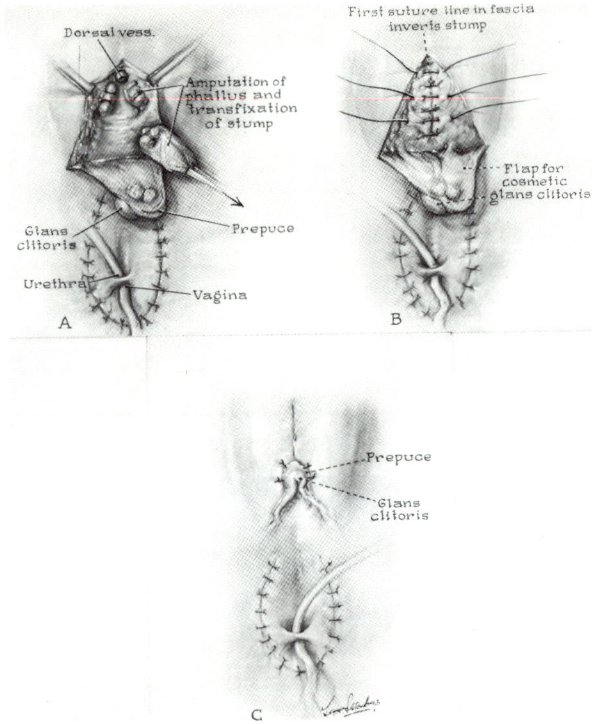

Fig. 2. With the flap turned down, the shaft of the phallus can be amputated, and the flap, including the shell of glans, then sutured into the defect.

Two of the secondary procedures were for excision of a painful clitoral stump. The remaining procedures were to provide a vaginal orifice adequate for coitus (Table 3).

The Operative Technique for Secondary Operations at the Vaginal Outlet

The problems at the vaginal orifice are basically of two kinds, although the operative procedure required for these is more or less the same.

TABLE 3. Surgical Procedures (84 patients)

Type	Surgical procedures		
	First	Second	Third
Reconstruction	84	–	–
Exteriorization of vagina	–	15	–
Relaxation of orifice	–	11	5
Other	–	2	–

The first kind of problem arises from failure to adequately exteriorize the vaginal orifice at the first operation. They may result from the fact that in some instances of marked external deformity the vagina communicates with the urogenital sinus deep in the perineum and, therefore, primary exteriorization is technically difficult, or from simple failure at the first operative procedure to carry the original midline incision far enough posteriorly to expose the vaginal orifice.

Fortunately, secondary exteriorization when the child is older is easier and facilitated by the feminization at puberty. In fact, it seems advisable not to attempt secondary exteriorization, especially that due to deep communication, until pubertal feminization has been well established.

In either circumstance, exteriorization can be accomplished by a midline incision of adequate length – 1 or 3 cm is usually sufficient – to expose the orifice. Suture of the urogenital sinus epithelium immediately peripheral to the orifice (vulvar vestibule) to the skin completes the procedure [4].

A second kind of problem at the outlet arises from linear contraction of the original incision between the epithelium of the urogenital sinus and the perineal skin. This may result in a contraction of the orifice if the original suture line is just at the orifice, ie at the hymenal ring where there is already a natural construction.

In most instances at the primary operation vaginal orifice exteriorization can be accomplished without actually invading the orifice per se. The exteriorization is best accomplished by utilizing the epithelium immediately surrounding the orifice, ie the epithelium of the vestibule, to suture to the skin. As noted above, it seems that when the original midline incision involves the orifice itself (hymen) that contraction of the orifice tends to occur by simple linear contraction of the scar.

The correction of this problem is very simple and can often be done on an outpatient basis. A vertical midline incision repaired horizontally should solve the problem, although it is interesting that it was necessary to repeat the procedure on 5 patients [3].

Even though about one-fourth of the patients with the adrenogenital syndrome require a secondary procedure at the vaginal outlet prior to vaginal function, it is nevertheless thought desirable to reconstruct these genitalia at a very early age to achieve a feminine appearance through the formative years. Furthermore, three-fourths of all patients will require no additional procedure, even when the operation is carried out during infancy.

Reconstruction of the masculinized genitalia of female pseudohermaphrodites of types other than those due to the adrenogenital syndrome can be done with this same technique. Although no formal follow-up of the nonadrenogenital female pseudohermaphrodites has been carried out, it is our clinical impression that such a follow-up would show a substantially lower necessity for secondary procedures, as compared with the adrenogenital syndrome.

MALE PSEUDOHERMAPHRODITES

The surgical approach to male pseudohermaphrodites reared as females differs from that to female pseudohermaphrodites in that it is necessary to remove the contradictory gonads, and in male pseudohermaphrodites without a vagina to modify the operation on the external genitalia accordingly. In some cases it is necessary to lengthen or widen an inadequate vagina. While, for academic and therapeutic purpose, it is desirable to use an etiologic and pathogenetic classification of male pseudohermaphroditism [4], for strictly surgical purposes, in those patients reared as females, a simpler grouping into 3 categories is useful: A) ambiguous external genitalia: 1) with a vagina; 2) with or without a uterus; 3) without a vagina; B) normal female external genitalia.

In general, most, but not all, of those with ambiguous external genitalia will masculinize at puberty. Those with normal female external genitalia are clearly of the feminizing variety (Table 1).

As with female pseudohermaphroditism, definitive surgery works very well when carried out prior to 18 months of age, except in those in the feminizing group when it can be deferred until after puberty, so that the patient will experience spontaneous feminization at puberty, and thus further imprint the feminine role.

Ambiguous External Genitalia With a Vagina

In the group with ambiguous genitalia, laparotomy may be necessary in some cases to remove the gonads and ascertain the status of the müllerian ducts. In other cases where the status of the müllerian duct development is clear, inguinal or labial incisions may suffice to excise the gonads and their adnexa.

Inguinal hernias are not uncommon in this group of patients and should be repaired at this time.

The operation on the external genitalia in those patients with ambiguous external genitalia and with a vagina, with or without a uterus, is essentially the same as for the female pseudohermaphrodite group.

Ambiguous External Genitalia Without a Vagina

For those with ambiguous external genitalia without a vagina, gonadectomy by whatever route is necessary. It is to be noted that this group includes some patients who might properly be classified as micropenis.

Many times in this group the phallus is quite small. There may be perineal hypospadias, and the posterior limits of the scrotolabial folds are located so far forward that a normally placed vaginal orifice would not be buttressed laterally by normally placed labia.

In infancy operation on the external genitalia consists of lowering the urethral meatus to a position relatively normal for the female, so that voiding in a sitting position presents no problem. In addition, scrotolabial folds can be moved posteriorly to some degree by a bilateral Y-plasty. There is a limit to the extent to which the scrotolabial folds can be moved posteriorly in some cases, because sloughing will occur at the tip of the flap if it is too long in relation to its base. Finally, the clitoris can be reduced to size by the technique used for female pseudohermaphroditism (Figs. 3 and 4).

At a much later date, when the child has attained full growth and vaginal function is desirable, a vagina may be created by split-thickness graft with an orifice in the more normal position. At that time, if the labia are not sufficiently posterior to be lateral to the vaginal orifice, a second-stage Y-plasty can be accomplished.

Normal Female External Genitalia

For patients with entirely female genitalia, gonadectomy can be deferred until after spontaneous puberty takes place, as there seems to be some delay in the rise

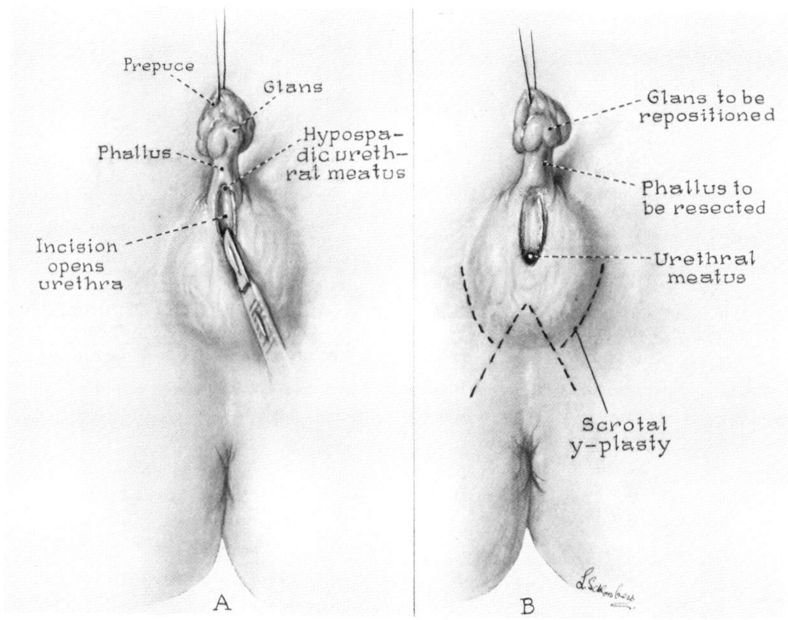

Fig. 3. In the male pseudohermaphrodite without a vagina the urethral meatus may be lowered in the perineum by simply incising the urethra to place it in a location which will enable the patient to void satisfactorily in a sitting position. It is necessary that the incision be carried only as far posteriorly as necessary, because if it meets with one of the limbs of the Y-plasty, the lower end of each of the lateral flaps will have seriously impaired circulation.

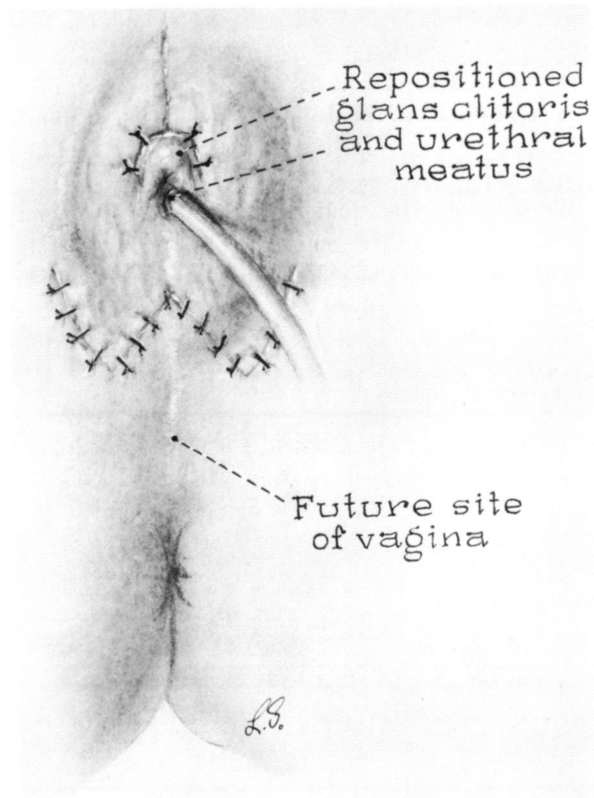

Fig. 4. The completed operation showing the bilateral Y-plasty to bring the scrotolabial folds as far posteriorly as possible, so that the subsequent vaginal orifice will be surrounded laterally by the scrotolabial folds. It is usually difficult to get the scrotolabial folds far enough posteriorly at this first stage. If they are placed under too much tension, they are apt to retract, especially if the incision is not kept dry. For this reason, an indwelling catheter for 5 days or so is exceedingly helpful.

of the curve of expectancy of tumors in these testes, as compared with the virilizing male pseudohermaphrodites [5]. The delay has the advantage of allowing the patient to experience spontaneous pubertal development, and thereby reinforce her feminine gender role. However, gonadectomy by age 20 at the latest is desirable to avoid the appreciable risk of tumor.

No operation is required on the external genitalia, but in some patients the vagina may be quite inadequate for intercourse due to shortness or narrowness, and in either case this difficulty can be corrected by a split-thickness graft using the McIndoe technique.

TRUE HERMAPHRODITISM

In the true hermaphrodite reared as a female, the gonadal portion of the operation will be limited to the excision of the testicular portion. Usually, there is a gross dissimilarity between the testicular and ovarian portion of the gonads. However, this is not always the case and liberal use of frozen section at laparotomy is essential to prevent mistakes. However, within the last few years, several hermaphrodites in various clinics have been successfully operated on with excision of the testicular portion of the gonadal maldevelopment, so that there is every reason to believe that fertility should be possible in at least some of these individuals. Certainly, normal menstruation has occurred in some patients of this kind.

Essentially all of the true hermaphrodites reared as females will have vaginas with ambiguity of the external genitalia, and the operation thereon corresponds precisely to that of the female pseudohermaphrodites.

REFERENCES

1. Jones HW Jr, Jones GES: The gynecological aspects of adrenal hyperplasia and allied disorders. Am J Obstet Gynecol 68:1330, 1954.
2. Jones HW Jr, Wilkins L: Gynecological surgery in 94 patients with intersexuality – Implications concerning the endocrine theory of sexual differentiation. Am J Obstet Gynecol 82:1142, 1961.
3. Jones HW Jr, Garcia SC, Klingensmith GJ: Secondary surgical treatment of the masculinized external genitalia of patients with virilizing adrenal hyperplasia. Obstet Gynecol 48:73, 1976.
4. Park IJ, Aimakhu VE, Jones HW Jr: An etiologic and pathogenetic classification of male hermaphroditism. Am J Obstet Gynecol 123:505, 1975.
5. Manuel M, Katayama KP, Jones HW Jr: The age of occurrence of gonadal tumors in intersex patients with a Y chromosome. Am J Obstet Gynecol 124:293, 1976.

EDITORIAL COMMENT

The editor has, for purposes of uniformity among papers in this volume, substituted the term, "male pseudohermaphrodite," for the author's terms "male intersex" and "male hermaphrodite," and has substituted "female pseudohermaphrodite" for "female intersex" and "female hermaphrodite."

Robert L. Summitt

The Nosology of Male Pseudohermaphroditism Due to Androgen Insensitivity *

Leonard Pinsky

The purposes of this paper are to provide a brief overview of male sexual dysmorphogenesis in man and a substantive discussion of congenital androgen insensitivity under the following headings: the main molecular varieties; their usual and variant clinical features that have diagnostic or pathogenetic significance; nomenclature and terminology; and delineation of certain priorities in development of the field.

MALE SEXUAL DYSMORPHOGENESIS

It is useful to classify male sexual dysmorphogenesis broadly as shown in Table 1 [1]. Defects at the earliest (determinative), or at a subsequent, step in testicular differentiation, and faulty biosynthesis of testosterone or perhaps of the müllerian inhibiting substance by well-differentiated testes are responsible for the first class. The second class comprises those situations in which the supply of androgenizing substances is normal in time and amount, but the ability of their respective embryonic target structures to respond is permanently defective. The third class is conceptually important in trying to understand certain mendelian forms of male pseudohermaphroditism in which external genitalia that are severely undermasculinized at birth begin to virilize impressively at puberty; it is best exemplified from the gonadal viewpoint by two of the prototypic sex chromosome disorders, mixed gonadal dysgenesis and the XXY Klinefelter syn-

*This review was prepared during tenure of a grant (MT-2830) from the Medical Research Council of Canada.

Birth Defects: Original Article Series, Volume XIV, Number 6C, pages 73–95

drome. In the former [2], gonads that were capable of inducing only minimal masculinization before birth often appear to acquire the ability to impose marked degrees of virilization at puberty; in the latter, the reciprocal is true. The frequent coexistence of sexual and somatic anomalies in males with various malformation syndromes [3] implies that differentiation of male sexual structures is not under the exclusive control of the androgenizing substances; in other words, they are subject to the same dysmorphogenetic influences that regularly affect somatic structures.

A PROTOTYPE ANDROGEN TARGET CELL

The main elements of an androgen target cell response are depicted according to current dogma in Figure 1. Molecules of testosterone (T) or 5α-dihydrotestosterone (DHT) enter the cytoplasm perhaps by a process of carrier-mediated transport [4] and bind with high affinity and low capacity to specific protein molecules – the receptors. There appears to be a single species of receptor for both of these main androgens. The androgen-receptor complex apparently needs to be "activated" before it can translocate into the nucleus.* Once in the nucleus, the complexes bind to sites on the chromatin called acceptors. These sites appear to be functionally encoded by certain nonhistone proteins as well as by their underlying nucleotide sequences; the relative contributions of these two chromatin components to "acceptor" activity is controversial [5]. The numbers of such sites, their differential binding affinity within and among various cell types, and how they are involved in initiating the chain of events that effect the characteristic responses of various androgen target organs remain largely to be determined. Selective uncoiling of the DNA duplex to expose the transcribing strand, and an increased number of "preinitiation" sites at which RNA polymerase can bind [6, 85], may be important steps in the chain. In any event, new molecules of mRNA that encode the synthesis of target organ-specific proteins are ultimately presented to the cytoplasm for translation.

THE NORMAL EVENTS IN MALE SEXUAL MORPHOGENESIS

In the presence of a normal Y chromosome – containing the gene(s) for the H-Y antigen(s) [7] – and in the absence of certain rare non-Y-linked mutations that can cause gonadal or testicular dysgenesis [8–14], the neutral gonads become testes. The embryonic testes have four spheres of influence (Figs. 2 and 3).

First, they promote differentiation of each wolffian or mesonephric duct to yield a vas deferens, a seminal vesicle, and an epididymis. This is accomplished by

*There appear to be some receptor-dependent androgen effects that are consummated within the cytoplasm [86]; these can be disregarded for the purposes of this review.

TABLE 1. Classification of Male Sexual Dysmorphogenesis

1. Permanent gonadal defect
2. Permanent target structure defect
3. Impermanent gonadal or target structure defect
4. Nonspecific gonadal and/or target structure defects in syndromal association with somatic anomalies

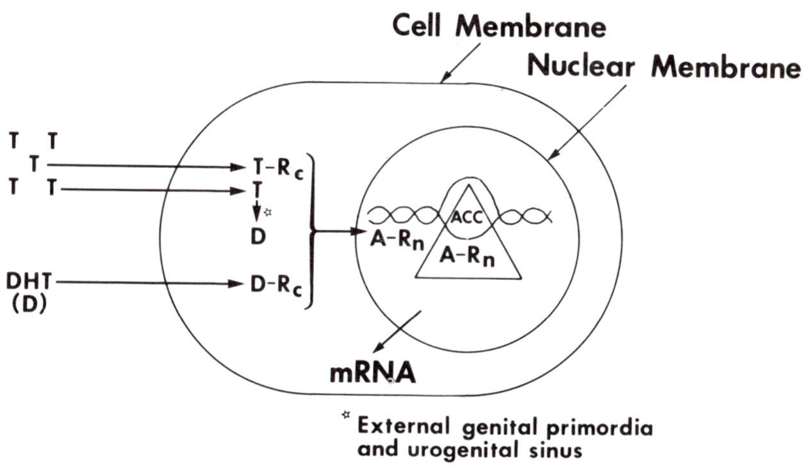

Fig. 1. An androgen target cell. T, testosterone; DHT (D), 5α-dihydrotestosterone; A, androgen; R_c cytoplasmic form of the receptor; ★, activation of the cytoplasmic hormone-receptor complex; R_n, nuclear form of the receptor; ACC, acceptor site. Uncoiling of the DNA duplex in the region of the A-R_n binding site is presumed to expose an RNA polymerase binding site. A more sophisticated version of this scheme [83] postulates two types of chromatin binding sites – one, with relatively low affinity, that serves to localize A-R_n complexes regionally, and a second, with higher affinity, that serves to effect transcription of an adjacent structural gene.

means of testosterone itself, and to some extent it is a local (ipsilateral) action of each testis rather than simply a product of their combined humoral actions [15].

Second, they repress the müllerian or paramesonephric ducts; this is certainly an ipsilateral action of each testis that is enforced by a polypeptide known as the müllerian inhibitory substance or the antimüllerian hormone [16].

The third and fourth roles of the embryonic testes depend upon the conversion of testosterone to DHT within the initially indifferent target structures of the urogenital (UG) sinus and the external genitalia. Under the influence of DHT the proximal UG sinus generates a prostate gland, while its distal portion becomes

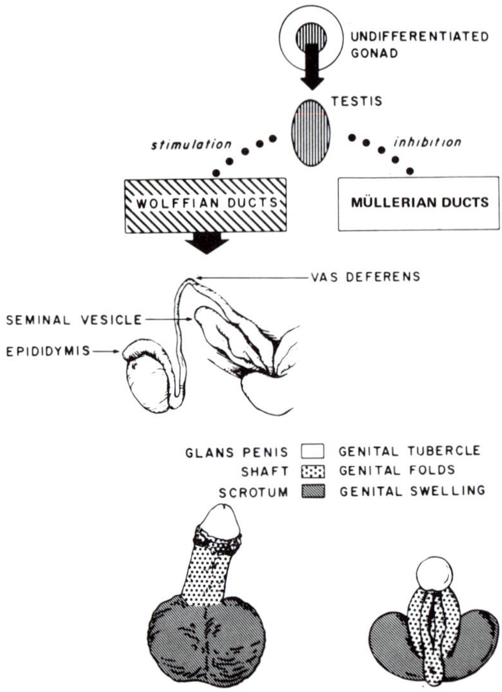

Fig. 2. The roles of the embryonic testes in development of the mesonephric (wolffian) and paramesonephric (müllerian) ducts and of the undifferentiated external genital primordia (adapted from Federman [87]).

assimilated into the penis without generating the inferior portion of the vagina. Similarly, the neutral external genitalia become the usual male external genital apparatus.

CONGENITAL ANDROGEN INSENSITIVITY

The consequences of a state of partial androgen insensitivity (AI) are depicted in Figure 4. The wolffian derivatives have appeared, but the prostate is small, the urethra has not been oriented towards the phallus, the latter has remained clitoral in size and shape, and a portion of the UG sinus has been permitted to bud-off a blind vagina. Müllerian suppression occurs normally because this is an androgen-independent process. It is easy to imagine how this figure would be altered in the presence of complete androgen insensitivity: There would be no derivatives of the wolffian duct and no prostate, the separation between the urethra and the blind vagina would be complete so that each would have its own perineal orifice, and the external genitalia would be unambiguously female.

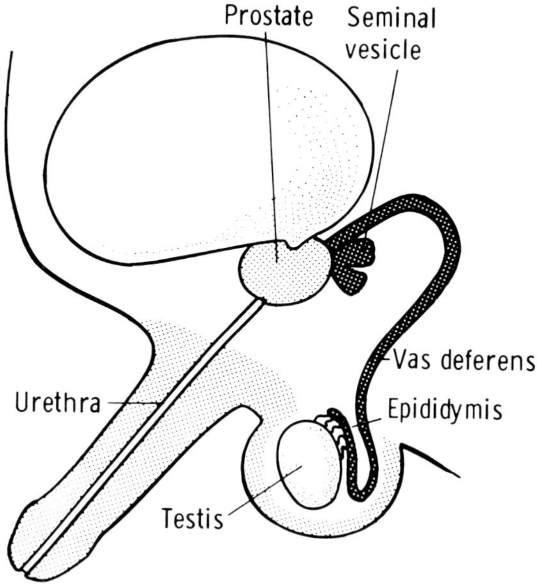

Fig. 3. The role of the testes in differentiation of the urogenital sinus (from Imperato-McGinley and Peterson [88]).

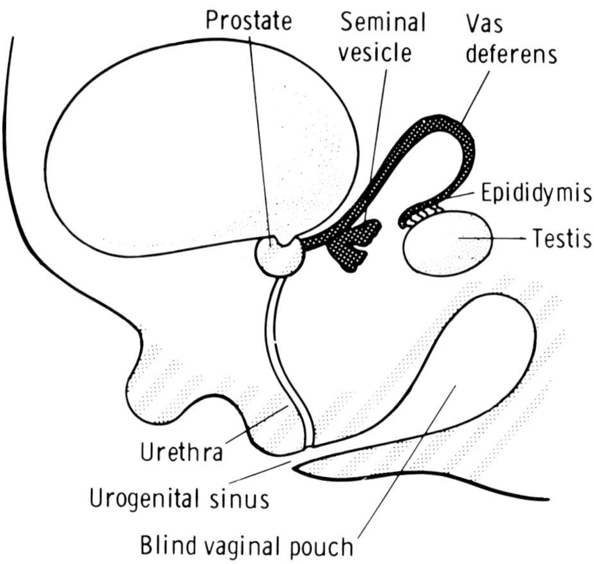

Fig. 4. One possible set of consequences resulting from a state of partial androgen insensitivity. Compare with Fig. 3.

TABLE 2: Classificatory Features of Congenital Androgen Insensitivity

Inheritance:	X-linked recessive (or autosomal dominant with sex-limited expression)
	Autosomal recessive
Distribution:	Universal
	Focal
Severity:	Complete
	Incomplete
Chronology:	Permanent
	Transient

TERMINOLOGY AND NOMENCLATURE

Some general terms that should be retained as guideposts whenever one thinks about androgen insensitivity are given in Table 2. Thus, under the heading of distribution one must ask whether the AI affects all target organs or only some; in regard to severity, whether the target tissues are totally insensitive or only partially; and, in regard to chronology, whether the targets are permanently insensitive or only transiently.

The term male pseudohermaphroditism (MP) can be applied to any XY individual whose internal and/or external genitalia are female, ambiguous, or ambivalent. The appellation has been adapted for classifying the various known types of congenital androgen insensitivity (CAI) in a generic sense (Table 3). The type of CAI that is essentially total by all the criteria listed above – the one that has been known as complete testicular feminization (TF) – is designated a form of complete MP. In contrast, the types of CAI that are incomplete, in one way or another, can be considered forms of incomplete MP.

Despite contrary opinion [17], I favor the retention of "MP" at least in everyday clinical parlance because, when qualified by "complete" or "incomplete," it has generic value as one step in differential diagnosis. For instance, there are some prepubertal males with the 17α-hydroxylase-deficient [18–22], 17, 20-desmolase-deficient [23], or 17-ketosteroid reductase-deficient [24–27] forms of testosterone

TABLE 3. Classification and Glossary of "Male Pseudohermaphroditism" According to Type of Androgen Insensitivity

1. "Complete": Testicular feminization
    a) Receptor-negative form
    b) Receptor-positive form(s)
2. "Incomplete":
    a) Type 1 (partial receptor deficiency)
    b) Type 2 (deficient 5α-reductase activity)
    c) "Type 3" (partial receptor deficiency; "Incomplete testicular feminization" [30])

TABLE 4. Core Features of Total Congenital Androgen Insensitivity

Classical Name:	Complete testicular feminization
Distribution:	Universal
Severity:	Complete
Chronology:	Permanent
Cause:	X-linked gene
Mechanism:	a) Absence of receptor activity
	b) Undefined receptor-positive form(s)

undersynthesis that, because their external genital phenotypes are unambiguously female, merit the designation of complete MP until successful pursuit of their correct diagnosis allows them to be differentiated from those with the same phenotype that are due to complete androgen insensitivity. Similarly, there are some patients with apparently partial degrees of the same enzyme deficiencies [28, 29] whose external and internal genital phenotypes are indistinguishable from those with various forms of incomplete androgen insensitivity. Such cases can be labeled incomplete MP until a precise diagnosis can be made.

The one term the retention of which I do not favor is "incomplete testicular feminization" (ITF). It has been used recently to name a variety of CAI whose definition is unclear [30]. "ITF" had currency until about 5 years ago, when it was used as a nosologic umbrella for all patients with incomplete MP assumed to have partial androgen insensitivity. There is clearly no need to resurrect its use at this time.

COMPLETE TESTICULAR FEMINIZATION

The core features of complete testicular feminization (CTF) are given in Table 4. There are two pathogenetic types – one with a severe deficiency of cytoplasmic androgen receptor activity [31–33], and a receptor-positive "variety" the specific defect(s) of which has (have) not yet been discovered [32, 34]. I am not aware of any clinical difference between these two laboratory types. These patients are born with unambiguously female external genitalia, and if they lack a positive family history (see below) and have intraabdominal rather than inguinal or labial testes, then they go unrecognized until puberty, when they fail to develop a normal amount of sexual hair or menses.

A word about their sexual hair is in order. They seldom, if ever, have axillary hair, but they often have a small amount of immature (soft, straight) pubic hair. We have not found any laboratory distinction between those that have no pubic hair and those that have some of it (Pinsky, unpublished). Indeed, patients with and without scant pubic hair occur within the same families [35].

These patients do not (and cannot be made to) menstruate, of course, because müllerian duct suppression is an androgen-independent process. Occasionally, however, a müllerian derivative does persist in a hypoplastic or vestigial form [36–38*, and unpublished observation of Dr. A. Ferenczy, Montreal], and this can be diagnostically disconcerting. Such müllerian remnants may occur no more often than in normal males [39], but I am unaware of formal control data on this point. In any event, the diagnosis of complete TF in a patient with otherwise classic signs should not be discarded just because endoscopy or surgery discloses a rudimentary oviduct or a hypoplastic uterus. Comments in a similar vein are in order with regard to the wolffian system. In theory, none of the major wolffian derivatives should be found in the pelvis, or attached to the testes, of a patient with CAI. But, not uncommonly, an epididymis or hypoplastic vas deferens will be found [40–43; and unpublished observation of Dr. A. Ferenczy, Montreal].

At puberty, a firm diagnosis of complete TF can be made clinically when a female develops breasts and female body contours in the absence of menses, normal sexual hair, or any signs of virilization. Normal (or elevated) adult levels of luteinizing hormone, solidify the diagnostic suspicion. Before puberty, patients with female external genitalia who come to clinical attention because of externally visible testes can be suspected of having complete TF if they have a normal plasma testosterone response after a short course of stimulation by human chorionic gonadotropin [44].

In patients of either age, but particularly in those who are prepubertal or lack a positive family history, the diagnosis should be proven – and for academic purposes molecularly classified – by assaying the integrity of the receptor-acceptor system within the patient's androgen target cells. The most convenient cells to use for this purpose are cultured skin fibroblasts, preferably those derived from genital skin. A simple way to do this is to expose cells in monolayer culture to medium containing ^{3}H-labeled T or DHT, and to characterize the process by which they are bound to the specific androgen receptor proteins in the cytoplasm and to the acceptor sites on the chromatin. If certain kinetic restrictions of the assay are obeyed, then estimates of the maximum specific binding capacity and the specific binding affinity for each level of binding can be obtained with relatively few cells. Representative results of the combined cytoplasmic and 0.4 M KCl-extractable nuclear receptor activity in skin fibroblasts cultured from the prepuce of a normal male and the labium majus of a patient with complete testicular feminization are shown in Figure 5.

The results obtained when the assay described in Figure 5 was applied to labium majus and nongenital skin fibroblast cell strains from controls and patients referred with the diagnosis of complete TF are shown in Figure 6. Notice that the labial strain results are plotted on a scale that is double that of the nongenital strains, and that our range of normals for the nongenital strains extends close to the limits of sensitivity of the assay. Furthermore, the lowest

*The abnormal karyotype was almost certainly fortuitous in view of the patient's testicular histology and family history.

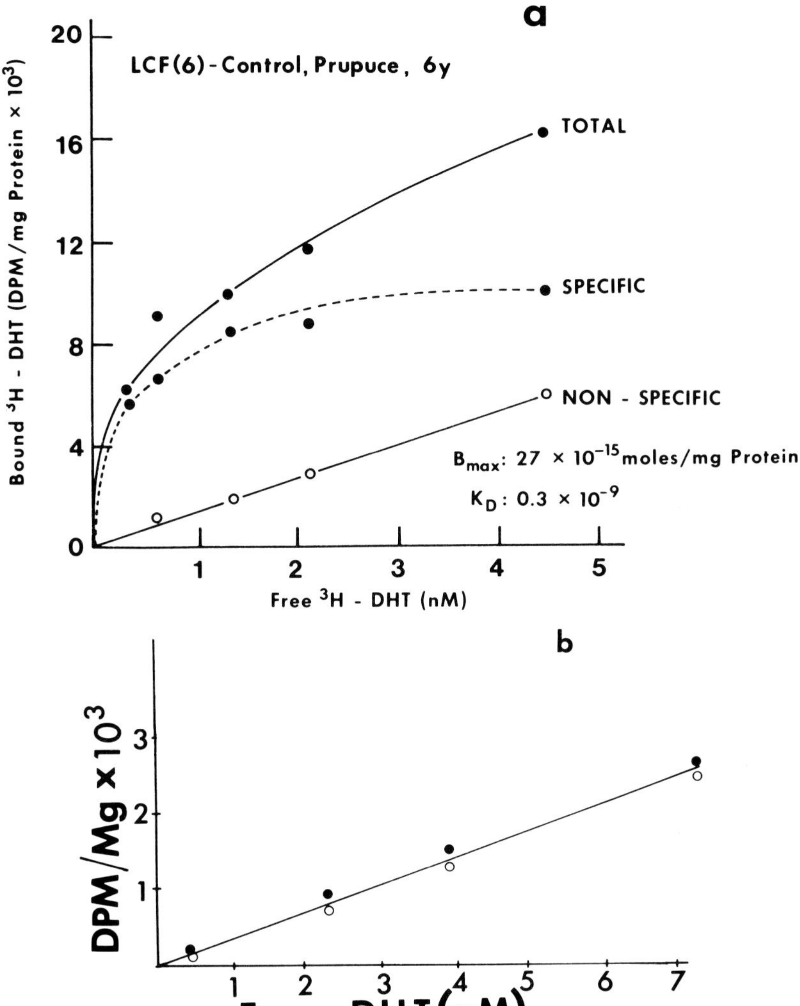

Fig. 5. a) Replicate cultures are incubated with serial concentrations of ^{3}H-labeled DHT in the presence or absence of excess (200-fold) unlabeled hormone. In the absence of excess of unlabeled hormone (top line) the high-affinity, low-capacity, specific cytoplasmic receptor proteins are complexed first, accounting for the steep early portion of the curve, and when these sites are filled, the nonspecific binding continues to increase in a linear, concentration-dependent, unsaturable way. In the presence of excess unlabeled DHT (bottom line) all the high-affinity sites are immediately loaded with unlabeled DHT molecules, and the only sites available for binding with the minority of ^{3}H-DHT molecules are the nonspecific ones. Subtraction of the nonspecific from the total binding values yields a saturable curve (dashed line) that defines the extent and affinity of specific binding as a function of the ligand concentration. b) These are the results on a labium majus skin strain from a patient with complete androgen insensitivity. The "total" and "nonspecific" lines are almost superimposable signifying an almost complete absence of specific androgen receptor activity (from "Clinical Genetics" [84]).

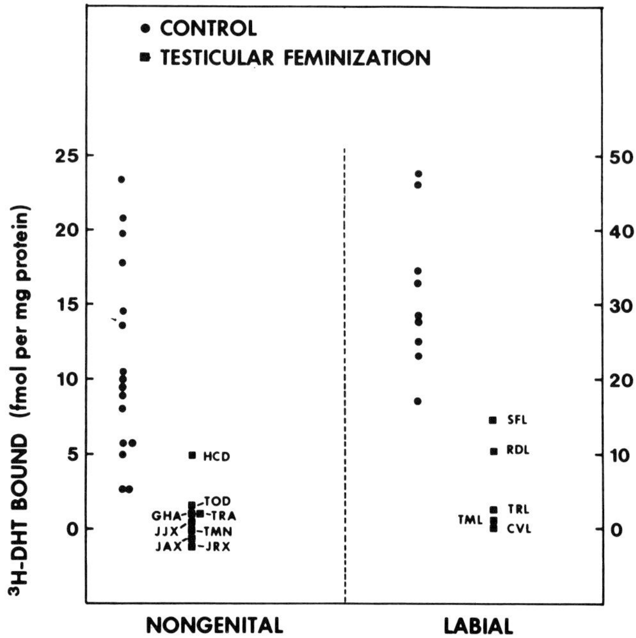

Fig. 6. Specific DHT receptor activity in nongenital and labium majus skin fibroblast strains of controls and patients referred with the diagnosis of complete testicular feminization. (From the Journal of Clinical Investigation [33]).

normal nongenital strain results are juxtaposed to the data obtained on seven of the eight patient strains. The outstanding result (denoted HCD) represents a patient, orchiectomized at 2 years of age, who had unambiguously female external genitalia when we saw her at 10 years of age. We had previously considered her to have complete TF [33] because her maternal grandmother had never had axillary hair. (It is well known that XX females carrying the gene for complete TF not uncommonly have sparse sexual hair and/or delayed menarche [35, 41, 42, 45]). Initially, we interpreted her low but normal androgen-binding activity to mean that she had a receptor-positive form of complete androgen insensitivity. But we have recently learned that at the time of her orchiectomy, the surgery recorded that she had a "somewhat larger uterus" than normal, and we suspect that he meant "clitoris," not uterus. If the clitoris was indeed larger than normal, then she does not merit the label of complete TF and her low-normal nongenital skin fibroblast result raises the question whether she may have a partial deficiency of androgen receptor activity in her genital skin fibroblasts. Furthermore, in the light of this new information, the fact that her maternal grandmother never had axillary hair takes on importance as a diagnostic element in the differential diagnosis of incomplete MP (see below).

The lower limit of the control labial results offers plenty of room for making the diagnosis of complete as well as partial androgen receptor deficiency states. Two of the five patient labial strains came from individuals with unassailable diagnoses of complete TF; yet they had levels of specific androgen receptor activity just below the normal range. The other three had very low levels. We are trying to acquire nongenital strains from the two "atypical" patients to see where these results will fall. If they fall in the normal range, then it will become apparent that some patients with the phenotype of complete TF associated with a partial quantitative deficiency of cytoplasmic androgen receptor activity can only be diagnosed with certainty by means of a study of their genital skin fibroblasts. Furthermore, the quantitative disparity between the clinical and receptor phenotypes of such patients would suggest that their receptor activity is qualitatively abnormal beyond its degree of quantitative deficiency.

The threefold difference in androgen receptor activity between labial and nonlabial skin fibroblast strains (which persists even when they are serially subcultured to near-senescence) deserves additional discussion from the academic viewpoint [46]. Ohno has argued [47, 48], that since androgen receptors seem to be a constitutive property of all cells (including those that seem to have no need for them), there is thus no need to postulate a level of regulation that controls the amount or specificity of the androgen receptors among and within different androgen target cells. In other words, he visualizes that the strength and specificity of androgen target cell responses reside entirely at the postreceptor level of regulation. The difference in androgen receptor activity between genital and nongenital strains counters this argument.

"VARIABLE" CONGENITAL ANDROGEN INSENSITIVITY

The core data on this form of incomplete AI are shown in Table 5. The remarkable thing about this disorder is its intrafamilial variability [49, 50]. Now, as those in the field of birth defects know too well, such variability among mendelian malformation syndromes is the rule rather than the exception. It is seldom, however, that we can ask whether such variability represents leakiness at the genotypic rather than at the phenotypic level, and here we can at least try. Indeed, the first data on the androgen receptor activity in cultured labial skin fibroblasts from three members of one such family are at least compatible with the premise that there is variable expression of the mutant gene at a level close to its gene product: One of the three patients had a result like that from patients with complete TF, whereas the other two fell in an intermediate position between the normal results and those of patients with complete TF [31].

Three grades of intrafamilial variability have been defined among the 11 affected members of this family (Table 6). Notice that those in the most under-

TABLE 5. Core Features of Familial-Variable Congenital Androgen Insensitivity

Previous Names:	Familial incomplete male pseudohermaphroditism, type 1: Reifenstein syndrome
Distribution:	Variable
Severity:	Variable
Chronology:	Variable
Cause:	X-linked gene (?)
Mechanism:	Variable (?) deficit of receptor activity

masculinized category nearly merit the diagnosis of complete TF. Indeed, in a second family [51] (whose androgen insensitivity is admittedly presumptive), the authors assigned that label (albeit inaccurately) to three affected individuals, and reserved the term "pseudohermaphroditimus masculinus" for a fourth who had greater masculinization. In a third such family [52], two sibs were identified as having the Reifenstein syndrome (a form of incomplete MP), while two of their maternal aunts merited the clinical diagnosis of complete TF.

FEMALE CARRIER EXPRESSION IN "VARIABLE" ANDROGEN INSENSITIVITY

It is not well known that XX female relatives of patients known or presumed to have "variable" AI may have sparse sexual hair. This has been seen in the families described by Temtamy et al [52], Williamson and Dewhurst [53], and Verma et al [54], and possibly, in that of our own patient (HCD) discussed earlier. The importance of such expression is dual: It somewhat favors X-linkage of the mutation rather than autosomal dominant inheritance with sex-modified expression; and it can serve as an early clue to the differential diagnosis of incomplete MP because females heterozygous for the autosomal recessive mutations that cause deficient testosterone biosynthesis or decreased T to DHT conversion are not known to have any clinical expression of their carrier states. In fact, there does not appear to be any clinical expression even in females who are homozygous for the mutation causing deficient 5α-reductase activity [55].

TESTOSTERONE 5α-REDUCTASE DEFICIENCY

This disorder is due to homozygosity for an autosomal recessive gene that results in a defect in the conversion of testosterone to DHT within androgen target cells. The enzyme deficiency yields a form of androgen insensitivity* that is specific for testosterone, focal in distribution, and inconstant in time; reciprocally, it can be seen to yield a state of DHT dependency.

*The designation of this disorder as a form of androgen insensitivity may be questioned. It is so designated on the grounds that a defect at any level in the normal physiology of a target cell may confer androgen insensitivity on it. For instance, it is possible that insensitivity could result from a defect in the carrier-mediated transport process that appears to govern the entry of androgen into at least some types of target cell [4].

TABLE 6. Familial-Variable Congenital Androgen Insensitivity. Phenotypic Spectrum in One Family [49]

Degree of defect	Phenotype
Minimal	Micropenis Scrotum (± small, ± prominent raphe, otherwise normal) No gynecomastia Normal wolffian derivatives
Moderate	Perineoscrotal hypospadias Single (common) urogenital orifice Gynecomastia Hypoplastic wolffian derivatives
Severe	Perineoscrotal hypospadias Separate vaginal and urethral orifices Gynecomastia Absent wolffian derivates

The external genital features of prepubertal patients are on the whole feminine: Their phallus is clitoral, their labioscrotal folds are primarily labial, and they usually have a single urogenital opening that encloses a urethral orifice anteriorly and a vaginal orifice posteriorly. Occasionally, there are separate vaginal and urethral orifices on the perineum, or there is a single urethral orifice with a blind vagina entering the urethra posteriorly, just proximal to its tip.

The focality of the prenatal androgen-insensitive phenotype in these patients is striking: Their wolffian derivatives differentiate normally but their external genital primordia and UG sinus remain feminine, with the exception that there is some posterior fusion of the labioscrotal folds and that a small prostate buds off the UG sinus. In other words, differentiation of the male internal genitalia can occur in the presence of T alone, but external genital and UG sinus differentiation requires DHT. This is entirely in keeping with the facts that wolffian duct derivatives acquire the ability to convert T to DHT after they differentiate, while the external genital primordia and the UG sinus acquire this ability before they differentiate [56].

At puberty dramatic, and equally focal, virilization occurs: Male muscularity appears, the voice deepens and the external genitalia enlarge impressively, but there is little or no beard growth, no temporal recession of the hair line, little or no acne, and the prostate remains small [55]. Inferentially, the former functions are T-dependent, while the latter are DHT-dependent. The pubertal "reactivation" of the external genitalia is incompletely explained: It may simply reflect the fact that external genital growth, as distinct from its differentiation is T-dependent; nevertheless, acquisition of increased 5α-reductase activity and the extent to which pubertal levels of luteinizing hormone can overcome a constant enzyme deficiency deserve consideration.

The important clinical point to make here is that this postpubertal picture and its prepubertal antecedent are not unique: A very similar course can occur in patients with a deficiency of 17-ketosteroid reductase [24–27], the enzyme responsible for the final step in T biosynthesis. It used to be thought that the absence of gynecomastia in those with 5α-reductase deficiency could differentiate these two forms of incomplete MP, but even this distinction has not been sustained [57].

The quickest way to support a suspicion of 5α-reductase deficiency is to measure the plasma levels of T and DHT simultaneously, and to do so in prepubertal patients only after a course of testicular stimulation by human chorionic gonadotropin [44]. The normal ratio of T to DHT in plasma is less than 20:1; in patients with 5α-reductase deficiency it is more than 30:1 [55]. The failure of one patient with sound evidence for 5α-reductase deficiency to display a diagnostic ratio is unexplained [58]. Another reason for evaluating the suspicion of 5α-reductase deficiency before orchiectomy is that at puberty the testes of these patients may exhibit advanced stages of spermatogenesis [55] in contrast to patients with similar phenotypes due either to deficient testosterone biosynthesis or a defective receptor-acceptor system in whose seminiferous tubules only spermatogonia and Sertoli cells can be seen.

If this disorder is suspected in a patient who has already been orchiectomized (unfortunately, still a much too frequent occurrence) then it seems that the next best way to document the enzyme deficiency is to incubate a fresh slice of labial skin with testosterone and then measure the rate of production of its various 5α-reduced derivatives compared to that of control labial skin. In one study, the reduced activity of the patient's tissue was dramatic [58].

The reason these particular diagnostic approaches have been stressed is that I am somewhat insecure about making the diagnosis by means of cultured skin fibroblasts, even when they are derived from genital skin.

THE DIAGNOSIS OF 5α-REDUCTASE DEFICIENCY IN CULTURED SKIN FIBROBLASTS

Most of the work in this area has been done by Jean Wilson and his colleagues [59–61]. They have confirmed previous data [62] indicating that, as a group, genital (prepuce, labium majus, scrotum) skin fibroblasts have more 5α-reductase activity than fibroblasts from various nongenital skin sites, and they have emphasized the remarkable variability in this activity among a large series of control genital strains, even when the cell culture conditions prior to and/or during the assay are rigidly standardized. In two of their studies [59, 60] foreskin strains varied 40- and 80-fold, respectively, and the labium majus strains varied even more [60]. The foreskin strains of two sibs with 5α-reductase deficiency could barely be distinguished from the lowest values obtained on controls, and, expectedly, neither of the mutant nongenital strains could be distinguished from the controls [59–61]. Discrimination of the mutant genital strains remained just as difficult even when the assay used was able to separate a background 5α-reductase activity with a pH optimum of 9 from the genital skin-specific form of the enzyme whose pH optimum is 5.5 [61]. Clearly, it is easy to exclude 5α-reductase deficiency using cultured genital skin fibroblasts, but it is considerably more difficult to affirm it. Furthermore, to affirm the diagnosis, it appears both in our experience [unpublished observation], and in others' [61], to be crucial that phallic skin fibroblasts be assayed, since control labial skin fibroblasts have a lower level of variability that is at the limit of sensitivity of the assay. The ability of Peterson and co-workers [63] to recognize 5α-reductase deficiency in cultured nongenital skin fibroblasts is unexplained; it may reflect an inadequate experience with control nongenital strains.

The enormous variability in 5α-reductase activity among control foreskin strains appears to stem from an equally large amount of variability that has been seen among clones of individual fibroblasts isolated from the earliest monolayer outgrowth of a single 1-mm^3 explant of foreskin [64]. Such constitutive-clonal heterogeneity in vitro either reflects a significant degree of cellular nonuniformity in the in situ differentiative process that distinguishes most genital skin fibroblasts from their nongenital skin counterparts, or it reflects a dedifferentiative behavior of genital skin fibroblasts in vitro that has an equal degree of clonal heterogeneity [46].

THE FUTURE OF CONGENITAL ANDROGEN INSENSITIVITY

Certain comments can be made about the future of CAI in man based on current knowledge about the subject as well as on various models of steroid insensitivity that have been described in a variety of heteroploid cell lines.

INCOMPLETE ANDROGEN INSENSITIVITY DUE TO RECEPTOR-RELATED DEFECTS: VARIABLE EXPRESSIVITY VERSUS GENETIC HETEROGENEITY

The variable expressivity within families with "variable" AI circumscribes that seen in a series of prototypic families in the older literature that illustrate maternal inheritance and whose phenotypes form a gradation of undermasculinization presumably on the basis of incomplete receptor-related androgen insensitivity [65–68]. It has been suggested, with some justification, that these more or less discrete family phenotypes merely represent different slices in the spectrum of expression of the same mutation that is responsible for "variable" AI. In order to evaluate this suggestion, particular attention should be paid to the question of phenotypic constancy when reporting on new families with incomplete AI [69, 70]. It will also be important to evaluate the receptor status of genital skin fibroblasts derived from such patients for two reasons: First, because partial deficiencies of receptor-related activities, even if universally distributed, are more likely to be recognizable in them than in nongenital skin fibroblasts; and second, because it is conceivable that some patients have receptor-related androgen insensitivity that is restricted to their genital tissues. Patients in the latter (hypothetic) category could be viewed as suffering from a fault in the normal differentiative (epigenetic) process, which determines that the average cultured genital skin fibroblasts should have more cytoplasmic androgen receptor and 5α-reductase activity than its nongenital counterpart [46].

COORDINATE VERSUS INCOORDINATE MASCULINIZATION IN "VARIABLE" AI

A feature of the phenotypic variability within the family with "variable" AI described in Table 6 is that, within any indivdual, the distribution of masculinzation is coordinate; for instance, the patient with the greatest degree of masculinization did not develop gynecomastia. Close examination of the available literature, however, reveals several instances of incoordinate masculinization, as documented in Table 7. Such apparent target organ control over the expression of certain mutations causing AI deserves close scrutiny in the future.

MODELS OF STEROID INSENSITIVITY

Insensitivity to the cytotoxicity of glucocorticoid has been observed in the lymphoblasts from patients with acute lymphoblastic leukemia [71, 72], and in various mouse cell lines [73, 74], including those developed from lymphomas [75] and leukemias [76]. In some of these situations, insensitivity has been attributable to quantitatively deficient cytoplasmic receptor activity (R^-); in others, receptor activity has been quantitatively normal (R^+) [77], implying that there are faults in the quality of the cytoplasmic receptor or in other components of the receptor-initiated process responsible for normal glucocorticoid sensitivity. Defects of the former type have been found in several dexamethasone-resistant mouse lymphoma cell lines [78], but not in any of the dexamethasone-resistant HTC rat hepatoma clones [79].

Insensitivity to the growth-promoting effect of androgenic steroids has been described in variant lines derived from the original androgen-dependent mouse mammary adenocarcinoma, Shionogi carcinoma 115. As described above for glucocorticoid insensitivity, autonomous (androgen-insensitive) Shionogi cell lines have been found with and without quantitatively normal levels of cytoplasmic androgen receptor activity [80–82]. Furthermore, the autonomous variants

TABLE 7. Some Examples of Incoordinate Masculinization in Patients Known* or Assumed† to Have Familial Incomplete Androgen Insensitivity Due to a Defect in the "Receptor" System

	Keenan et al [70]	Temtamy et al [52]	Verma et al [54]	Williamson and Dewhurst [53]
Patient number	Case 1	Case 2 – family 2	III-1	Case 1
Age (yr)	20	14	16	15
Habitus	Male	Female	Male	Female
Voice	High	Masculine	Masculine	Husky
Breast development	Mastectomy	Yes	No	Full development
Hair-axillary	Yes	No	No	No
-pubic	Yes (♀ pattern)	Scanty	No	Sparse
External genitals	Perineoscrotal hypospadias	Third-degree hypospadias	Clitoral phallus (2 cm) Labia majora No vaginal orifice	Moderate clitoromegaly Labia majora Very short vagina

*The family described by Keenan et al was shown to have androgen insensitivity in vivo and supranormal testosterone 5α-reductase activity within cultured inguinal skin fibroblasts.

†Each of the families described by Temtamy et al, Verma et al, and Williamson and Dewhurst contains affecteds in more than one generation linked maternally and each contains at least one fertile female with scanty sexual hair, a carrier phenotype indicating X-linkage or sex-limited, autosomal dominant inheritance. The form of androgen insensitivity due to deficient 5α-reductase activity is determined by an autosomal recessive gene and is not known to find clinical expression in carrier females whether they are heterozygous or homozygous for the mutant gene.

appear spontaneously in vitro during serial subculture of dependent clones, so that a variety of androgen-insensitivity mutations should be available from this source for comparative analysis with cell strains derived from patients with CAI.

It is important to underline in all such work that steroid insensitivity in a target cell with quantitatively normal cytoplasmic receptor activity cannot be taken to indicate a "postreceptor" defect until all the functional domains of the receptor have been verified: In addition to a normal capacity and affinity for cytoplasmic binding of the steroid, these include the abilities to transfer the steroid-receptor complex to the nucleus and to bind the complex to chromatin and, probably, the subsequent ability to initiate the sequence of events that culminates in the transcription of target organ-specific mRNA molecules.

REFERENCES

1. Pinsky L: Human male sexual maldevelopment: Teratogenic classification of monogenic forms. Teratology 10:193–204, 1974.
2. Davidoff F, Federman DD: Mixed gonadal dysgenesis. Pediatrics 52:725–742, 1973.
3. Smith DW: "Recognizable Patterns of Human Malformation." Philadelphia: WB Saunders, 1976.
4. Georgi EP: Studies on androgen transport into canine prostate in vitro. J Endocrinol 68:109–119, 1976.
5. Mainwaring WIP, Symes EK, Higgins SJ: Nuclear components responsible for the retention of steroid-receptor complexes, especially from the standpoint of the specificity of hormonal responses. Biochem J 156:129–141, 1976.
6. Tsai SY, Tsai MJ, Schwartz R, Kalimi M, Clark JH, O'Malley BW: Effects of estrogen on gene expression in chick oviduct: Nuclear receptor levels and initiation of transcription. Proc Natl Acad Sci USA 72:4228–4232, 1975.
7. Silvers WK, Wachtel SS: H-Y antigen: Behavior and function. Science 195:956–960, 1977.
8. Espiner EA, Veale AMO, Sands VE, Fitzgerald PH: Familial syndrome of streak gonads and normal male karyotype in five phenotypic females. N Engl J Med 283:6–11, 1970.
9. Sternberg WH, Barclay DL, Kloepfer HW: Familial XY gonadal dysgenesis. N Engl J Med 278:695–700, 1968.
10. Baron J, Rucki T, Sim S: Familial gonadal malformations. Gynecologia 153:298–308, 1962.
11. Barr ML, Carr DH, Plunkett ER, Soltan HC, Weins RG: Male pseudohermaphroditism and pure gonadal dysgenesis in sisters. Am J Obstet Gynecol 99:1047–1055, 1967.
12. Bartlett DJ, Grant JK, Pugh MA, Aherne W: A familial feminizing syndrome. J Obstet Gynacol Br Commun 75:199–210, 1968.
13. Chemke J, Carmichael R, Stewart JM, Freer RH, Robinson A: Familial XY gonadal dysgenesis. J Med Genet 7:105–111, 1970.
14. Jirasek JE: Testicular dysgenesis syndromes. In Bergsma D (ed): Part X. "The Endocrine System." Baltimore: Williams & Wilkins for The National Foundation – March of Dimes, BD:OAS VII(6):159–165, 1971.
15. Winter JSD, Faiman C, Reyes FI: Sex steroid production by the human fetus: Its role in morphogenesis and control by gonadotropins. In Blandau RJ, Bergsma D (eds): "Morphogenesis and Malformation of the Genital System." New York: Alan R Liss for The National Foundation–March of Dimes, BD:OAS XIII (2):41–58, 1977.

16. Josso N, Picard J-Y, Tran D: The anti-müllerian hormone. Ibid pp 59–84.
17. Smith DW: Discussion following: Classification of hereditary disorders of sexual development. In Bergsma D (ed): "Genetic Forms of Hypogonadism." Miami: Symposia Specialists for The National Foundation–March of Dimes, BD:OAS XI(4):15, 1975.
18. Gobel P: 17-Hydroxylation deficiency with mineralocorticoid and testicular feminization syndrome. In Bergsma D (ed): Part XVI. "Urinary System and Others." Baltimore: Williams & Wilkins for The National Foundation–March of Dimes, BD:OAS X(4):301–303, 1974.
19. Heremans GFP, Moolenaar AJ, van Gelderen HH: Female phenotype in a male child due to 17α-hydroxylase deficiency. Arch Dis Child 51:721–723, 1976.
20. Mantero F, Busnardo B, Riondel A, Veyrat R, Austoni M: Hypertension artérielle, alcalose hypokaliémique et pseudohermaphrodisme mâle par deficit en 17α-hydroxylase. Schweiz Med Wochenschr 101:38–43, 1971.
21. Bricaire H, Luton JP, Laudat P, Legrand JC, Turpin G, Corvol PM, Lemmer M: A new male pseudohermaphroditism associated with hypertension due to a block of 17α-hydroxylation. J Clin Endocrinol Metab 35:67–72, 1972.
22. Tourniaire J, Audi-Parera L, Loras B, Blum J, Castelnovo P, Forest MG: Male pseudohermaphroditism with hypertension due to a 17α-hydroxylation deficiency. Clin Endocrinol 5:53–61, 1976.
23. Goebelsman U, Zachmann M, Davajan V, Israel R, Mestman JH, Mishell DR: Male pseudohermaphroditism consistent with 17,20-desmolase deficiency. Gynecol Invest 7:138–156, 1976.
24. Saez JM, de Peretti E, Morera AM, David M, Bertrand J: Familial male pseudohermaphroditism with gynecomastia due to a testicular 17-ketosteroid reductase defect. I. Studies in vitro. J Clin Endocrinol Metab 32:604–610, 1971.
25. Saez JM, Morera AM, de Peretti E, Bertrand J: Further in vivo studies in male pseudohermaphroditism with gynecomastia due to a testicular 17-ketosteroid reductase defect (compared to a case of testicular feminization). J Clin Endrocinol Metab 34:598–600, 1972.
26. Goebelsman U, Horton R, Mestman JH, Arce JJ, Nagata Y, Nakamura RM, Thorneycroft IH, Mishell DR Jr: Male pseudohermaphroditism due to testicular 17β-hydroxysteroid dehydrogenase deficiency. J Clin Endocrinol Metab 36:867–879, 1973.
27. Zurbrugg RP: Inborn errors in testosterone biosynthesis with special reference to 17-oxosteroid reductase deficiency. Helv Paediatr Acta (Suppl) 34:63–77, 1974.
28. Zachmann M, Vollmin JA, Hamilton W, Prader A: Steroid 17,20-desmolase deficiency: A new cause of male pseudohermaphroditism. Clin Endocrinol 1:369–385, 1972.
29. New MI: Male pseudohermaphroditism due to 17α-hydroxylase deficiency. J Clin Invest 49:1930–1941, 1970.
30. Madden JD, Walsh PC, MacDonald PC: Clinical and endocrinologic characterization of a patient with the syndrome of incomplete testicular feminization. J Clin Endocrinol Metab 41:751, 1975.
31. Griffin JE, Punyashthiti K, Wilson JD: Dihydrotestosterone binding by cultured human fibroblasts. Comparison of cells from control subjects and from patients with hereditary male pseudohermaphroditism due to androgen resistance. J Clin Invest 57:1342–1351, 1976.
32. Keenan BS, Meyer WJ III, Hadjian AJ, Jones HW, Migeon CJ: Syndrome of androgen insensitivity in man: Absence of 5α-dihydrotestosterone binding protein in skin fibroblasts. J Clin Endocrinol Metab 38:1143–1146, 1974.
33. Kaufman M, Straisfeld C, Pinksy L: Male pseudohermaphroditism presumably due to target organ unresponsiveness to androgen. Deficient 5α-dihydrotestosterone binding in cultured fibroblasts. J Clin Invest 58:345–350, 1976.

34. Amrhein JA, Meyer WJ III, Jones HW Jr, Migeon CJ: Androgen insensitivity in man: Evidence for genetic heterogeneity. Proc Natl Acad Sci USA 73:891–894, 1976.
35. Pion RJ, Dignan WJ, Lamp EJ, Moore JG, Frankland MV, Simmer HH: Testicular feminization. Am J Obst Gynecol 93:1067–1075, 1965.
36. Sawetawan V: Testicular feminization syndrome. A report on 4 siblings. J Med Assoc Thai 54:751–757, 1971.
37. Gordon DL, Krmpotic E, Zelinger BB, Rosenblum LE: Sex chromsome mosaicism in the testicular feminization syndrome. Obstet Gynecol 34:779–782, 1969.
38. Gordon DL, Miller LR, Bensch KG: Electron miscroscopic observations of the gonad in the testicular feminization syndrome. Lab Invest 13:152–160, 1964.
39. Naftolin F, Judd HL: Testicular feminization. In Wynn RM (ed): "Obstetrics and Gynecology Annual." New York: Appleton-Century-Crofts, vol 2, 1973, pp 25–53.
40. Lockwood E, Ghosh PC, Pennington GW, Tipton R: Steroid excretion and metabolism by gonadal tissue from a subject with testicular feminization syndrome. J Clin Pathol 27:135–139, 1974.
41. Turksoy RN, Mitchell DW, Safaii HS: Testicular response to exogenous gonadotropins in the syndrome of feminizing testes. Fertil Steril 27:670–676, 1976.
42. Laurian L, Finkelstein JO, Soferman N, Loewenthal M, Diligdisch L: Testicular feminization syndrome. Isr J Med Sci 7:668–673, 1971.
43. Faiman C, Winter JSD: The control of gonadotropin secretion in complete testicular feminization. J Clin Endocrinol Metab 39:631–638, 1974.
44. Grant RB, Laurance BM, Atherden SM, Ryness J: HCG stimulation test in children with abnormal sexual development. Arch Dis Child 51:596–601, 1976.
45. Morris JM, Mahesh VB: Further observations on the syndrome "testicular feminization." Am J Obstet Gynecol 87:731–745, 1963.
46. Kaufman M, Straisfeld C, Pinsky L: Expression of androgen-responsive properties in human skin fibroblast strains of genital and nongenital origin. Somat Cell Genet 3:17–25, 1977.
47. Ohno S: Major regulatory genes for mammalian sexual development. Cell 7:315–321, 1976.
48. Ohno S: Testosterone and cellular response. In Blandau RJ, Bergsma D (eds): "Morphogenesis and Malformation of the Genital System." New York: Alan R Liss for The National Foundation–March of Dimes, BD:OAS XIII(2):99–106, 1976.
49. Wilson JD, Harrod MJ, Goldstein JL, Hemsell BL, MacDonald PC: Familial incomplete male pseudohermaphroditism, type 1: Evidence for androgen resistance and variable clinical manifestations in a family with the Reifenstein syndrome. N Engl J Med 290:1098–1103, 1974.
50. Walker AC, Stack EM, Horsfall WA: Familial male pseudohermaphroditism. Med J Aust 1:156–160, 1970.
51. Gardo S, Papp Z: Clinical variations of testicular intersexuality in a family. J Med Genet 11:267–270, 1974.
52. Temtamy SA, Loutfy AH, Hetta F, Raafat M, Atteya OM, Boulos SY: Familial male hermaphroditism. In Bergsma D (ed): Part XVI. "Urinary System and Others." Baltimore: Williams & Wilkins for The National Foundation –March of Dimes, BD:OAS X(4): 243–247, 1974.
53. Williamson JA, Dewhurst CJ: Testicular feminization with clitoral enlargement. J Obstet Gynecol Br Commun 77:356–358, 1970.
54. Verma IC, Sansi PK, Kumar V, Ahuja MMS: Familial male pseudohermaphroditism with female external genitalia, presence of labial testes, male habitus and voice, and complete absence of müllerian structures and lack of breast development. In Bergsma D (ed): "Genetic Forms of Hypogonadism." Miami: Symposia Specialists for The National Foundation–March of Dimes, BD:OAS XI(4):145–152, 1975.

55. Peterson RE, Imperato-McGinley J, Gautier T, Sturla E: Male pseudohermaphroditism due to steroid 5α-reductase deficiency. Am J Med 62:170–191, 1977.
56. Siiteri PK, Wilson JD: Testosterone formation and metabolism during male sexual differentiation in the human embryo. J Clin Endocrinol Metab 38:113–125, 1974.
57. Givens JR, Wiser WL, Summitt RL, Kerber IJ, Andersen RN, Pittaway DE, Fish SA: Familial male pseudohermaphroditism without gynecomastia due to deficient testicular 17-ketosteroid reductase activity. N Engl J Med 291:938–944, 1974.
58. Walsh PC, Madden JD, Harrod MJ, Goldstein JL, MacDonald PC, Wilson JD: Familial incomplete male pseudohermaphroditism, type 2: Decreased dihydrotestosterone formation in pseudovaginal perineoscrotal hypospadias. N Engl J Med 291:944–949, 1974.
59. Wilson JD: Dihydrotestosterone formation in cultured human fibroblasts. Comparison of cells from normal subjects and patients with familial incomplete male pseudohermaphroditism, type 2. J Biol Chem 250:3498–3504, 1975.
60. Moore RJ, Griffin JE, Wilson JD: Diminished 5α-reductase activity in extracts of fibroblasts cultured from patients with familial incomplete pseudohermaphroditism, type 2. J Biol Chem 250:7168–7172, 1975.
61. Moore RJ, Wilson JD: Steroid 5α-reductase in cultured human fibroblasts. J Biol Chem 251:5895–5900, 1976.
62. Pinsky L, Finkelberg R, Straisfeld C, Zilahi B, Kaufman M, Hall G: Testosterone metabolism by serially subcultured fibroblasts from genital and nongenital skin of individual human donors. Biochem Biophys Res Commun 46:364–369, 1972.
63. Imperato-McGinley J, Guerrero L, Gautier T, German JL, Peterson RE: Steroid 5α-reductase deficiency in man. An inherited form of male pseudohermaphroditism. In Bergsma D (ed): "Genetic Forms of Hypogonadism." Miami: Symposia Specialists for The National Foundation–March of Dimes, BD:OAS XI(4):91–103, 1975.
64. Kaufman M, Pinsky L, Straisfeld C, Shanfield B, Zilahi B: Qualitative differences in testosterone metabolism as an indication of cellular heterogeneity in fibroblast monolayers derived from human preputial skin. Exp Cell Res 96:31–36, 1975.
65. Lubs HA Jr, Vilar O, Bergenstal DM: Familial male pseudohermaphroditism with labial testes and partial feminization: Endocrine studies and genetic aspects. J Clin Endocrinol Metab 19:1110–1120, 1959.
66. Gilbert-Dreyfus S, Sebaoun CIA, Belaisch J: Etude d'un cas familial d'androgynoidisme avec hypospadias grave, gynecomastie et hyperoestrogenie. Ann Endocrinol (Paris) 18:93–101, 1957.
67. Bowen P, Lee CSN, Migeon CG, Kaplan NM, Whalley PG, McKusick VA, Reifenstein EC: Hereditary male pseudohermaphroditism with hypogonadism, hypospadias, and gynecomastia (Reifenstein's syndrome). Ann Intern Med 62:252–270, 1965.
68. Rosewater S, Gwinup G, Hamwi GJ: Familial gynecomastia. Ann Intern Med 63:377–385, 1965.
69. Perez-Palacios G, Ortiz S, Lopez-Amor E, Morato T, Febres F, Lisker R, Scaglia H: Familial incomplete virilization due to partial and organ sensitivity to androgens. J Clin Endocrinol Metab 41:946–952, 1975.
70. Keenan PS, Kirkland JL, Kirkland RT, Clayton GW: Male pseudohermaphroditism with partial androgen insensitivity. Pediatrics 59:224–231, 1977.
71. Lippman ME, Halterman RH, Leventhal BG, Perry S, Thompson EB: Glucocorticoid binding proteins in human acute lymphoblastic leukemic blast cells. J Clin Invest 52:1715–1725, 1973.
72. Bird CC, Waddell AW, Robertson AMG, Currie AR, Steel CM, Evans J: Cytoplasmic receptor levels and glucocorticoid response in human lymphoblastoid cell lines. Br J Cancer 32:700–707, 1975.

73. Lippman ME, Thomspon EB: Steroid receptors and the mechanism of the specificity of glucocorticoid responsiveness of somatic cell hybrids between hepatoma tissue culture cells and mouse fibroblasts. J Biol Chem 249:2483–2488, 1974.
74. Garroway NW, Orth DN, Harrison RW: Binding of cytosol receptor-glucocorticoid complexes by isolated nuclei of glucocorticoid-responsive and non-responsive cultured cells. Endocrinology 98:1092–1100, 1976.
75. Sibley CH, Tomkins GM: Mechanisms of steroid resistance. Cell 2:221–227, 1974.
76. Lippman ME, Perry S, Thompson EB: Cytoplasmic glucocorticoid-binding proteins in glucocorticoid-unresponsive human and mouse leukemic cell lines. Cancer Res 34:1572–1576, 1974.
77. Yamamoto KR, Stampfer MR, Tomkins GM: Receptors from glucocorticoid-sensitive lymphoma cells and two classes of insensitive clones: Physical and DNA binding properties. Proc Natl Acad Sci USA 71:3901–3905, 1974.
78. Gehring U, Tomkins GM: A new mechanism for steroid unresponsiveness: Loss of nuclear binding activity of a steroid hormone receptor. Cell 3:301–306, 1974.
79. Thompson EB, Aviv D, Lippman ME: Variants of HTC cells with low tyrosine amino transferase inducibility and apparently normal glucocorticoid receptors. Endocrinology 100:406–419, 1977.
80. Bruchovsky N, Meakin JW: The metabolism and binding of testosterone in androgen-dependent and autonomous transplantable mouse mammary tumors. Cancer Res 33:1689–1695, 1973.
81. Bruchovsky N, Sutherland DJA, Meakin JW, Minesita T: Androgen receptors: Relationship to growth response and to intracellular androgen transport in 9 variant lines of the Shionogi mouse mammary carcinoma. Biochim Biophys Acta 381:61–71, 1975.
82. Stanley ER, Palmer RE, Sohn U: Development of methods for the quantitative in vitro analysis of androgen-dependent and autonomous Shionogi carcinoma 115 cells. Cell 10:35–44, 1977.
83. Buller RE, O'Malley BW: The biology and mechanism of steroid hormone receptor interaction with the eukaryotic nucleus. Biochem Pharmacol 25:1–12, 1976.
84. Kaufman M, Straisfeld C, Pinsky L: Specific 5α-dihydrotestosterone binding in labial skin fibroblasts cultured from patients with male pseudohermaphroditism. Clin Genet 9:567–574, 1976.
85. Mainwaring WIP: "The Mechanisms of Action of Androgens." Monographs on Endocrinology. Philadelphia: Springer-Verlag, 1977, vol 10, pp 92–94.
86. Mainwaring WIP: Ibid pp 19 and 52.
87. Federman DD: "Abnormal Sexual Development." Philadelphia: WB Saunders, 1968.
88. Imperato-McGinley J, Peterson RE: Male pseudohermaphroditism: The complexities of male phenotypic development. Am J Med 61:251–272, 1976.

EDITORIAL COMMENT

I agree wholeheartedly with the author in this plea that the term, "incomplete testicular feminization" be abandoned. I agree also that the term, "male pseudohermaphroditism," is useful and should be retained.

However, I would further plead that the use of the terms "complete male pseudohermaphroditism" and "incomplete male pseudohermaphroditism," be abandoned. The terms are, in my opinion, conceptually unsound and quite confusing. I believe that the terms, "male pseudohermaphroditism with female external genital phenotype" and "male pseudohermaphroditism with intersex genital phenotype" are conceptually far more sound. In addition, the use of the terms, "Type 1," "Type 2," and "Type 3" to designate, respectively, partial receptor deficiency, deficient 5α-reductase activity and, again, partial receptor deficiency, is unsound. First, Types 1 and 3 are confusing in that they both designate partial receptor deficiency. No scientific basis exists for the separation of Type 3 from Type 1 as originally proposed by Wilson and colleagues. Second, as the author points out, there are many forms of "incomplete male pseudohermaphroditism" not due to receptor deficiency, in *addition to* deficient 5α-reductase activity. This second point also emphasizes another problem surrounding Types 1, 2, and 3. While the author makes clear that their designation in his paper refers only to types of androgen insensitivity, their usage in general may suggest to the reader, particularly in view of the way they are used by Wilson and colleagues who coined them, that they are the *only* types of "incomplete male pseudohermaphroditism" known – such is not the case. I suggest, instead, for the purpose of designating a form of male pseudohermaphroditism with intersex genitalia, the use of the following: Male pseudohermaphroditism with intersex genitalia due to . . . (eg deficient 5α-reductase activity, deficient 17-ketosteroid reductase activity, etc), or, if the defect is not known, male pseudohermaphroditism with intersex genitalia, etiology undetermined.

I would make one additional plea: Since we would abandon the term, "incomplete testicular feminization," why not also abandon "complete testicular feminization?" I suggest, instead, eg "male pseudohermaphroditism with female phenotype due to complete androgen insensitivity," or, even preferrably, simply, "complete androgen insensitivity," the latter assuming that it is understood that complete androgen insensitivity produces a female phenotype. These changes in terminology require complete delineation of the etiology of each case of male pseudohermaphroditism as stressed by Dr. Pinsky.

Robert L. Summitt

Cytogenetic and Clinical Notes on a Girl With a 46,X,i(Yq) Karyotype, H-Y Antigen-Negative, and a Gonadoblastoma

Jack M. Rary, Doris K. Cummings, Howard W. Jones, Jr., John A. Rock, and Conrad G. Julian

INTRODUCTION

Isochromosomes for the long arm of the Y chromosome are rare and only a few cases have been reported in the literature [1–3]. The isochromosome for the long arm of the Y chromosome serves as a unique opportunity for looking at factors relating to the genetic function of the Y chromosome. Localization of the gene(s) responsible for differentiation of the human gonad on either the long or short arm of the Y chromosome has been a long-standing question [2, 4–7]. This case is also pertinent to the reports of increased risks of gonadal tumors in gonadal dysgenesis patients with either a normal or abnormal Y chromosome [8–10] present in their karyotype. Fibroblasts from this individual were determined to be H-Y antigen-negative. This evidence strongly suggests that the H-Y antigen gene is located on the short arm of the Y chromosome in man.

CASE REPORT

An 11-year-old white female was seen for the first time in September 1975 with the chief complaint of fever.

The family history revealed one brother, age 23, who was mentally retarded from birth. That sib was born with bilateral clubfeet. A sister, age 18, is normal. There are no other discernible family genetic abnormalities.

Birth Defects: Original Article Series, Volume XIV, Number 6C, pages 97–107

Until the time of the present illness the patient had unremarkable prepubertal female development. In August 1975 she received a diphtheria-tetanus booster in her arm and an oral polio vaccine. Her arm became swollen and fever developed. She received no relief from Tylenol and she was brought to her community hospital for evaluation. On physical examination she appeared to be a perfectly normal prepubertal female with a height of 143 cm and a weight of 39 kg. She had slight axillary and pubic hair growth but no breast development. External genitalia were normal for a prepubertal girl of this age. The only positive finding was a large smooth semisolid, semicystic mass arising out of the pelvis to about the level of the umbilicus. Preoperative studies included routine blood work which showed a mild anemia and a normal chest x-ray examination, barium enema, and intravenous pyelogram. Sonography confirmed the findings of a pelvic mass.

Exploratory laparotomy performed on September 11, 1975 revealed a large, soft, thin-walled, cystic lesion of the right gonad. The mass was adherent to the posterior aspects of the broad ligament and a tiny, but otherwise normal, uterus. The right tube and tumor were removed and there was some spillage of tumor. The left gonad appeared to be a streak and was not biopsied or removed. The appendix was removed routinely. The pathology showed a poorly differentiated malignant teratoma (see Figs. 1–3) with prominent dysgerminomatous components.

She was admitted for the first time in September 1975 to The Johns Hopkins Hospital, at which time an extensive metastatic workup was performed and found to be negative. Peripheral blood was obtained for chromosomal analysis and this demonstrated a 46,X,i(Yq) karyotype. She was started on triple chemotherapy on a monthly, pulse fashion. The 3 agents used were vincristine, actinomycin-D, and cytoxan. On October 1, 1975 she received 15 mCi of intraperitoneal radioactive phosphorous. She was readmitted to the hospital in February 1976 for reexploration and removal of the streak. Abdominal and pelvic reexploration revealed obvious signs of peritoneal irradiation and multiple adhesions between the terminal ileum and the uterus. The streak gonad (Fig. 4) was removed along with the left tube. There was no suggestion of recurrent tumor on exhaustive peritoneal exploration. It was elected not to remove the uterus because of the risk of injuring the irradiated small bowel. Multiple peritoneal biopsies, the removed omentum, and peritoneal washings were all negative for tumor. Multiple sections along the length of the streak revealed no evidence of neoplasm (Fig. 4). Sections through the left tube demonstrated typical müllerian architecture. There was no evidence of testicular development. The patient's postoperative course was uneventful. She has continued to receive triple pulse chemotherapy with little morbidity, and at 11 months following the initial diagnosis is free of any evidence of recurrent or metastatic tumor.

Fig. 1. Right dysgenetic gonadal tumor. Area of dysgerminoma prominent in upper portion of photo. (H & E, × 155).

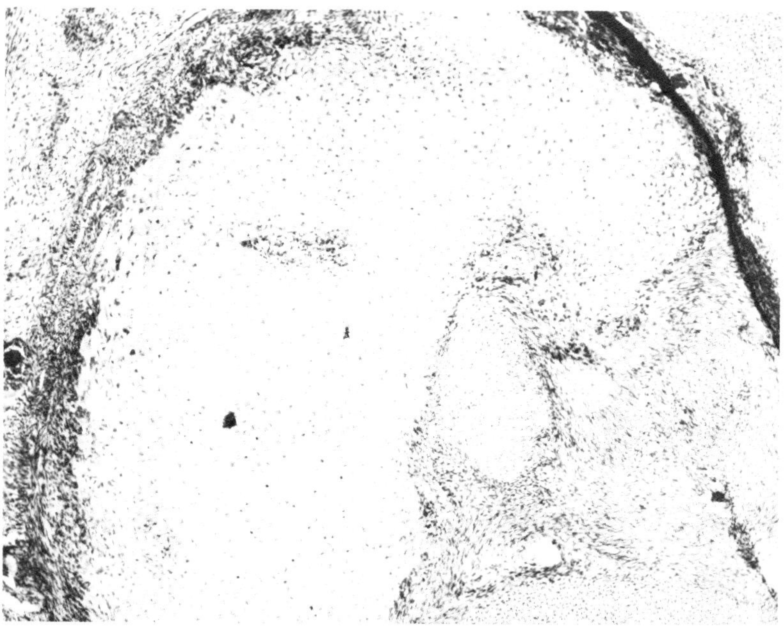

Fig. 2. Dysgenetic right gonadal tumor showing immature cartilage (H & E, × 80).

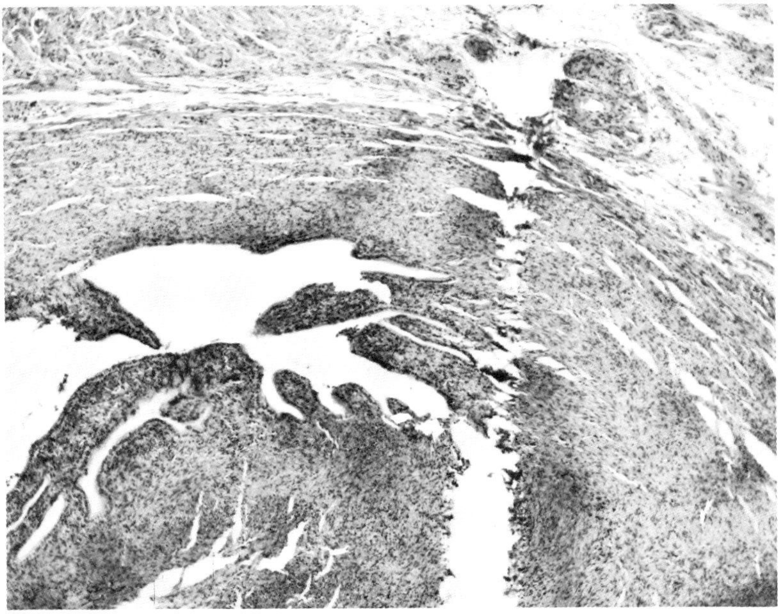

Fig. 3. Right dysgenetic gonadal tumor. Areas of müllerian epithelium in center and left side of figure. (H & E, × 80).

MATERIALS AND METHODS

Metaphase chromosome preparations were obtained by culturing whole blood, skin, and gonadal streak fibroblasts with subsequent harvesting of each by conventional methods. Metaphase figures were then stained with quinacrine mustard and photographed with a Zeiss 50W fluorescent microscope. "C"-Banding and urea-Giemsa banding were also performed on both leukocyte and fibroblast chromosomes.

The serologic test for H-Y antigen was a modification of that published by Wachtel [11]. The major modifications were as follows: Fibroblasts were used instead of leukocytes. After the serologic test, permanent slides were made of sperm for live–dead determination [13].

The H-Y antigen antiserum was obtained from C57BL/6 mice that had been injected weekly for 3–6 weeks with spleen and testis cells from male C57BL/6 mice. The females were then bled (from the eye or tail) 1 week after the last injection. The serum was then stored in liquid nitrogen until testing or use. The BALB mice sperm suspensions were prepared by the method of Goldberg et al [12]. The fibroblasts were grown in culture flasks and prior to confluency they were removed by trypsinization. The fibroblasts were then washed at least 4

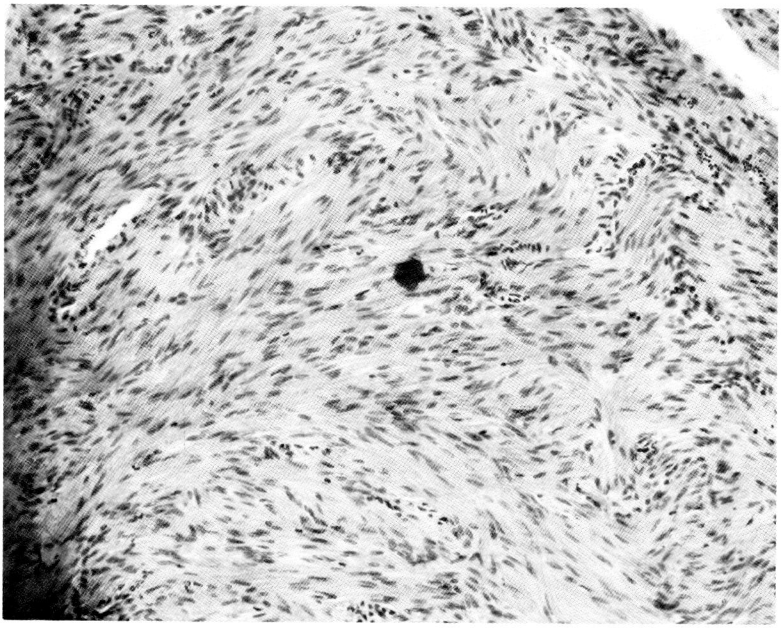

Fig. 4. Left streak gonadal tissue. Areas of interlacing stromal cells. No epithelial, ovarian, or testicular elements are present. (H & E, × 110).

times in Hanks' balanced salt solution prior to absorption. The fibroblasts were then used to absorb the antisera. After absorption the serum was removed from the fibroblasts by centrifugation. Equal volumes of the absorbed antisera, absorbed rabbit complement, and BALB sperm suspension were incubated at 37°C for 40 minutes. Trypan blue was added and incubation continued for 20 minutes. The sperm were then prepared by the method of Dooher and Bennett [13] for live-dead determination. All live-dead sperm counts were done blindly. A detailed account of the H-Y antigen testing technique used in this paper is in preparation and will be published (Cummings) in the near future.

RESULTS

Quinacrine [14], urea-Giemsa [15], and "C"-Banding [16] techniques were performed on metaphase chromosomes of leukocytes and fibroblasts of the proband. A nonmosaic 46,X,i(Yq) karyotype (Fig. 5) was found in repeated leukocyte, skin, and streak gonadal fibroblast cultures from the proband.

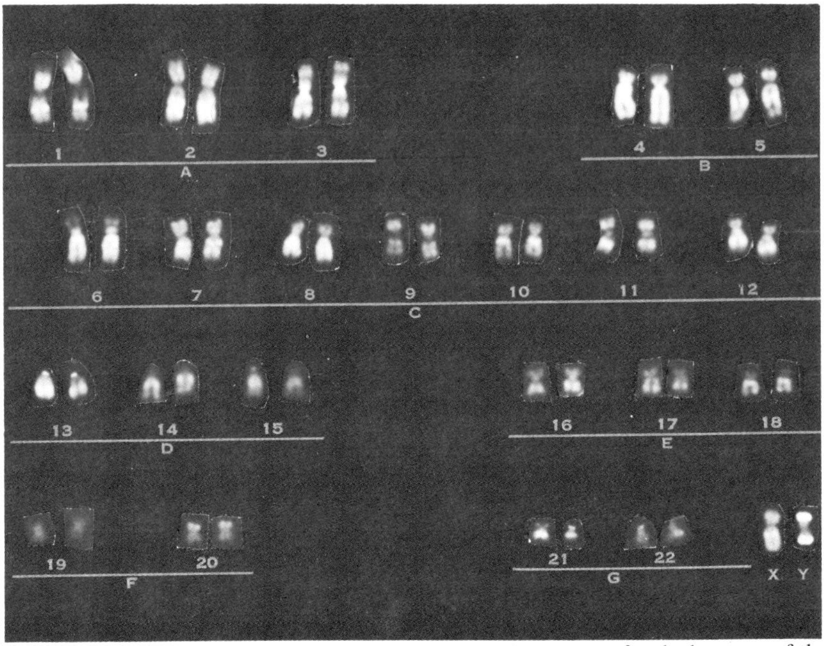

Fig. 5. A quinacrine-banded karyotype showing the isochromosome for the long arm of the Y chromosome.

Barr bodies were absent in the nuclei of cells observed from a buccal smear. However, when the buccal cells were stained with quinacrine, a single large fluorescent body was observed in the interphase nuclei (Fig. 6).

The isochromosome for the long arm of the Y was observed to be metacentric with only one centromeric constriction. Diagrammatic schemes of the normal Y, the isochromosome for the long arm of the Y, and a quinacrine-banded i(Yq) from the proband are shown in Figure 7.

Leukocyte cultures and chromosome analyses were also performed on the parents and sibs of the proband, and in each case the karyotype was found to be normal.

Fibroblast cultures from the streak gonad are available from the Genetic Mutant Cell Repository, Institute for Medical Research, Camden, New Jersey (GM-1709).

Three different fibroblast cell lines were used for the H-Y antigen determination. These were: (1) a normal 46,XY male, (2) a normal 46,XX female, and (3) the fibroblasts from the proband which were 46,X,i(Yq) karyotypically. Two separate H-Y tests of the fibroblast lines were conducted. In each experiment the 3 fibroblast lines were coded and the complete analysis was conducted blindly.

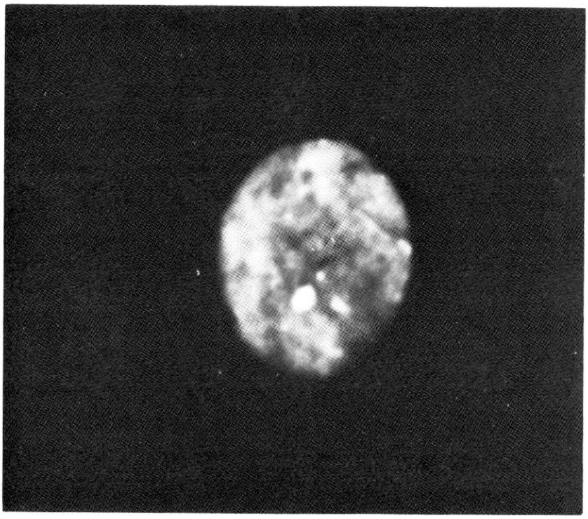

Fig. 6. Interphase nucleus showing large fluorescent "Y body." The smaller fluorescent bodies in the nucleus were determined to be the bright centromeric areas of the number 3 chromosomes and the number 13 chromosome.

The 46,X,i(Yq) fibroblast line was found to be H-Y antigen-negative in each experiment. Table 1 shows the percentage of mouse sperm killed for the antisera absorbed with 46,X,i(Yq), 46,XY and 46,XX fibroblasts for the 2 experiments. Antisera absorbed with the 46,X,i(Yq) and the 46,XX female in each experiment showed a similar sperm kill, with the 46,XY always showing a much lower sperm kill, thus indicating that the 46,X,i (Yq) fibroblasts are H-Y antigen-negative.

DISCUSSION

A phenotypic comparison of 4 females with an isochromosome for the long arm of the Y chromosome is presented in Table 2. The phenotypes of all 4 patients were strikingly similar, each being female with no Turner stigmata, with undifferentiated gonadal streaks, and apparently normal external genitalia with no breast development.

The fact that none of the 4 patients exhibited any stigmata of Turner syndrome is good evidence that genetic information to prevent such stigmata is located on the long arm of the Y chromosome, since Turner syndrome, 45,X, and the 46,X,i(Yq) individuals only differ karyotypically by the presence of 2 Y chromosome long arms in the latter. It therefore appears that the long arm of the Y chromosome does indeed have enough genetic information to prevent the basic stigmata of Turner syndrome.

The genetic material for determining whether or not a 46,XX female attains normal height has been assigned to the short arm of the X chromosome [14].

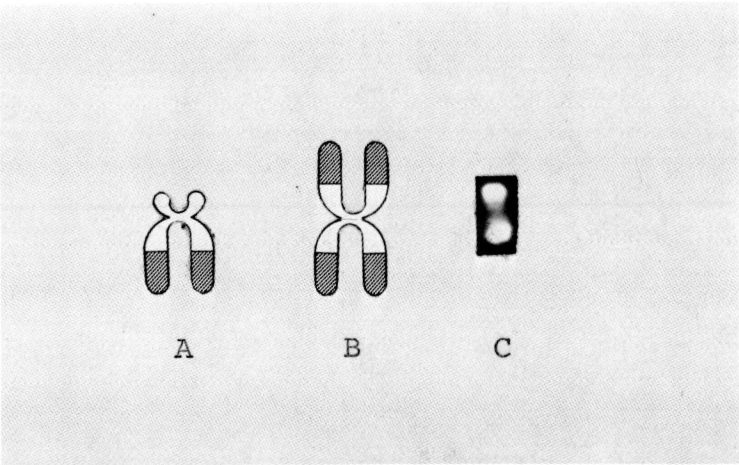

Fig. 7. A diagrammatic scheme of a normal banded Y chromosome (A), an isochromosome for the long arm of the Y chromosome (C) and the quinacrine-banded isochromosome Y of the proband (B).

TABLE 1. Cytotoxicity Test Results From Two Separate Experiments, a and b, on Mouse Epididymal Sperm With Mouse H-Y Antiserum Absorbed With Fibroblasts From a Normal Female 46,XX, a Normal Male 46,XY, and Fibroblasts From the Proband 46,X,i(Yq)*

Experiment	Karyotype of fibroblasts absorbed with antiserum	Percentage of sperm dead
a	46,X,i(Yq)	46
	46,XX	45
	46,XY	31
	Sperm control	12
b	46,X,i(Yq)	42
	46,XX	33
	46,XY	18
	Sperm control	8

*The percentage of dead sperm was derived from averaging blind counts made by 2 different people. At least 200 sperm were scored for each point.

TABLE 2. Phenotypic Comparison of Four Females With Isochromosome for the Long Arm of the Y Chromosome

Reference	Turner stigmata	Gonadal differentiation	Age	Height (cm)	Gonadotropins	Uterus	Tumor	Breast development
Book et al [1]	No	Streak[a]	26	152	Elevated	Rudimentary	No	None
Jacobs and Ross [2] *Case 1*	Some suggestion of webbed neck	Streaks[a]	34	155	Unknown	Underdeveloped	No	Infantile
Case 2	No	Unknown	62	163	Unknown	Unknown	No	Infantile
Present case	No	Streak[a]	11	150	Elevated	Small	Yes	Prepubertal

[a] Gonadal agenesis without ovarian or testicular elements.

Individuals with only one short arm of the X chromosome and no Y chromosome are short as, for example, individuals with a 45,X or 46,X,i(Yq) karyotype. On the other hand individuals who have 2 X chromosome short arms and only 1 long arm, for example, 46,X,i(Xp) or 46,X,Xq–, are of normal stature. The presence of a Y chromosome with any combination of X chromosomes of course provides the needed genetic material for attaining normal stature.

From this we could then conclude that there are at least 2 locations of genetic material, 1 on the Y chromosome and 1 on the short arm of the X, that, if present with a normal X, can enable the individual to attain normal height. The fact the 46,X,i(Yq) (with only 1 X chromosome short arm and 2 long arms of the Y chromosome) individuals are not classified as short leads us to agree with Jacobs and Ross [2] that this is good evidence to indicate that the Y height gene(s) is located on the long arm of the Y and not the short arm.

Using the same information that was used for assigning the height gene to the long arm of the Y chromosome one could additionally assign the gene(s) for gonadal differentiation and testicular formation to the short arm of the Y chromosome. Neither in the present case nor in 2 of the 3 cases (the fourth was unknown) of i(Yq) reported in the literature was there any gonadal differentiation, testicular or ovarian elements formed in the gonadal streak (see Fig. 4). Apparent normal testicular development in the 47,XYY individuals probably precludes the presence of 2 long arms of the Y chromosome in 46,X,i(Yq) individuals as being a causal factor of gonadal agenesis in the 46,X,i(Yq) individuals. Therefore, we feel that the lack of gonadal differentiation and testicular formation in the 46,X,i(Yq) individuals is good supportive evidence for previous reports [16, 17] which have assigned the gonadal differentiation or testicular forming genes to the short arm of the Y chromosome.

The high incidence and pathologic data of tumors developing in dysgenetic gonads of patients with either a normal or abnormal Y chromosome have been well documented [8–10, 15]. The present case is the only 1 of the 4 cases of isochromosome for the long arm of the Y that had a tumor. The tumor in this case was an embryonal dysgerminoma which originated in the right gonadal streak. Figures 5–7 show typical histologic sections of the tumor. Areas of dysgerminoma, immature cartilage, and müllerian epithelium were common.

The findings of 1 tumor in the dysgenetic gonads of 4 different 46,X,i(Yq) individuals is numerically not sufficient to predict statistically the likelihood or frequency of gonadoblastoma occurrence in such individuals. The fact that a gonadoblastoma has indeed occurred in a patient with an isochromosome for the long arm of the Y and also in patients with other Y chromosome abnormalities is sufficient evidence to indicate that gonadal streaks in such patients should be removed. It is significant that a tumor developed in a gonadal streak in the absence of that part of the Y chromosome that is thought to contain the genes for testicular formation.

We feel that our determination that the 46,X,i(Yq) fibroblasts are H-Y antigen-negative is good evidence that the genetic determinant of the H-Y antigen is located on the short arm of the Y chromosome. Kunkel et al [18] have recently reported that these same 46,X,i(Yq) fibroblasts have essentially a double dose of the Y-chromosome-specific reiterated DNA. Furthermore they additionally localized the Y-chromosome-specific reiterated DNA to the long arm of the Y chromosome.

REFERENCES

1. Book JA, Eilon B, Halbrecht I, Komlos L, Shabtay F: Isochromosome Y[46,Xi(Yq)] and female phenotype. Clin Genet 4:410–414, 1973.
2. Jacobs PA, Ross A: Structural abnormalities of the Y chromosome in man. Nature 210:352–355, 1966.
3. Wachtel SS, Koo GC, Breg WR, Elias S, Boyse EA, Miller OJ: Expression of H-Y antigen in human males with two Y chromosomes. N Engl J Med 293:1070–1072, 1975.
4. Boczkowski K: Male and female differentiation of the human gonad. Clin Genet 4: 213–219, 1973.
5. Ferguson-Smith UA, Boyd E, Ferguson-Smith UE, Yresuf AFM, Gray B: Isochromosome for long arm of Y chromosome in a patient with Turner's syndrome and six chromosome mosaicism 45,X/46,XYqi. J Med Genet 6:422, 1969.
6. Ferguson-Smith UA: Karyotype phenotype correlations in gonadal dysgenesis and their bearing on the pathogenesis of malformation. Humangenetik 14:83–84, 1971.
7. Siebers JW, Vogel W, Hepp H, Bolze H, Dittrich A: Structural aberrations of the Y chromosome and the corresponding phenotype. Humangenetik 19:57–66, 1973.
8. Manuel M, Katayama KP, Jones HW Jr: The age of occurrence of gonadal tumors in intersex patients with a Y chromosome. Am J Obstet Gynecol 124(3):293–300, 1976.
9. Teter J, Boczkowski K: Occurrence of tumors in dysgenetic gonads. Cancer 20:1301–1310, 1967.
10. Scully RE: Gonadoblastoma. Cancer 25:1340–1356, 1970.
11. Wachtel SS, Koo GC, Zuckerman EE, Hammerling U, Scheid MP, Boyse EA: Serological crossreactivity between H-Y (male) antigens of mouse and man. Proc Natl Acad Sci USA 71:1215–1218, 1974.
12. Goldberg EH, Boyse EA, Bennett D, Scheid M, Carswell EA: Serological demonstration of H-Y (male) antigen on mouse sperm. Nature 232:478–480, 1971.
13. Dooher GB, Bennett D: A simple technique for preparing easy-to-read, permanent cytotoxicity tests on mouse spermatozoa. Transplantation 23(4):381, 1977.
14. Caspersson T, Zech L, Johannson C: Analysis of the human metaphase set by aid of DNA-binding fluorescent agents. Exp Cell Res 62:490–492, 1970.
15. Berger R: Une novelle technique d'analyse de caryotype. C R Acad Sci [D] (Paris) 273:2620–2622, 1971.
16. Arrighi FE, Hsu TC: Localization of heterochromatin in human chromosomes. Cytogenetics 10:81–86, 1971.
17. Porter IH: The clinical side of cytogenetics. J Reprod Med 17(1):3–18, 1976.
18. Kunkel LM, Smith KD, Boyer SH, Borgaonkar DS, Wachtel SS, Miller OJ, Breg WR, Jones HW Jr, Rary JM: Analysis of human Y-chromosome-specific reiterated DNA in chromosome variants. Proc Natl Acad Sci USA 74:1245–49, 1977.

Bilateral Gonadoblastoma in a Phenotypic Female With 45,X/46,X, Dicentric Iso Y [45,X/46,X,idic(Yq)] Mosaicism*

Charles R. King, MD and David M. Cook, MD

Increased knowledge of the endocrine and chromosomal aspects of sex differentiation have yielded new aspects to the clinical delineation of problems of sex development. A key element in this determination is the presence of the Y chromosome. Generally a Y chromosome foretells male sex differentiation. Numerous examples of XY gonadal dysgenesis with varying degrees of masculinization and intersexuality have been reported. The present case is an example of XY gonadal dysgenesis in a phenotypic female with masculinization, but who had bilateral gonadoblastomata and a 45,X/46,X, dicentric iso Y [45,X/46,X, idic (Yq)] karyotype.

CASE REPORT

An 18-year-old white female (Fig. 1) presented with amenorrhea and short stature (146.5 cm). She had never menstruated. A left nephrectomy had been performed at age 6 because of chronic pyelonephritis secondary to urethral stenosis. The past medical history was otherwise noncontributory. No other family members were identified with short stature, sexual ambiguity, or amenorrhea.

*This work was supported by NIH grant 5 R01 H007997, a grant for a Genetics Study Center (NIH 10 H 820004 08), and a grant for a Mental Retardation Training Program (920).

Birth Defects: Original Article Series, Volume XIV, Number 6C, pages 109–122

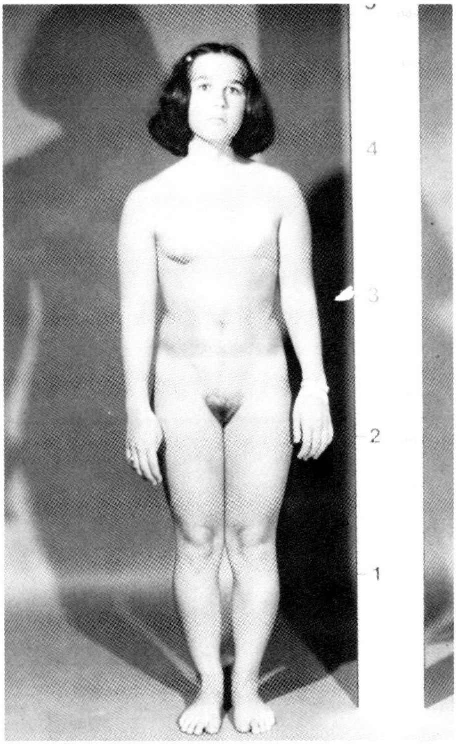

Fig. 1. The patient at age 18.

Physical Examination

Blood pressure was 120/70mm Hg. The ear canals were straight. No neck webbing was present. One observer thought the posterior hairline was low placed. The chest was clear to auscultation and percussion. No heart murmurs were heard. A normal sinus rhythm was present. Moderate breast development with rather immature nipples was present. No abdominal masses were palpable. The neurologic system was grossly intact. The 4th metacarpals were not short. The 2nd, 3rd, and 4th toes were bilaterally elongated. The external genitalia were normal. No clitoromegaly was present. The pubic hair was normal in amount and distribution. A nulliparous cervix was visualized in a normal vagina. The uterine body was small in size. No adnexal masses were palpable.

Laboratory Studies

Routine complete blood count, urinalysis, and SMA-20 were normal. Serum FSH (136 IU/ml) and LH (40 IU/ml) measured by radioimmunoassay were elevated to the postmenopausal range. The serum testosterone (37 ng/100 ml) was well within the normal range for adult females. Serum estradiol (20 pg/ml) registered at the extreme lower limit of normal for the follicular phase of the menstrual cycle. Radiographs confirmed a bone age of 14 years, or about 4 years behind the patient's chronologic age. Peripheral lymphocytes were studied for H-Y antigen expression by the techniques reported by Wachtel [1, 2]. Preliminary results suggest the patient is H-Y antigen negative. Further studies are in progress to confirm this observation.

Cytogenetics

A 45,X/46,X, idic (Yq) karyotype was identified with phytohemagglutinin stimulated peripheral lymphocytes banded with quinacrine (Fig. 2) [3]. Sixty percent of lymphocytes contained the anomalous Y chromosome. A similar karyotype was confirmed in fibroblasts cultured from skin and from both gonads. These cultures contained a smaller percentage of cells with the isodicentric Y chromosome (Table 1). The Y chromosome was mitotically unstable. Occasional cells were encountered with an XY karyotype. Likewise cells were seen with 2 isodicentric Y chromosomes. Observations of anaphase cells demonstrated frequent anaphase lag of the dicentric Y chromosome. Interphase nuclei were not X-chromatin (Barr body) positive. Approximately 50% of lymphocytes were "Y Body" positive. Most of these cells demonstrated only a single "Y Body," but some contained 2 distinct "Y Bodies." C-Banding with a modification of the technique reported by Arrighi and Hsu [4] confirmed that the Y chromosome contained 2 distinct centromeres (Fig. 3). The long arm heterochromatin was symmetric in both arms of the chromosome. Short arm material was definitely present between the centromeres. Chromosome analysis of lymphocytes from the patient's parents and her younger brother were normal. Her father and brother both possessed a Y chromosome morphologically indistinguishable from half of the proband's isodicentric Y chromosome.

Pathology

Bilateral gonadectomy and hysterectomy were performed uneventfully. The fallopian tubes and uterus were normal in gross appearance. Histologically the cervix and endometrium were prepubertal in development. The ovaries were

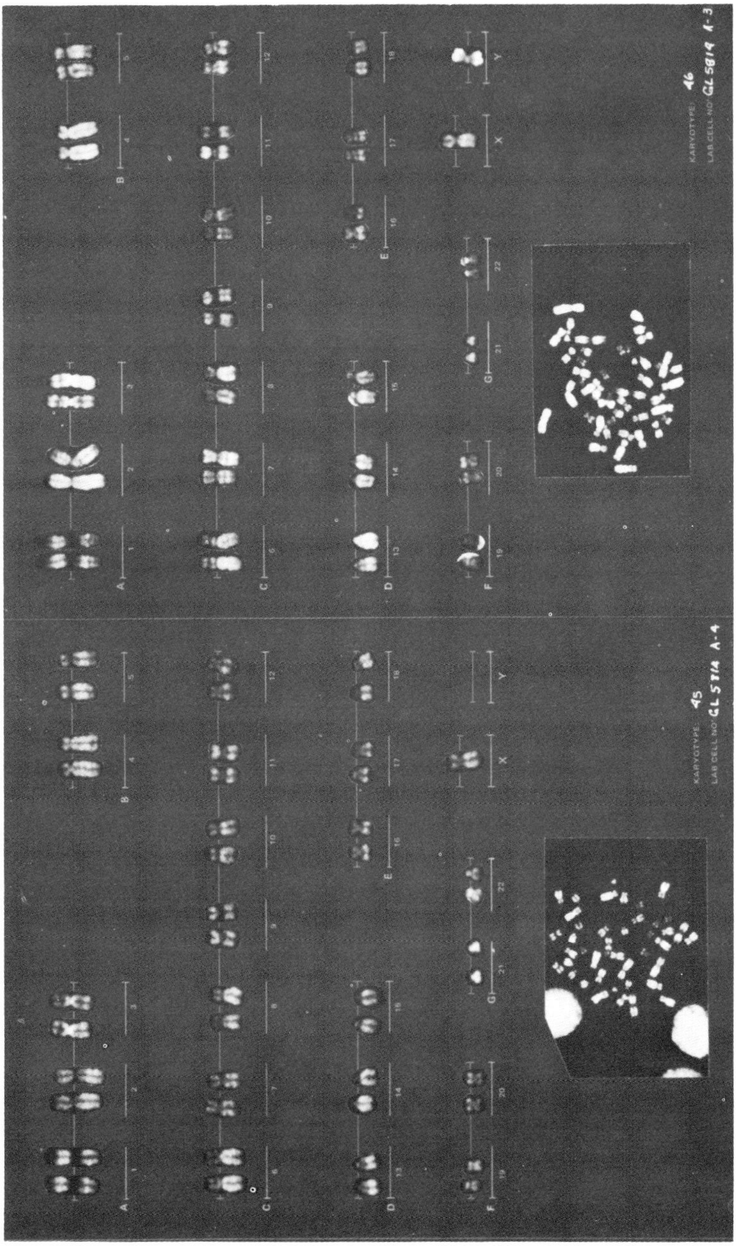

Fig. 2. Quinacrine-banded karyotypes demonstrating 45,X/46,X,idic (Yq) mosaicism.

TABLE 1. Distribution of 45,X and 46,X, dic(iYq) Cells in Studied Tissues

Tissue	45,X	46,X, dic(iYq)
Peripheral lymphocytes	42	58
Skin fibroblasts	75	25
Right gonad	80	20
Left gonad	60	40

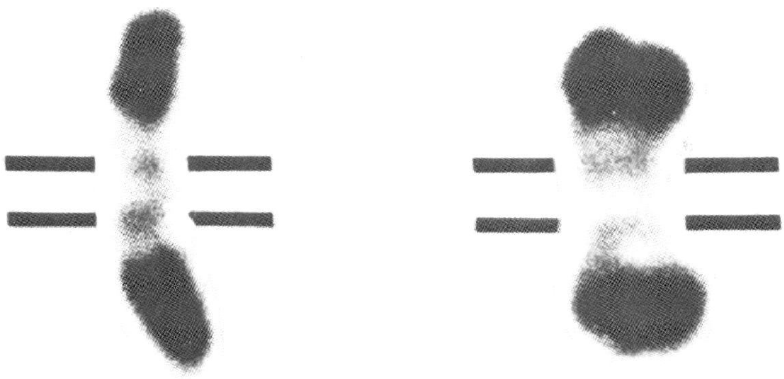

Fig. 3. C-banded Y chromosome demonstrating 2 centromeres. (X 10,000)

small, measuring 2.3 X 1.2 X 1 cm (right) and 2.6 X 1 X 1 cm (left). Both ovaries contained firm yellow tumor masses. Histologically follicular structures were present, but no ova were seen. Much of the ovarian tissues was fibromuscular in character as described for "streak" gonads. Sheets of Leydig-like cells with no crystalloids of Reinke were present in both ovaries. Scattered tubular structures with a basement membrane lined by Sertoli-like cells were seen. These frequently assumed an acinar pattern around pockets of colloid. No calcification was present. Only rare mitotic activity was seen. Both tumors were compatible with a diagnosis of gonadoblastoma (Fig. 4). The patient's postoperative course was uneventful. She was given oral estrogen therapy.

DISCUSSION

Gonadal dysgenesis is not a single entity. As cytogenetic and biochemical studies have become more precise, greater heterogeneity has been demonstrated.

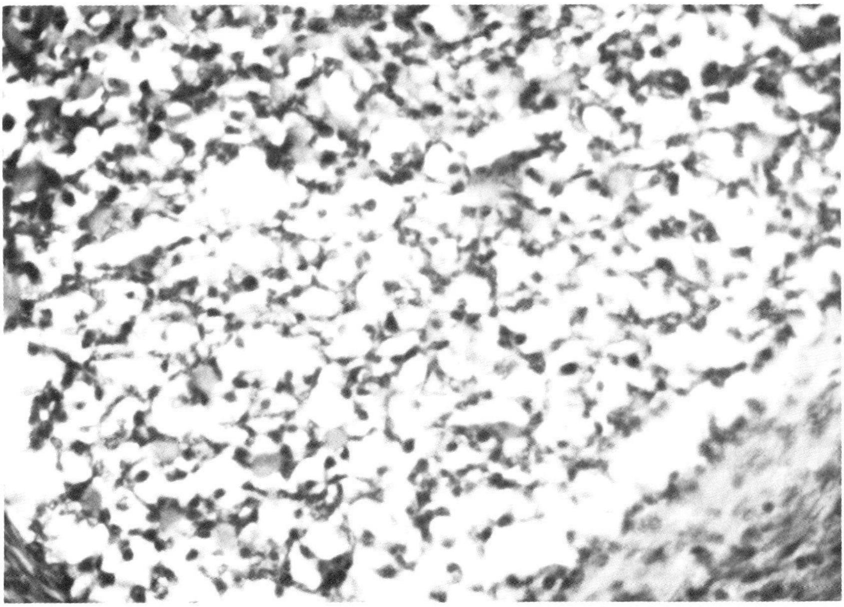

Fig. 4. Gonadoblastoma from the right gonad (H & E original magnification × 160).

Two large categories, one with and another without the Y chromosome, can be identified. A much smaller subgroup of patients with XY gonadal dysgenesis is composed of individuals with an abnormal Y chromosome. Within this last category 21 patients have been reported with a dicentric Y chromosome (Tables 2 and 3). Even this much smaller group of patients does not represent a single phenotype. These patients demonstrate a spectrum from a normal male with azoospermia [5] to the present patient who is a phenotypic female of short stature. A patient has not been reported with normal menses, fertility, or pregnancy.

Seven patients have been reported as phenotypic males (Table 2) [5–11]. All of these patients except the patient noted above of McIlree [5] have had ambiguous genitalia. Fertility has not been documented in these patients, but many were reported in early childhood. All of the patients have had at least 1 testis; in most it was reported as histologically normal. One patient developed a gonadoblastoma [10]. All of the patients have been mosaic for the anomalous Y chromosome.

Fourteen phenotypic female patients have been described (Table 3) [10, 12–21]. Five of these patients have had evidence of masculinization of the external

TABLE 2. Reported Cases of Dicentric Iso Y Chromosome: Phenotypic Males

Karyotype	Age	Stature (cm)	External genitalia	Gonadal pathology	Reference
45,X/46,X,dic(Y)	24	---	Male	Testes Azoospermia	McIlree 1966 [5]
45,X/46,X,dic(Y)/ 46,XXq–	7	110	Ambiguous	Testes No tubular lumens	Boschetti 1968 [6]
45,X/46,X,dic(Y)	13	142	Ambiguous	Streak Normal testis	Ferrier 1968 [8]
45,X/46,XY/ 46,XX/46,X,dic(Y)/ 47,XX,dic(Y)	8	---	Ambiguous	Prepubertal testes	Jalbert 1969 [9]
45,X/46,XY/ 46,XXq–/ 46,X,dic(Y)	22	---	Ambiguous	Testis Gonadoblastoma Ovarian stroma	Morillo-Cucci 1971 [10]
45,X/46,X,dic(Y)	2	---	Ambiguous	Streak Dysgenetic testis	Ying 1971 [11]
45,X/46,X,dic(Y)	4	---	Ambiguous	Dysgenetic testis Normal testis	Buchanan 1976 [7]

TABLE 3. Reported Cases of Dicentric Iso Y Chromosome: Phenotypic Females

Karyotype	Age	Stature (cm)	External genitalia	Gonadal pathology	Reference
46,X,i(Yq)	34	155	Infantile	Streak	Jacobs 1966 [18]
46,X,i(Yq)	62	163	Infantile	---	Jacobs 1966 [18]
45,X/46,X,i(Yq)	66	147	Infantile Clitoromegaly	Streak	Ferguson-Smith 1969 [16]
45,X/46,X,dic(Yq)	14	138	Infantile	Streak	Angell 1970 [12]
45,X/46,X,dic(Yq)	36	147	Infantile	Streak	Angell 1970 [12]
45,X/46,X,dic(Yp)	11	124	Infantile	---	Angell 1970 [12]
45,X/46,XY/ 46,X,dic(Y)/ 47,X,dic(Y)	49	---	Clitoromegaly	Dysgenetic testis Streak	Morillo-Cucci 1971 [10]
46,X,dic(Y)	16	138	Infantile clitoromegaly	Streak gonadoblastoma	Armendares 1972 [13]
46,X,i(Y)	26	152	Female	Streak	Book 1973 [14]
45,X/46,X,i(Yp)	20	138	Infantile	Streak	Siebers 1973 [21]
45,X/46,X,dic(Y)	11	---	Ambiguous	Streak Dysgenetic testis	Cohen 1973 [15]
45,X/46,X,dic(Y)	22	156	Female	Streak	Johnston 1974 [19]
45,X/46,XY−/ 46,X,idic(Yp)	19	148	Infantile	Streak Gonadoblastoma	Malkova 1975 [20]
45,X/46,X,dic(Y)	15	134	Ambiguous	Dysgenetic testis Gonadoblastoma	Hayek 1975 [17]
45,X/46,X,dic(Y)	19	146	Female	Streak Gonadoblastoma	Present case

genitalia. The remaining patients have had no masculinization, but breast development and normal postpubertal sex development have not occurred. The external genitalia have been infantile in development. None has experienced menses or pregnancy. Short stature and features of the Turner phenotype have been frequently noted. This is not an unexpected finding since only 4 patients did not demonstrate a 45,X cell line. This cell line when present would be expected to affect the phenotypic expression of the patient. Gonadal pathology has not been reported with normal ovarian histology, but isolated islands of ovarian stroma have been observed. Three patients have had gonadoblastoma [13, 17, 20]. Two demonstrated masculinization of the external genitalia. All but 4 have had a mosaic chromosomal constitution. Three had only peripheral blood specimens examined and thus another cell population may not have been identified. The fourth patient did not demonstrate mosaicism in bone marrow, blood, or skin fibroblasts [13]. Gonadal tissue failed to grow and was thus not analyzed.

The development of quinacrine- and Giemsa-banding techniques have added tremendously to the knowledge of chromosome structure [3, 22]. Prior to the introduction of these methods to human cytogenetics in 1968, only a few chromosomes could be identified with certainty [22]. About one-half of the cases of dicentric Y chromosome were reported without the use of banding analysis. The abnormal chromosome was identified by autoradiography in a few of these instances, however [8, 11, 18]. Since banding studies were not performed in all cases, the anomalous Y chromosome may not actually be dicentric in structure in all the reported patients. Some of the published karyotypes do not confirm the presence of 2 centromeres. In only 3 cases including the present case was C-banding performed to further confirm the presence of 2 centromeres [7, 14]. One would expect that, because of the phenotypic similarities of the published cases, if appropriate studies were done, many of the other reported cases would demonstrate similar chromosomal findings.

The mitotic behavior of the anomalous Y chromosome has not been specifically examined in most studies; however, all but 4 cases represent mosaicism. This in itself suggests that the anomalous Y chromosome may be unstable. Several cases have demonstrated multiple cell lines. The case reported by Jalbert [9] exhibited 5 different cell populations. Anaphase nuclei were specifically observed for a lagging Y chromosome in the present case, the case reported by Buchanan [7], and that reported by Ying [11]. In all 3 instances the anomalous Y chromosome exhibited delayed migration to daughter cells. This phenomenon has previously been suggested as important for the etiology of other chromosomal errors. Fitzgerald [23] has proposed that premature centromere division could provide a mechanism for nondisjunction of the X chromosome in normal females. He suggests that this mechanism is operative in the increased incidence of monosomy X in the normal female as she ages. A similar means, ie premature centromere division, may be responsible for meiotic or mitotic errors leading to nondisjunction and finally the anomalous Y chromosome.

Interpretation of "Y Body" studies in patients mosaic for the Y chromosome is at best hazardous. In the present case the 45,X cells obviously lacked a "Y Body." Cells with the abnormal Y usually demonstrated a single "Y Body" but many had 2 distinct "Y Bodies." Similar observations have previously been noted [7]. The important implication to draw from these cytogenetic observations is that complete karyotype analysis should be performed, by banding techniques, in any patient with sexual ambiguity. Errors can easily arise in sex assignment or patient management when short cuts are mistakenly followed.

The chromosome anomaly in these patients might arise either by meiotic or mitotic error. A meiotic error during spermatogenesis could produce a sperm bearing a dicentric Y chromosome. All normal males produce abnormal sperm. Cytologic abnormalities occur frequently in sperm from normal men. Cytogenetic studies of normal males [24] have demonstrated sperm bearing 2 Y chromosomes and occasionally a dicentric Y chromosome. Such a sperm, if it fertilized a normal X bearing ovum, would produce a zygote with a 46,X, dic (Y) chromosome complement.

An alternate mechanism would be by way of a mitotic error. A normal male zygote arising from a Y bearing sperm and an X bearing ovum could by an error early in mitosis produce the anomalous Y chromosome. The resultant dicentric Y chromosome would produce the abnormal cell line observed. The loss of this unstable Y chromosome would yield the 45,X cell line. The mitotic studies reported in this case confirm mitotic instability of the anomalous Y chromosome. Although this does not explain whether the developmental error is of meiotic or mitotic origin, it does provide evidence for the occurrence of mosaicism. Multiple errors must be postulated in a patient like that reported by Jalbert [9] with multiple cell lines. At present it is not possible by cytogenetic or biochemical means to identify the meiotic or mitotic origin of the dicentric Y chromosome.

The gonadoblastoma is an unusual tumor of the gonad [25, 26]. Histologically it is characterized by germ cells and cells resembling immature Sertoli cells [25, 27]. Leydig-type cells are usually present as well. Calcification is frequently also seen. These tumors are thought to be hamartomatous lesions of the dysgenetic testis [28]. Ultrastructural studies have demonstrated the apparent testicular origin of these tumors [27, 29]. Ishida [29] has reported that the germ cells present in this tumor are identical to fetal germ cells by ultrastructural studies. The Sertoli-like cells contain Charcot-Bottcher crystalloids suggesting their origin is actually as Sertoli cells. The Leydig-like cells contain crystalloids typical of fetal Leydig cells. The basal laminae frequently found in gonadoblastoma are identical ultrastructurally to the seminiferous tubules. Histochemically the Leydig-like cells have been demonstrated to be active steroid producers. These cells, particularly at the lysosomal membranes, contain Δ5-3β hydroxysteroid dehydrogenase activity.

As previously proposed, hormonal activity of the gonadoblastoma may account for some of the feminization of these patients. The normal serum testosterone and marginal estradiol levels in this case may well be of gonadoblastoma origin. As suggested by Cantu [30] this steroid production either directly as estrogens or indirectly as androgens metabolized to estrogens could produce the breast development and other feminine features of this patient. Even if this is correct the patient's tumor did not produce enough estrogen to achieve mature breast development or to stimulate endometrial or cervical mucosal maturity.

These histologic and more particularly histochemical and ultrastructural studies support a testicular origin of the gonadoblastoma. No case has been reported of metastatic gonadoblastoma. Certainly malignancy such as dysgerminoma may occur in addition to gonadoblastoma in a dysgenetic gonad and may even metastasize, but not the gonadoblastoma itself. Mulvihill [31] has nicely illustrated the occurrence of gonadoblastoma in patients with dysgenetic testes. Most patients with gonadoblastoma have demonstrated some aspect of sexual ambiguity or intersexuality. The majority of these patients have had a Y chromosome, whether normal or abnormal. Only rarely has a patient been evaluated as was the present case with cytogenetic analysis of gonadal tissue. Such cases have demonstrated a Y-containing cell line in the gonadal tumor [26]. A reportedly normal female with normal menses and an ectopic pregnancy with gonadoblastoma [32] did demonstrate Y material [26] on cytogenetic study of the gonads even though lymphocytes demonstrated a normal 46,XX karyotype. Talerman [33] has reported the second case of a phenotypically normal male with a gonadoblastoma in normally descended testes.

A review of the Johns Hopkins experience and the literature on patients with tumors arising in dysgenetic gonads illustrates the striking increase in tumor incidence with the patient's advancing age [34]. The significant increase in incidence occurs at about the time of puberty. Tumors have not been observed before age 5. The precise incidence of tumors is difficult to determine due to such factors as bias of ascertainment and prophylactic surgery. As an approximation, 50% of patients with XY gonadal dysgenesis and masculinization will have developed tumors by middle adult life. For this reason, gonadectomy should be performed as soon as the diagnosis is made, even if this is at a very young age. The only exception to this is patients with the testicular feminization syndrome who some feel should be allowed to progress through puberty and achieve feminization before gonadectomy. The incidence of tumors in this group of patients is about 3% before age 20.

An additional avenue of study is the assay of H-Y antigen in patients with gonadal dysgenesis. This antigen has been important in the study of male sex development. It was originally described in male mice [35], and study has now been extended to other mammalian species including man [1, 35]. It correlates strongly with male phenotype and/or the presence of the Y chromosome. It is present in a wide variety of intersex conditions possessing the Y chromosome. Phenotypic males with a 46,XX karyotype have been demonstrated by Wachtel and associates to exhibit the H-Y antigen [35]. For this reason and a multitude of phylogenetic data it has been proposed that the H-Y antigen is either essential for testicular differentiation or is closely linked to the gene or genes necessary for this to occur. These genes are thought to be quite close to the centromere in the short arm of the Y chromosome. Preliminary study in the present case does not demonstrate H-Y antigen expression on peripheral lymphocytes. Wachtel has noted a similar lack of expression of antigen activity in some mosaic patients (personal communication). Further study is in progress to confirm this initial observation. In addition cultured fibroblasts from the gonadoblastoma will be assayed for H-Y antigen activity and reported separately. It would be expected that if H-Y antigen is as closely linked to male sex development as present data would indicate, and if the gonadoblastoma truly arises in a dysgenetic testis, then the H-Y antigen will be demonstrated in this patient. The H-Y antigen has not been previously studied in tissue from such tumors.

Previous patients with a dicentric Y chromosome have not been studied for the presence of H-Y antigen. Because this chromosome is dicentric and the locus for H-Y antigen is close to the centromere on the Y short arm, it might be possible to demonstrate H-Y antigen levels above normal male levels in some patients with a dicentric Y chromosome. A dosage effect has previously been demonstrated in patients with 2 separate Y chromosomes. Males with either a 47,XYY or 48,XXYY karyotype have produced increased amounts of H-Y antigen as compared to normal 46,XY males [2]. If this observation is extended to the anomalous dicentric Y chromosome, it would be important not only for what it tells us about the structure of this particular chromosome, but also because it has much wider implications for the study of gene mapping and dosage effects.

Patients with errors of sex development are of interest not only because of what they tell us about abnormal sex development, but also because of what they tell us about normal sex development. These patients provide an experiment of nature comparable to laboratory manipulation of sex development in lower animals. Precise cytogenetic and biochemical study of these developmental errors will enable further elucidation of the mechanism of normal sex development. The importance of the Y chromosome in this process is essential. The H-Y antigen provides a biochemical marker of this development. These observations should be extended to other patients with errors of sex development. Only by complete study – cytogenetic, biochemical, anatomic, psychologic, etc – will complete understanding of the process of sex development occur. Only with a similar complete evaluation will each individual patient receive the appropriate diagnosis, therapy, and prognosis.

ACKNOWLEDGMENTS

We thank Cathy Olson and Tim Donlon for cytogenetic laboratory assistance.

REFERENCES

1. Wachtel SS, Ohno S, Koo GC, Boyse EA: Possible role for H-Y antigen in the primary determination of sex. Nature 257:235–236, 1975.
2. Wachtel SS, Koo GC, Breg WR, Elias S, Boyse EA, Miller OJ: Expression of H-Y antigen in human males with two Y chromosomes. N Engl J Med 293:1070–1072, 1975.
3. Caspersson T, Lomakka G, Zech L: The 24 fluorescence patterns of the human metaphase chromosomes – distinguishing characters and variability. Hereditas 67:89–102, 1971.
4. Arrighi FE, Hsu TC: Localization of heterochromatin in human chromosomes. Cytogenetics 10:81–86, 1971.
5. McIlree ME, Price WH, Court-Brown WM, Tullock WS, Newsome JE, Maclean N: Chromosome studies on testicular cells from 50 subfertile men. Lancet 2:69–71, 1966.
6. Boschetti R, Gilbertas B, Noel B, Quack B: Pseudohermaphrodisme masculin et mosaique 45,X/46,XY dic/46,XXq–. Ann Genet 11:62–65, 1968.
7. Buchanan PD, Wyandt HE, D'Ercole AJ, Rao KW, Hartsell ML: A mitotically unstable human dicentric Y chromosome in a male pseudohermaphrodite. Cytogenet Cell Genet 17:42–50, 1976.
8. Ferrier PE, Ferrier SA, Bill AH: A male pseudohermaphrodite with a dicentric Y chromosome. Humangenetik 6:131–141, 1968.
9. Jalbert P, Convert A, Mouriquand C, Malka J: A propos d'une observation d'Y dicentrique chez un pseudohermaphrodite masculin avec mosaique gonosomale complexe. Ann Genet 12:253–258, 1969.
10. Morillo-Cucci G, German J: Abnormal Y chromosomes and monosomy 45,X: A concept derived from the study of three patients. In Bergsma D (ed): Part X. "The Endocrine System." Baltimore: Williams & Wilkins for The National Foundation–March of Dimes, BD:OAS VII(6):210, 1971.
11. Ying KL, Ives EJ: Mitotic behavior of a human dicentric Y chromosome. Cytogenetics 10:208–218, 1971.
12. Angell R, Giannelli F, Polani PE: Three dicentric Y chromosomes. Ann Hum Genet 34: 39–50, 1970.
13. Armendares S, Buentello L, Salamanca F, Cantu-Garza JM: A dicentric Y chromosome without evidence of sex chromosomal mosaicism, 46,XYq dic, in a patient with features of Turner's syndrome. J Med Genet 9:96–100, 1972.
14. Book JA, Eilon B, Halbrecht I, Komlos L, Shabtay F: Isochromosome Y (46,X,iYq) and female phenotype. Clin Genet 4:410–414, 1973.
15. Cohen MM, MacGillivray MH, Capraro VJ, Aceto TA: Human dicentric Y chromosomes. J Med Genet 10:74-79, 1973.
16. Ferguson-Smith MA, Boyd E, Ferguson-Smith ME, Pritchard JG, Yusuf AFM, Gray B:

Isochromosome for long arm of Y chromosome in patient with Turner's syndrome and sex chromosome mosaicism (45,X/46,XYqi). J Med Genet 6:422–425, 1969.
17. Hayek A, Yunis E: Dicentric Y chromosome in mixed gonadal dysgenesis. J Med Genet 12:210–212, 1975.
18. Jacobs PA, Ross A: Structural abnormalities of the Y chromosome in man. Nature 210:352–354, 1966.
19. Johnston AW, Speed RM, Klopper A, Robinson JA: A patient with a dicentric Y chromosome. Clin Genet 6:326–331, 1974.
20. Malkova J, Michalova K, Chrz R, Kobilkova J, Motlik K, Starka L: Dicentric Yp chromosome in a patient with the gonadal dysgenesis and gonadoblastoma. Humangenetik 27:251–253, 1975.
21. Siebers JW, Vogel W: Structural aberrations of the Y chromosome and the corresponding phenotype. Humangenetik 19:57–66, 1973.
22. Caspersson T, Farber S, Foley GE, Kudynowski J, Modest EJ, Simonsson E, Wagh V, Zech L: Chemical differentiation along metaphase chromosomes. Exp Cell Res 49: 219–224, 1968.
23. Fitzgerald PH, Pickering AF, Mercer JM, Miethke PM: Premature centromere division: A mechanism of nondisjunction causing X chromosome aneuploidy in somatic cells of man. Ann Hum Genet 38:417–428, 1975.
24. Sumner AT, Robinson JA, Evans HJ: Distinguishing between X, Y and YY-bearing human spermatozoa by fluorescence and DNA content. Nature (New Biol) 229:231–233, 1971.
25. Scully RE: Gonadoblastoma: A review of 74 cases. Cancer 25:1340–1356, 1970.
26. Williamson HO, Underwood PB, Kreutner A, Rogers JF, Mathur RS, Pratt-Thomas HR: Gonadoblastoma: Clinicopathologic correlation in six patients. Am J Obstet Gynecol 126:579–584, 1976.
27. Garvin AJ, Pratt-Thomas HR, Spector M, Spicer SS, Williamson HO: Gonadoblastoma: Histologic, ultrastructural, and histochemical observations in five cases. Am J Obstet Gynecol 125:459–471, 1976.
28. Schellhas HF: Malignant potential of the dysgenetic gonad. Parts I and II. Obstet Gynecol 44:298–309, 455–462, 1974.
29. Ishida T, Tagatz GE, Okagaki T: Gonadoblastoma: Ultrastructural evidence for testicular origin. Cancer 37:1770–1781, 1976.
30. Cantu JM, Olivera E, Gonzalez D, Hidalgo F, Gonzales M, Gallegos AJ: A possible explanation for spontaneous breast development in XY gonadal dysgenesis. Ann Endocrinol 33:517–522, 1972.
31. Mulvihill JJ, Wade WM, Miller RW: Gonadoblastoma in dysgenetic gonads with a Y chromosome. Lancet 1:863, 1975.
32. Pratt-Thomas HR, Cooper JM: Gonadoblastoma with tubal pregnancy. Am J Clin Pathol 65:121–125, 1976.
33. Talerman A, Delemarre JFM: Gonadoblastoma associated with embryonal carcinoma in an anatomically normal man. J Urol 113:355–359, 1975.
34. Manuel M, Katayama KP, Jones HW: The age of occurrence of gonadal tumors in intersex patients with a Y chromosome. Am J Obstet Gynecol 124:293–300, 1976.
35. Silvers WK, Wachtel SS: H-Y antigen: Behavior and function. Science 195:956–960, 1977.

The Turner Syndrome and the Y Chromosome: Mechanisms of Diminished Y-Determined Maleness

Vincent M. Riccardi, MD, Stephen Duck, MD,
and Paul Katayama, MD

INTRODUCTION

The Turner syndrome (TS) is characterized clinically by short stature, webbed neck, cubitus valgus, short metacarpals, "shield chest," aortic coarctation, renal and urinary tract anomalies, multiple pigmented nevi, distal axial triradii, a predominance of the fine-ridged, vertically oriented ulnar loops on the fingertips, primary amenorrhea, immature breasts and genitalia, other deficiencies of secondary sex characteristics, and streak (dysgenetic) gonads [1–3]. Most often the TS is associated with a 45,X chromosome constitution, although in some instances there may be an X-chromosome deletion [eg del(Xp)], an isochromosome [ie i(Xq)], or mosaicism of a 45,X cell line with 46,XX, 46,XY, and/or other cell lines (eg 47,XXX) [1–5]. When 45,X monosomy is mosaic with other karyotypes, the severity of the TS is often tempered or the clinical picture is distinct from the usual TS (eg mixed gonadal dysgenesis) [6]. When the full-blown TS is present in the face of mosaicism with either a 46,XX or 46,XY chromosome pattern, the lack of influence of the apparently normal sex chromosome compliment has been attributed to small numbers of normal cells in the involved tissues, although this has not been directly demonstrated. The present case is of special interest because a relatively large proportion of the proposita's cells appear to have a 46,XY chromosome pattern (the rest being 45,X), and because the patient is H-Y antigen-positive [7, 8] and yet she has not manifested Y-chromosome expression except by the presence of a gonadoblastoma. This in turn suggests that additional factors might be at play in masking or suppressing Y-chromosome function.

Birth Defects: Original Article Series, Volume XIV, Number 6C, pages 123–132

CASE REPORT

The proposita presented at age 13 years, 8 months for evaluation of short stature and delayed sexual development. She was born at 8 months gestation, the first of dizygotic twins, with a birthweight of 2.61 kg and length of 43 cm; no neonatal abnormalities were noted. Her motor and psychosocial development and overall performance were normal and comparable with her twin sister's, though she has always been the shortest child in her school class. She experienced no sexual development or pubertal growth spurt, both of which had been apparent in her twin for at least a year. She admitted to shortness of breath and mild exercise intolerance relative to her peers; her habits were sedentary. A heart murmur was noted when she was a young child but it was not evaluated. Her school work was above average. Her family history was unremarkable for similarly affected persons; 15- and 12-year-old sisters were normal.

On physical examination her general appearance was typical of the Turner syndrome. Her weight was 39.4 kg (10th percentile), her height 135 cm (height age: 9 years, 4 months), arm span 133 cm, and head circumference 52.1 cm. She was verbal and cooperative, with apparently normal intelligence. Her stature was proportionately and symmetrically short, and her habitus was neuter with generalized obesity. There were excessive numbers of pigmented nevi, distributed diffusely. There was mild webbing of the neck associated with a low, trident nuchal hairline. The eyes were normal with no ptosis. The chest was broad with wide-spread nipples and immature breasts (Tanner stage I). The cardiovascular findings were limited to a grade 2/6 midsystolic, left-sided murmur and nonpalpable femoral, dorsalis pedis, and posterior tibial pulses bilaterally. Blood pressures in both upper limbs averaged 145/80, but with some diastolic values at 99 mm Hg. There was no mature pubic hair (Tanner stage I). The external genitalia were those of a prepubertal female, with underdeveloped labia majora and minora; the clitoris was unremarkable. Cubitus valgus was present, the 4th and 5th metacarpals were short bilaterally, and both 5th fingers were short. The toenails were hypoplastic. The axial triradius was in a t″ position bilaterally and fingerprint patterns were characterized by fine-ridged, vertically oriented ulnar loops.

Radiographic studies revealed a bone age of 12 years, 8 months, evidence of aortic coarctation, an apparent uterine cervical os (genitogram), and nonvisualization of the uterine fundus and broad ligament structures (pneumogynegram); the sella turcica and urinary tract were normal. Routine laboratory results were within normal limits, including a serum thyroxine level of 8.4 μg%. Three consecutive 24-hour urine collections revealed prepubertal excretion levels of 17-hydroxysteroids (average 6.5 mg/24 hours) and 17-ketosteroids (average 4.9 mg/24 hours). Morning and evening serum levels of follicle-stimulating hormone (FSH), luteinizing hormone (LH), and estradiol were consistent with nonfunctioning ovaries, the former two being markedly increased, and the latter markedly decreased. Serum testosterone levels on two consecutive days, a short time later, were consistent with the absence of functional testes (23 and 20 ng%). An insulin tolerance test did indicate a positive response in serum levels of both growth hormone and cortisol, but the lowest blood sugar was only 44 mg% (at 15 minutes). A buccal smear was X-chromatin negative and an initial karyotype of 75 peripheral blood cells was uniformly 46,XY by G-banding and C-banding techniques (Fig. 1) [9].

Eight weeks later the patient's coarctation was surgically repaired. Approximately 6 months after presentation an exploratory laparotomy revealed a normal vagina, normal cervix, and small uterus. Bilateral streak gonads were removed, the left containing a small, circumscribed gonadoblastoma. Q-banded chromosome analyses at this time revealed mosaicism for 45,X and 46,XY cell lines: right streak ovary – 68 of 70 mitoses were 45,X; left streak ovary – 15 of 15 mitoses were 45,X (but 3 of 51 interphase cells showed a Y-chromatin mass); skin – 5 of 10 mitoses were 45,X. The Y chromosome showed normal Q-banding and C-banding.

Five and one-half months after surgery H-Y antigen assays (courtesy of Drs. S. S. Wachtel, G. C. Koo, and O. J. Miller), using peripheral blood samples, were performed on the proposita and her parents [7, 8]. She and her father were H-Y-positive, while her mother was H-Y-negative. Respecting the presence of a 46,XY chromosome constitution in 75 of 75 T-lymphocytes from circulating blood,

Fig. 1. C-banded metaphase spread from proposita. The arrow indicates the unequivocal Y chromosome in this peripheral lymphocyte metaphase. Except for the X/XY mosaicism the patient's chromosome constitution was normal.

these data are consistent with normal expression of the H-Y antigen locus of her Y chromosome. Her father's Y chromosome was indistinguishable from hers with C-banding techniques.

DISCUSSION

It is not yet known how chromosome abnormalities express themselves other than in relatively gross clinical terms. Although occasionally tissue-level abnormalities may be apparent, and gonadal endocrine and germ cells may not develop or function normally, on a cellular or intracellular level the immediate consequences of chromosome abnormalities are largely unrecognized. We present here a patient whose combination of clinical and laboratory features, including chromosomal mosaicism, provides an opportunity to evaluate expression of 45,X and 46,XY chromosome constitutions in their respective cell lines in vivo. The only apparent expression of the Y chromosome in the proposita was the presence of H-Y antigen [7, 8] and a gonadoblastoma [10, 11]. Otherwise it was as though all her cells were 45,X. The degree of mosaicism, although one element, may not be the entire mechanism to account for the lack of the Y chromosome's expression.

The TS is most often a result of a 45,X chromosome constitution in each of the affected individual's cells [1–3]. Among the alternative chromosomal mechanisms are X-chromosome structural aberrations and mosaicism of 45,X cells with one or more additional cell lines having either normal (ie 46,XX or 46, XY) or otherwise abnormal sex chromosome constitutions [eg 47,XXX, 46,X,i(Xq)]. A second cell line not infrequently accompanies 45,X monosomy [2, 4, 5] and the resultant clinical pictures are very variable. In the literature on 45,X/46,XY mosaicism, TS and gonadal dysgenesis [1–5, 12–22] two tacit, though rarely explicated, presumptions underlie our current rationalizations of this group of conditions: 1) that there is a continuum of phenotypes which can result from 45,X/46,XY mosaicism and the purported continuity from one phenotype to the other reflects different proportions of 45,X and 46,XY cells in various tissues [2]; 2) that the karyotype of each cell will determine whether that cell will contribute to a defective organ or tissue [2]. These two presumptions are the basis for the expectation that if the proportion of 45,X and 46,XY cells could be known for each tissue and organ, then the phenotype for the entire person could be predicted with accuracy. Conversely, if the phenotype is different from what is ordinarily associated with the demonstrated karyotype(s) in one or two tissues, mosaicism of different proportions is presumed for other, ostensibly more critical tissues.

The main purpose of this article is to question these two presumptions and to consider that the ultimate phenotype in a 45,X/46,XY individual may reflect more than the relative proportions of each cell type in the various tissues. The spectrum of phenotypes represents more than a quantitative variation; at certain times additional qualitatively distinct influences also must be considered.

The manifestations of a chromosome disorder are of two types, primary and secondary. The primary manifestations are those immediately and directly due to the chromosome aberration itself – eg failure of both germ cell formation and gonadal endocrine function. Secondary manifestations are those mediated by one or more factors beyond the chromosome aberration per se – eg failure to develop genitalia or to mature sexually due to sex hormone deficiency. Although it is yet unclear which of the TS features are primary and which are secondary manifestations of 45,X monosomy, we would presume that at least some are primary features and therefore in 45,X/46,XY mosaics we would expect them to be diminished or absent in proportion to the 46,XY cells present in that tissue or organ. Yet, at times it is as though cells with a 46,XY chromosome complement are not even present: Individuals with 45,X/46,XX or 45,X/46,XY mosaicism may be virtually indistinguishable, the overlapping phenotypes including the full-blown Turner syndrome, incomplete forms of the Turner syndrome, and females with primary amenorrhea with an external genital and somatic habitus almost indistinguishable from the normal (Table 1).

For a person with 45,X/46,XY mosaicism the resultant disorder should reflect the relative contributions of the 45,X and 46,XY cell lines in at least some of the tissues which may be abnormal (Tables 1 and 2). At one extreme, where the XY cells far outnumber the X cells, a basically male phentoype would be expected, and at the other extreme, where the X cells far outnumber the XY cells, the TS phenotype would be expected. In between, with more nearly equal representations of X and XY cells, a syndrome such as mixed gonadal dysgenesis, characterized by both male and female/TS features, would be expected [6]. In other instances a more normal female phenotype may be seen. The range of clinical pictures has suggested to some a continuum, from an almost normal male through mixed gonadal dysgenesis and the TS to an almost normal female, accounted for merely in terms of the relative proportions of 45,X and 46,XY cells in the affected individual. We would suggest that this schema is an oversimplification and that, in at least some instances, an additional set of factors must be considered to account for the full-blown Turner syndrome in the presence of a relatively large proportion of 46,XY cells. Our patient illustrates this approach to understanding the phenotypic expression of chromosome abnormalities. She has too many XY cells apparent to accept the presumption that they were outnumbered in each of the tissues or organs showing typical features of the TS. Rather, it is as though an additional influence superseded, or masked, the intracellular expression of the Y chromosome in her many 46,XY cells.

There are two compelling reasons why factors other than mosaicism percentages should be sought to account for the presence of the TS without apparent

TABLE 1. Karyotype-Phenotype Correlations for Some Sex Chromosome Abnormalities

Sex chromosome constitution	Resultant cell types	Clinical picture
X	X-monosomy	Turner syndrome[a]
XX	Female	Normal female[b]
XY	Male	Normal male[b]
X/XX	X-monosomy and female	Turner syndrome[a]
		Incomplete Turner syndrome[c]
		Primary amenorrhea[d]
X/XY	X-monosomy and male	Turner syndrome[a]
		Incomplete Turner syndrome[c]
		Primary amenorrhea[d]
		Mixed gonadal dysgenesis[e]
		Male genitalia[f]
XXX	Triple-X	Triplo-X female
XXY	Klinefelter	Klinefelter syndrome

[a]Including all or most of the classic features of this syndrome at or beyond the time of expected puberty.
[b]Exclusive of mendelian, teratogenic, autosomal, or other causes of abnormal sex differentiation.
[c]Definite, though partial, manifestations of the Turner syndrome per se.
[d]A phenotypic female, without the features of Turner syndrome, but with incomplete maturation of secondary sex characteristics and with absence of menarche.
[e]Ambiguous genitalia, usually with a unilateral testis and contralateral fallopian tube and streak ovary.
[f]Male genitalia, including testes and wolffian structures, though hypospadias may be present (pseudohermaphroditism).

influence of the 46,XY cell line in the context of 45,X/46,XY mosaicism: 1) Mosaicism of and by itself does not account for the marked variation in expression of these two cell admixtures, including the phenotypes designated in Tables 1 and 2. These clinical pictures, except for the differences between the TS and "incomplete TS" appear to represent qualitative differences rather than quantitative ones along a continuum. In addition, other types of X-chromosome aberrations, including 45,X/46,XX mosaicism, can lead to the same spectrum of phenotypes with the apparent exception of mixed gonadal dysgenesis. Thus, each clinical picture may, on the one hand, be associated with 45,X/46,XY mosaicism and, on the other hand, be associated with another chromosomal constitution. 2) Mosaicism does not intrinsically account for lack of expression of a normal 46,XY karyotype in at least some cells of those tissues whose differentiation and function presumably are not sex hormone-dependent. That is, why is the phenotypic picture in some 45,X/46,XY cases, such as the proposita, so completely that of the TS; why are some features not tempered, so to speak? Fifty percent of the proposita's cultured skin cells showed a 46,XY chromosome pattern; yet the pigmented nevi and dermatoglyphics were typical of 45,X TS.

TABLE 2. Karyotype-Phenotype Correlations as a Function of Mosaicism

Sex chromosome constitution	Resultant cell types	Clinical picture	Relative proportions of X-monosomy and diploid cells[a]
X/XX	X-monosomy and female	Turner syndrome	X >> XX
		Incomplete Turner syndrome	X > XX
		Primary amenorrhea	X < XX
X/XY	X-monosomy and male	Turner syndrome	X >> XY
		Incomplete Turner syndrome	X >> XY
		Primary amenorrhea	X ≅ XY
		Male genitalia	X < XY

[a]Relative proportions of respective cell types are estimated; notice that the proportions are expected to be different for the same clinical picture (eg primary amenorrhea) when the diploid cells are different.

One possibility deriving from these considerations is that the alternative phenotypes associated with 45,X/46,XY mosaicism do not represent a continuous spectrum and that indeed an additional factor or set of factors is a major determinant of the final phenotype. One such factor would be a structurally abnormal Y chromosome, well known to be present in some instances of X/XY mosaicism [2, 4], with resultant functional deficiencies. The proposita's Y chromosome was apparently normal and similar in appearance to her father's Y chromosome. Another possibility is that of a mendelian point mutation involving the functional, euchromatic proximal segment of the Y-chromosome long arm. In the proposita 2 indications of Y-chromosome function, H-Y antigen positiveness [7, 8] and the development of a gonadoblastoma [10, 11, 23], were present although mutations that would spare those functions obviously were not ruled out. It is possible, for example, that the complete TS will develop in the presence of X/XY mosaicism only if, additionally, there is a more or less specific Y mutation. Alternatively, if Y-chromosome expression involves interaction with an autosomal or X-linked locus [24, 25] and if that locus were mutant, then Y-chromosome expression would be obviated even if the Y chromosome itself were intact.

There are thus a number of schemes other than mosaicism itself that could account for the various phenotypes associated with X/XY mosaicism. We believe that it is precisely the distinctiveness of 4 of the phenotypes associated with X/XY mosaicism that warrants a search for another factor beyond the mosaicism itself. Although we cannot yet specify this additional influence, we would suggest that investigators should search for it in 45,X/46,XY TS patients rather than merely presume that cell admixtures account for the clinical picture and lack of Y-chromosome expression.

ACKNOWLEDGMENTS

Thanks to Dr. J. Frank for his help in patient follow-up, to Dr. S. S. Wachtel, Dr. G. C. Koo, and Dr. O. J. Miller for the H-Y antigen assay data, to Ms. K. Michalski for cytogenetic analyses, and to Mrs. C. Gibson for assistance with manuscript preparation.

REFERENCES

1. Engel E, Forbes AP: Cytogenetic and clinical findings in 48 patients with congenital defective or absent ovaries. Medicine 44:135–164, 1965.
2. Simpson JL: "Disorders of Sexual Differentiation. Etiology and Clinical Delineation." New York: Academic Press, 1976.
3. Ferguson-Smith MA: Karyotype-phenotype correlations in gonadal dysgenesis and their bearing on the pathogenesis of malformations. J Med Genet 2:142, 1965.
4. Fonatsch C, Flatz SD, Freymann R: Non-fluorescent Y chromosome in a male infant with Turner's symptoms and XO/XY mosaicism. Clin Genet 11:235–240, 1977.

5. Goldstein DE, Kelly TE, Johanson AJ, Blizzard RM: Gonadal dysgenesis with 45,X/46,XX mosaicism demonstrated only in a streak gonad. J Pediatr 90:604–605, 1977.
6. Davidoff F, Federman DD: Mixed gonadal dysgenesis. Pediatrics 52:725, 1973.
7. Wachtel SS, Koo GC, Breg WR, Elias S, Boyse EA, Miller OJ: Expression of H-Y antigen in human males with two Y-chromosomes. N Engl J Med 293:1070–1072, 1975.
8. Koo GC, Wachtel SS, Saenger P, New MI, Dosik H, Amarose AP, Dorus E, Ventruto V: H-Y antigen: Expression in human subjects with the testicular feminization syndrome. Science 196:655–656, 1977.
9. Miller OJ, Miller DA, Warburton D: Application of new staining techniques to the study of human chromosomes. Prog Med Genet 9:1–47, 1973.
10. Manuel M, Katayama KP, Jones HW Jr: The age of occurrence of gonadal tumors in intersex patients with a Y-chromosome. Am J Obstet Gynecol 124:293–300, 1976.
11. Psaroudakis A, Oettinger M, Byrd JR, Greenblatt RB: Cytogenetic studies in gonadal dysgenesis with dysgerminoma. Am J Obstet Gynecol 126:508–510, 1976.
12. Mittwoch U: "Genetics of Sex Differentiation." New York: Academic Press, 1973.
13. Federman DD: "Abnormal Sexual Development." Philadelphia: WB Saunders, 1967.
14. Wilroy RS, Summitt RL, Martens PR, Tharapel AT, Tipton RE, Givens JR: Phenotype-karyotype correlations in 81 patients with the Turner syndrome. (Abstract) Clin Res 25:74A, 1977.
15. Goldberg MB, Scully AL, Solomon IL, Steinbach HL: Gonadal dysgenesis in phenotypic female subjects. Am J Med 45:529–543, 1968.
16. Ferrier PE, Ferrier SA, Klein D, Fernex C: XX/XO mosaicism. Lancet 1:54, 1963.
17. DeGrouchy J, Lamy M, Aicardi J, Pellerin D: Mosaique XY/XO chez un pseudo-hermaphrodite masculin. Ann Pediatr 39:789, 1963.
18. DeGrouchy J, Blondet P, Royer P, Vermeil G, Lamy M: Syndrome de Turner et caryotype a 47 chromosomes. Ann Genet 6:25, 1963.
19. Sohval AR: Hermaphroditism with "atypical" or "mixed" gonadal dysgenesis. Am J Med 36:281, 1964.
20. Pfeiffer RA, Lambertz B, Friederiszick FK, Distel H, Pawlowitzki IH, Nicole R, Ober KG, Ruckes J: Die nosologische Sellung des XO/XY-Mosaizimus. Arch Gynaekol 206:369, 1968.
21. Morishima A, Grumbach MM: The interrelationship of sex chromosome constitution and phenotype in the syndrome of gonadal dysgenesis and its variants. Ann NY Acad Sci 155:695, 1968.
22. Srivastava K, Stoeckle H, Rogers WA, Townsend JF: XO/XY mosaicism in a 2-year-old phenotypic male. J Genet Hum 23:155–162, 1975.
23. Szokol M, Kondrai G, Papp Z: Gonadal malignancy and 46,XY karyotype in a true hermaphrodite. Obstet Gynecol 49:358–360, 1977.
24. Bjersing L: H-Y antigens and sexual differentiation. Lancet 1:311, 1977.
25. Kralova J, Demant P: Expression of the H-Y antigen on thymus cells and skin: Differential genetic control linked to the K end of H-2. Immunogenetics 3:583–594, 1976.

EDITORIAL COMMENT

The authors of this paper make a plea that factors other than cell proportions in various tissue in 45,X/46,XY mosaicism be searched for, particularly when such patients do not manifest or only minimally manifest phenotypic features attributable to the 46,XY cell line. The proportions of 45,X and 46,XY cells in

45,X/46,XY patients not ostensibly in keeping with the various phenotypes seen in this type of mosaicism concern the authors.

In general the plea and concern of the authors is intriguing. However, the reader is cautioned not to abandon totally the idea that the phenotypes in 45,X/46,XY mosaicism may be a function of proportions of 45,X and 46,XY cells in critical tissues at critical times in embryogenesis. Several points made by the authors deserve comment:

1) "It is impossible to study the chromosome complement of every cell in the body." This is a cliche often invoked to explain dichotomies between cell proportions and phenotype in mosaics. Yet the concept cannot be ignored. In addition, cell proportions are measured in a specific tissue at a specific point in time, both far removed from potential critical times and tissues in the developing embryos.
2) The authors imply that 45,X/46,XX mosaicism can produce all the phenotypes in the spectrum from an almost normal male on one end to an almost normal female on the other, as in 45,X/46,XY mosaicism, "with the apparent exception of mixed gonadal dysgenesis." This suggests that 45,X/46,XX mosaicism may be seen in association with an "almost normal male" phenotype. I know of no evidence to support this.
3) The authors offer two possible mechanisms other than cell proportions as explanations for apparent phenotype-cell proportion dichotomies in 45,X/46,XY mosaicism: Structural alteration of the Y chromosome present in the XY cells; and a mutant Y-linked gene. I would wonder if these two possibilities might be even less likely than currently favored explanations.

Robert L. Summitt

The KOP Translocation*

Philip D. Pallister, MD, and John M. Opitz, MD

INTRODUCTION

This is a belated clinical tidying-up of an X-autosome translocation which has been known for over 10 years, has been used extensively in somatic cell genetics by many investigators and has led to important advances in X chromosome and autosome mapping [1–9]. McKusick and Ruddle recently alluded to the history of this translocation in footnote 100 of their authoritative review of the status of the gene map of the human chromosomes [10].

In 1967 the patient was ascertained at 19 years by one of us (Dr. Pallister) at the Boulder River School and Hospital (BRSH) Boulder, Montana, to which he had been admitted for vocational rehabilitation. He was found to be borderline mentally retarded and a diagnosis of the Klinefelter syndrome was made on the basis of the clinical findings. A buccal smear showed many cells with a Barr body which was interpreted as larger than that formed by a normal single X chromosome. On a chromosome examination performed in the Cytogenetics Laboratory of BRSH, he was found to have a complex anomaly (Fig. 1): he had 47 chromosomes: 14C, 5D, a normal Y, three 17-like chromosomes, as well as two apparently identical, large, satellited acrocentric chromosomes. The patient's phenotypically normal mother had an apparently balanced chromosome abnormality which was interpreted as a Cq–; Dq+ translocation (Fig. 2). In view of the patient's clinical phenotype, his Barr body findings and these chromosome abnormalities, it was postulated 1) that his mother's chromosome constitution was 46,X,t(Xq–; Dq+)

* Studies of Malformation Syndromes of Man LI; Paper No. 1555 from the University of Wisconsin Genetics Laboratory. This study was supported by DHEW/PHS grants GM15422 and GM20130 from The National Institute of General Medical Sciences.

Birth Defects: Original Article Series, Volume XIV, Number 6C, pages 133–146

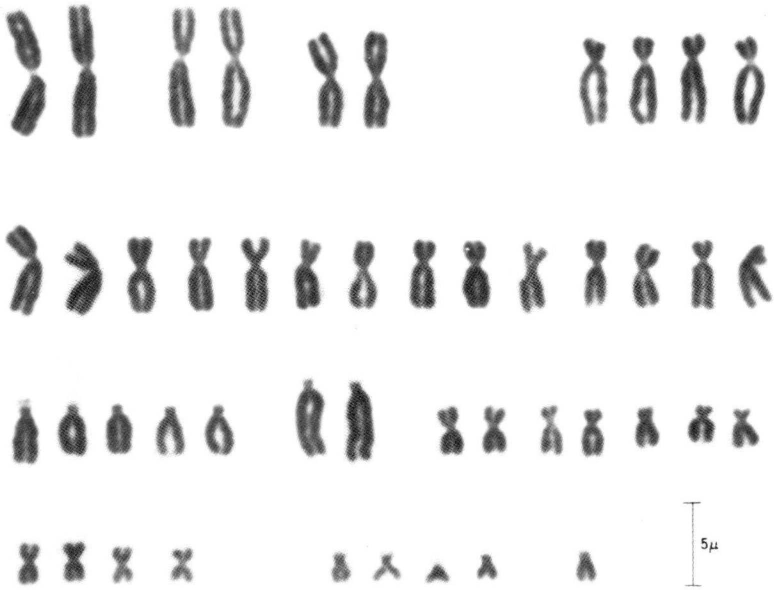

Fig. 1. Karyotype of the proband (KOP-2) prepared from peripheral lymphocyte cultured at the Wisconsin State Laboratory of Hygiene; the two X-D translocation chromosomes are placed between the D and E groups; the small X-D translocation chromosome cannot be distinguished from a 17. The Y is easily identified. Giemsa stain (original magnification × 3,000).

and 2) that the patient had no normal X chromosome, but that he was homozygous for the Dq+ ("X–D") translocation chromosome on the basis of second meiotic nondisjunction which gave him *one* X short arm (in the form of the Xq– chromosome), the Klinefelter syndrome due to *two* X-long arms, and nearly complete trisomy of a D chromosome. It was further postulated that if this hypothesis were correct, autoradiographic studies might show the mother's *normal* X chromosome to be consistently inactivated, and late replication of the DNA of most or all of *one* of the patient's two X–D chromosomes.

The phenotypic effects of trisomy 14 and 15 were and remain unknown. In order to facilitate scrutiny of the patient for known D-trisomy effects, it was hoped that the chromosome involved in this translocation would turn out to be 13. At the time of his initial ascertainment and study, it was also realized that this translocation might be of use in providing information on gene mapping, at least with respect to the X chromosome.

In view of these implications, intensive studies of the patient were begun in 1968 by several collaborating colleagues and laboratories.

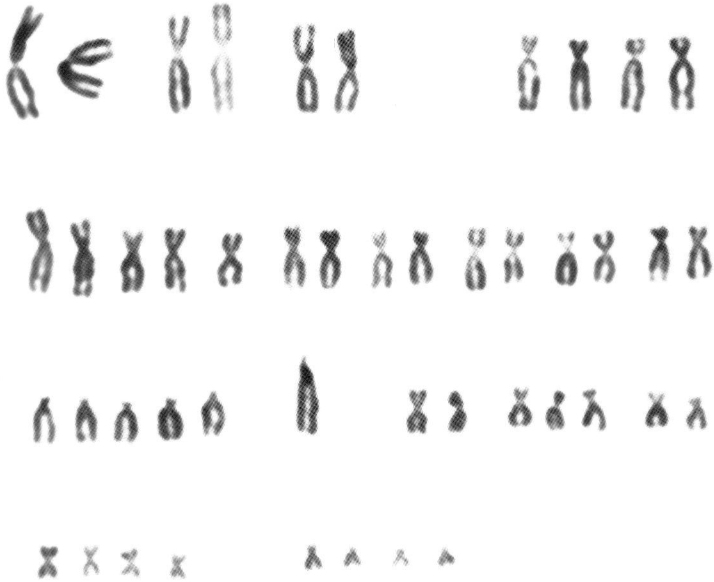

Fig. 2. Karyotype of the proband's mother (KOP-1) – note only 15C, 5D and three 17-like chromosomes, and the X-D chromosome; Giemsa stain (original magnification × 3,000).

CLINICAL STUDIES

In 1967 (at 19 years), the patient's height was 162.5 cm, weight 58.96 kg, occipito-frontal head circumference (OFC) 53 cm, span 157.5 cm, upper (UBS) and lower body segment (LBS) 77.5 and 85 cm, respectively. Over the ensuing four years his weight increased to 65.54 kg, his height to 164.8 cm, his span to 163 cm, UBS and LBS to 78.8 and 86 cm, respectively; his OFC remained the same. He had gynecomastia which enlarged over 2 years.

In addition to his eunuchoidal body proportions, the patient showed other manifestations of hypogonadism: His voice remained at an early pubertal pitch and he did not show prominence of the larynx or temporal recession of the hairline; he had no beard, minimal axillary hair, and a female escutcheon. The penis was somewhat small for age, the scrotum remained relatively hypoplastic and the testes were small and firm: The right testicle measured 2 × 1 × 1 cm, the left 2.5 × 1.5 × 1 cm. In addition he had brachycephaly with low posterior hairline, slight facial asymmetry and a relatively flat midface, slight anteversion of nostrils and hypoplasia of the nasal alae, micrognathia (shortness of mandible), slight posterior angulation of the left and anteversion of the right auricle with hypoplasia of the lobules and lower helices, Brushfield spots and upper epicanthic folds, slight malpositioning of several teeth with retention of left upper and lower lateral deciduous incisors; slight pes cavus, calcaneovalgus and metatarsus varus deformities; relatively short metatarsals 2–5 bilaterally, and genua valga and recurvata. The distal interphalangeal flexion creases of the 4th fingers were hypoplastic and on the left 5th finger he had only one flexion crease; both 5th fingers were short, and there were slight flexion contractures of both thumbs at the metacarpophalangeal joints. He had bridged creases on the right and a single (simian) crease on the left palm, *t′* palmar axial triradii bilaterally, a low arch dermatoglyphic pattern on all 20 digits, and a distal loop/distal pocket loop hallucal pattern on the right and left soles respectively. He had slight cubitus valgus bilaterally, a shallow sacral dimple and a right inguinal herniorrhaphy scar (Figs. 3–6).

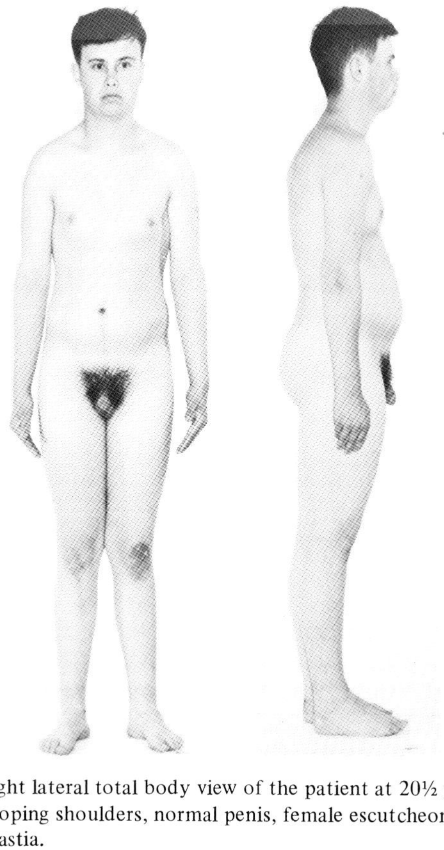

Fig. 3. Frontal and right lateral total body view of the patient at 20½ years. Note slight generalized obesity, sloping shoulders, normal penis, female escutcheon, hypoplastic scrotum, and minimal gynecomastia.

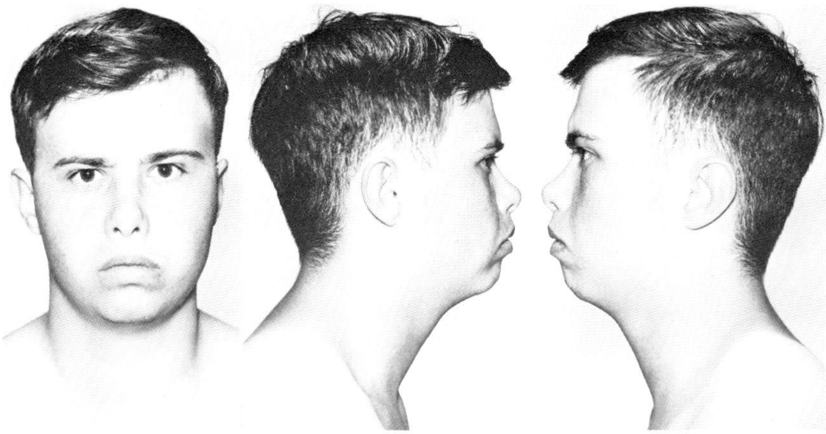

Fig. 4. Frontal right and left lateral profile views of head and face of the patient at 20½ years. Note slight facial asymmetry, brachycephaly, slight anteversion of right auricle, posterior angulation of left auricle, hypoplasia of lower helix and lobule bilaterally, epicanthic folds, relatively flat midface, hypoplastic nasal alae and micrognathia.

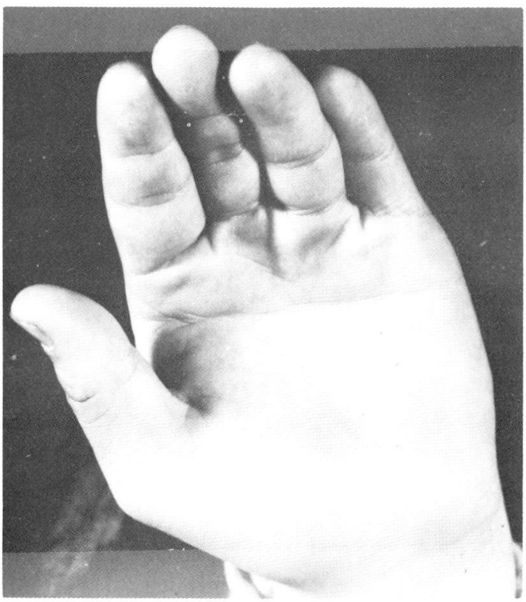

Fig. 5. Left hand showing camptodactyly of thumb at metacarpo-phalangeal joint, simian crease, shortness of 5th finger with single flexion crease of that digit, and hypoplastic distal IP flexion crease of the 4th finger.

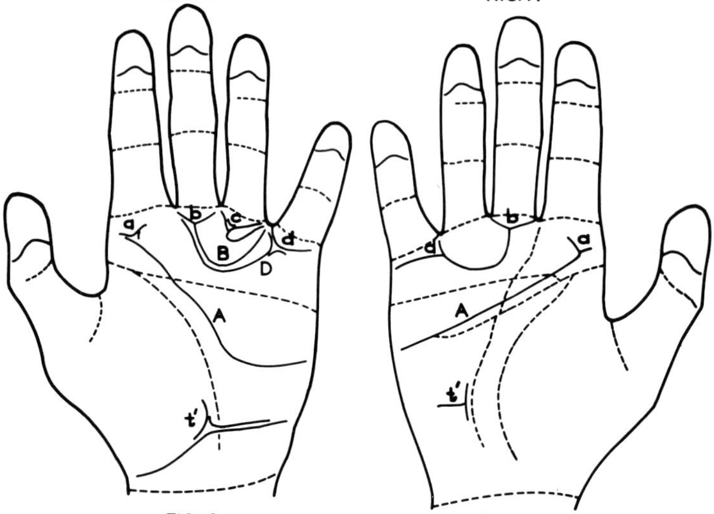

Fig. 6. Sketch of palmar and digital dermatoglyphic findings in the patient; note 10 low arches, *t'* triradius bilaterally, vestigial loop distal in left 4th interdigital area, left simian crease, single flexion crease of 5th fingers and absence of *c* triradius on the right.

The rest of the physical examination was unremarkable. The cardiovascular system, fundi, sight and hearing were essentially normal as was the neurologic status. The skeletal survey showed only those minor abnormalities expected on the basis of the physical examination (no radioulnar synostosis); at a chronologic age of 19 years, his bone age was estimated between 16 and 17 years.

Two electroencephalograms were normal. On psychologic examination (WAIS) he had an overall IQ of 73, and a relatively immature and unstable personality.

Past medical history was unremarkable, except for the usual childhood diseases, a right indirect inguinal hernia repaired at 12 years, slow psychomotor development, delay in the loss of certain deciduous teeth and in the onset of puberty.

Family history (Fig. 7). The patient's father was normal; he had apparently normal chromosomes, his family history was essentially unremarkable. Both maternal grandparents were deceased at the time of the study; at that time the mother's oldest brother (who had had 3 normal daughters) and her oldest sister (who had produced a normal son and a normal daughter) were also dead. The mother was the youngest of 5 sibs. Her surviving brother had a normal son and a normal daughter; he was examined and found to be normal, and his karyotype was normal. Her surviving sister is reportedly normal as are the sister's son and 2

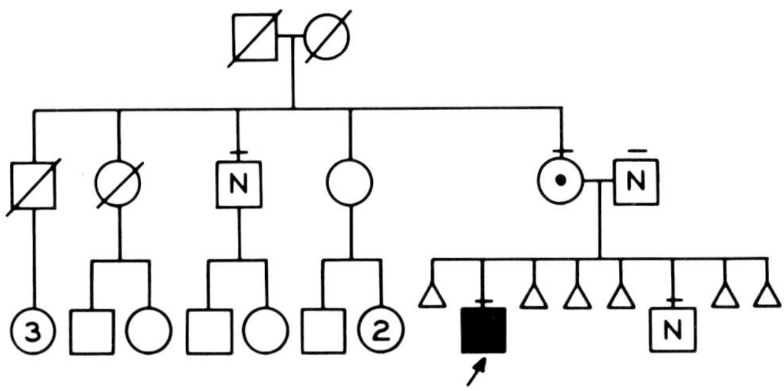

Fig. 7. Pedigree of the patient's family: see text. Bar over pedigree symbol – personally examined and studied cytogenetically (N = apparently normal chromosomes); small triangles: spontaneous abortions at 6–12 weeks of gestation.

daughters; to date it has not been possible to complete a chromosome examination on the sister. It is therefore not possible to conclude with certainty that the patient's mother was the first translocation carrier in her family. Altogether she has had 8 pregnancies: 6 early spontaneous abortions at 6 to 12 weeks, the patient, and a chromosomally and clinically normal son as pregnancies number 2 and 6, respectively.

SPECIAL STUDIES

The patient's color vision was normal and no X-linked disorders were present in the patient or his family. The results of determinations of hemoglobin binding capacity, haptoglobin, transferrin and Gc types performed by Dr. Eldon Sutton, Department of Zoology, the University of Texas at Austin (Table 1), of blood group typing (Table 2) performed by Drs. Ruth Sanger and R.R. Race of the MRC Blood Group Research Unit, the Lister Institute, London, England, and of

TABLE 1. Haptoglobin, Transferrin and Gc Types in Patient, His Mother, Father and Brother*

Person	Hemoglobin binding capacity (mg/100 ml)	Hapto-globin type	Trans-ferrin type	Gc type
Father	71	2–2	C	2–2
Mother	77	2–2	C	1–1
Patient	85	2–2	C	2–1
Brother	52	2–2	C	2–1

*Performed by Dr. H. Eldon Sutton

TABLE 2. Red Cell Antigen Types of Patient, His Mother, Father, and Brother*

Person	ABO	MNS	P	Rh	Lu^a	K	Le^a	Le^b	Fy^a	Fy^b	Xg^a
Father	O	MSMs	+	rr	−	−	−	+	+	+	+
Mother	O	MSMs	−	R_1R_2	−	−	+	−	+	−	−
Patient	O	MsMs	−	R_1r	−	−	−	+	+	+	−
Brother	O	MsMs	+	R_1r	−	−	−	+	+	−	−

*Performed by Dr. R. R. Race and Ruth Sanger
a, b are blood group type antigens

other polymorphic typing studies (Table 3) done by Dr. Elizabeth B. Robson of the MRC Human Biochemical Genetics Resarch Unit, the Galton Laboratory, University College, London, England, were not of help in autosomal or X-chromosome gene mapping.

Cytologic and autoradiographic studies performed on cultured fibroblasts and peripheral lymphocytes of the patient and his mother initially by the Wisconsin State Laboratory of Hygiene (Dr. Stanley L. Inhorn) confirmed 1) that the patient's X-chromatin mass appeared to be about twice as large as a normal X-chromatin mass; 2) that his mother's X-chromatin mass appeared to be of normal size (Fig. 8); 3) that his mother's normal X chromosome as consistently her latest replicating chromosome; 4) that the D chromosome involved in the

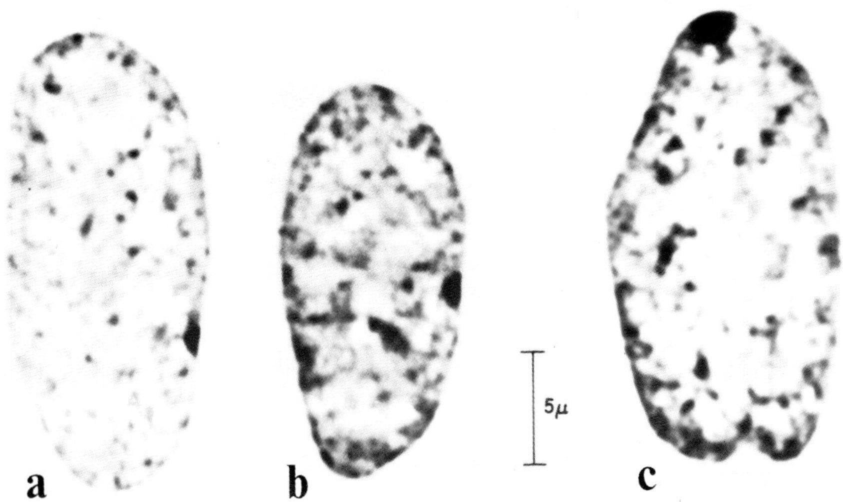

Fig. 8. Orcein stained X-chromatin masses in nuclei of fibroblasts cultured at the Wisconsin State Laboratory of Hygiene: a, b, are from the mother, c from the patient (original magnification × 3,000).

TABLE 3. Other Polymorphisms of Patient, His Mother, Father, and Brother*

Person	Acid phosphatase	PGM	AK	Peptidases A	B	C	D	PHI	ADA	Diaphorase
Father	BA	1	1	1		1	1	1	1	1
Mother	CB	2–1	1	1		1	1	1	1	1
Patient	CA	2–1	–	–		–	1	–	1	–
Brother	CA	2–1	–	–		–	1	–	1	–

*Performed by Dr. Elizabeth B. Robson

translocation was 14 and not 13 or 15; 5) that in many of the patient's cells one of the X–D translocation chromosomes replicated its DNA later than the other and that this apparently involved the entire chromosome and not just its distal, X-derived one-half (Fig. 9) and 6) that the small translocation chromosome presumed to consist of Xp, X centromere, possibly a tiny remnant of proximal Xq and a tiny (telomeric) piece of 14q could not be distinguished from chromosomes 17. Though suspected to be near the centromere of the Xq and near the tip of the 14q, the exact location and minimum number of breaks required to produce the original translocation remained unknown.

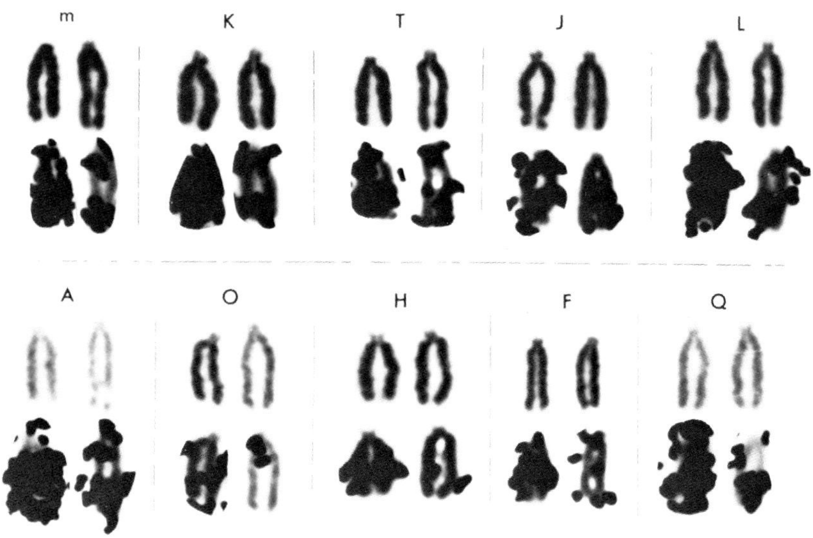

Fig. 9. Autoradiographic study of the two X-14 translocation chromosomes from 10 cells of the patient showing late DNA replication of one (the left) member of the pair in each case (Wisconsin State Laboratory of Hygiene and JMO).

In the fall of 1969 the patient, his mother and maternal uncle were studied by Drs. Klinger and Siniscalco at the Albert Einstein College of Medicine of Yeshiva University in New York; the above historic, genealogic and clinical data were confirmed and clinical photos were obtained. Cell (lymphocyte, skin and gingival fibroblast) cultures established at that time and on several subsequent occasions from the patient and his mother have been studied by several collaborating laboratories and investigators, including the Albert Einstein Medical Center (Drs. Klinger, Siniscalco and coworkers), the Department of Human Genetics, Leiden University, Holland (Dr. Siniscalco and collaborators), the Department of Biology of Yale University (Drs. F.H. Ruddle and Florence Ricciuti), and Columbia University (Dr. O.J. Miller) [3–9 and H.P. Klinger in a discussion of KOP translocation following paper of Mary Lyon at Schering Symposium on Intrinsic and Extrinsic Factors in Mammalian Development, Venice 1970].

Using man-mouse and man-syrian hamster somatic cell hybrids the HGPRT, G6PD and PGK loci were linked to the acrocentric long arm fragment of the two translocation chromosomes in the following sequence: centromere: PGK:HGPRT: G6PD [5–8]. That same fragment was of use in assigning the nucleoside phosphorylase (NP) locus to 14q [9]. The cell lines derived from mother and son in this case were designated KOP-1 and KOP-2 respectively, and were placed in the Human Genetic Mutant Cell Repository (Camden, NJ) by Dr. Ruddle.

Fluorescent and other banding studies by many laboratories led to the conclusion on the following designation of breaks in mother and son respectively: mother (KOP-1) – 46,X,t(X;14) (Xpter→Xq13: :14q32→14qter?;14pter→14q32: :Xq13→Xqter); son (KOP-2) – 47,Y,der(14), t(X;14) (Xpter→Xq13::

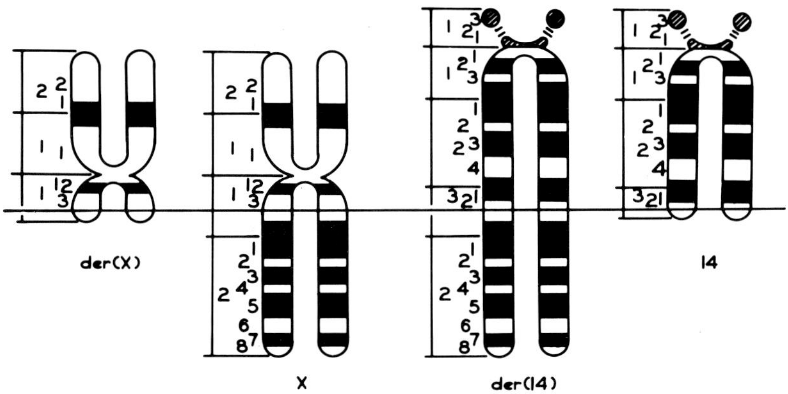

Fig. 10. Breakpoints Paris Nomenclature, based on published results of several investigations [1, 2]; t(X;14) (Xpter→Xq13::14q32→14qter?; 14pter→14q32::Xq13→Xqter).

14q32→14qter?; 14pter→14q32: :Xq13→qter)mat, + der(14), t(14pter→14q32: : Xq13→Xqter) mat [1, 2], (Fig. 10). Mouse-man somatic cell hybrids using these cell lines have been used to study linkage relationships of 17 human gene loci by Ruddle et al [5].

Further autoradiographic studies with tritiated thymidine were done by Dr. P.W. Allderdice at Columbia University, New York; these confirmed the earlier results.

The der(X) translocation chromosome was not preferentially inactivated in either the mother or the son and it was not possible to distinguish it from chromosomes 17 and 18 by terminal label. Quinacrine fluorescence studies agreed with autoradiographic analysis. Presumed meiotic pairing configuration of the translocated chromosomes in the mother is schematically represented in Fig. 11.

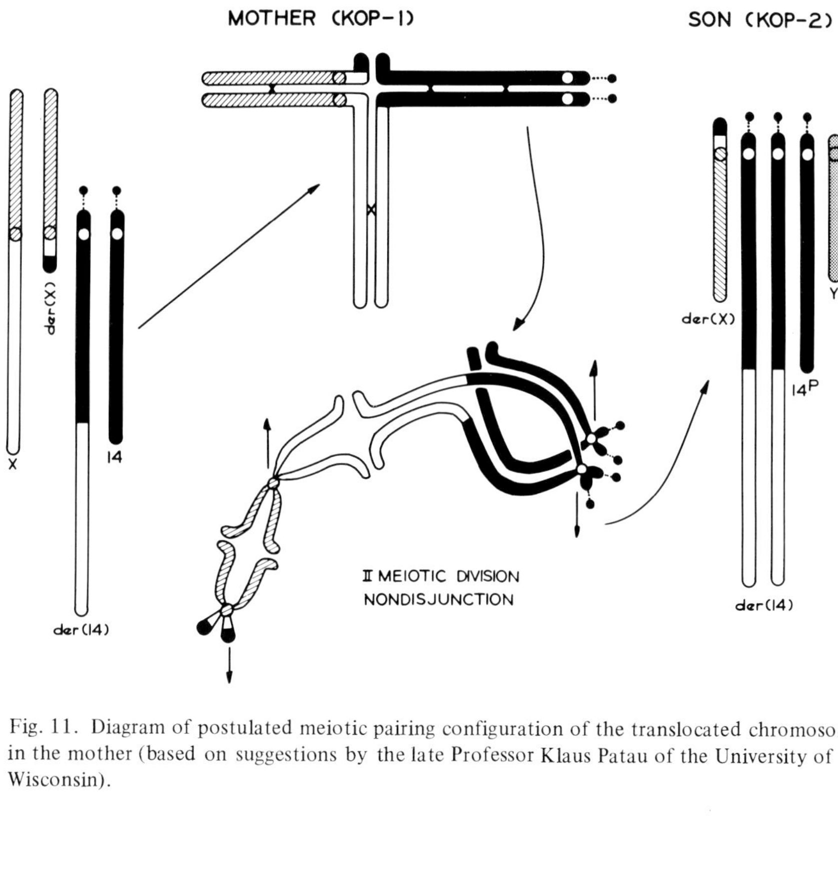

Fig. 11. Diagram of postulated meiotic pairing configuration of the translocated chromosome in the mother (based on suggestions by the late Professor Klaus Patau of the University of Wisconsin).

CONCLUSIONS

The KOP translocation

1) showed that careful phenotype analysis and the making and testing of pathogenetic and causal hypotheses in a clinical setting may permit use of the scientific method in its entirety and lead to discoveries of fundamental biologic importance;

2) showed that the Klinefelter syndrome can be produced in a Y bearing individual without normal X chromosome(s) and possessing only one Xp and two Xq's as translocation chromosomes;

3) in the same family confirmed the Lyon hypothesis and provided an important exception to it: In the Klinefelter male with two Xq's one was inactivated (in the sense that it was late labeling and formed a single larger than normal X-chromatin mass while the other and an Xp remained active. In the mother, with a balanced X-autosome translocation, random inactivation of the normal X and of the two translocation chromosomes bearing the other Xp and Xq respectively did not occur: The normal X was preferentially inactivated, leaving her with a functionally balanced set of autosomes. Whether selection plays a role in this cannot be answered from this investigation;

4) showed that additional (trisomic) autosomal material attached to a second, almost complete Xq will become functionally and physically inactivated in the sense of being without apparent phenotypic effect on physical development, enhancing the size of the X-chromatin mass and becoming as late labeling as the inactive X portion attached to it;

5) showed again, that in spite of the apparently "foolproof" inactivation of the attached large piece of autosomal material the Y + 2Xq aneuploidy effect (the Klinefelter syndrome) was not abolished – ie that inactivation of a second X is a quantitatively and possibly qualitatively different phenomenon than inactivation of autosomal by X-chromosomal material. [In other words, extra (more than one X chromosome) X-derived chromosomal material, even though genetically inactivated, obviously exerts a deleterious effect in embryogenesis to produce the Klinefelter phenotype, while extra autosomal material (14-derived in this instance), when inactivated by extension of inactivation from the attached X-derived segment, produces no discernible effect of the sort expected of autosomal trisomy. At the same time it is puzzling that, if all X-chromosomes (and all autosomes translocated thereto) are genetically active in early embryogenesis prior to formation of the X-chromatin mass, the autosomal trisomy is not in some way (some *very obvious* way) expressed. Editor's note.]

6) led to assignment of the loci PGK, HPRT and G6PD in proper order to the long arm of X;

7) led to assignment of the locus nucleoside phosphorylase to the long arm of 14;

8) showed again that the inactivated long arm of a normal second X makes up the bulk of the normal X-chromatin mass in view of the fact that inactivated Xq + attached inactive autosomal segment of about the same length as the Xq makes for an X-chromatin mass of about twice normal size, ie the same size as that produced by an inactivated i (Xq).

9) does not contradict the hypothesis of an X inactivation center Xq;

10) provides another example of a q13-break balanced X-autosome translocation with apparently normal gonadal function in the carrier female.

The authors are undecided whether they enjoyed more the making and testing in a clinical setting, of the hypotheses presented herein, or the early free sharing of this material with the investigators and laboratories most capable of putting it to best use.

This paper is to be read in conjunction with that by Summitt et al (this volume, pp 219–248) on X-autosome translocations.

Colophon (17 May 1977)

The proband son has married. His wife is blind in one eye, partially blind in the other, is retarded and has cerebral palsy and epilepsy as a result of an accident. They exist, with much social support, on medicaid, food stamps and welfare financing. Counseling has been given for frustration leading to occasional wife abuse. He no longer has problems with the police for molesting small girls. He has become obese, weighing over 200 pounds.

ACKNOWLEDGMENTS

The authors are most grateful to the following persons in assisting with these investigations: Drs S.L. Inhorn, L.F. Meisner, H. Eldon Sutton, Ruth Sanger, R.R. Race, Andrew L. Kosseff, Harold P. Klinger, Elizabeth B. Robson, Marcello Siniscalco, Frank H. Ruddle, Penelope W. Allderdice, Orlando J. Miller and Jürgen Herrmann. We are particularly grateful to the patient and his family for their patient and most gracious cooperation in being examined, donating blood and skin biopsies and in travelling to New York under Dr. Siniscalco's auspices to be studied there by our east coast colleagues.

REFERENCES

1. Opitz JM, Pallister PD, Ruddle FH: An (X;14) translocation, balanced, 46 chromosomes. (Human genetic mutant cell) repository identification No. GM-73 (KOP-1 cell line). Cytogenet Cell Genet 12:289–290, 1973.

2. Opitz JM, Pallister PD, Ruddle FH: An (X;14) translocation, unbalanced, 47 chromosomes. (Human genetic mutant cell) repository identification No. GM-74 (KOP-2 cell line). Cytogenet Cell Genet 12:291–292, 1973.
3. Siniscalco M: Somatic cell hybrids as tools for genetic studies in man. Symp Int Soc Cell Biol 9:205–231, 1970.
4. Allderdice PW, Miller OJ, Klinger HP, Pallister PD, Opitz JM: Demonstration of a spreading effect in an X-autosome translocation by combined autoradiographic and quinacrine-fluorescence studies. In "4th International Congress on Human Genetics." Paris: Excerpta Medica. Int Congr Series No 233, Abstract 39, 1971, p 14.
5. Ruddle FH, Chapman VM, Ricciuti F, Murnane M, Klebe R, Meera Khan P: Linkage relationships of 17 human gene loci as determined by man-mouse somatic cell hybrids. Nature [New Biol] 232:69–73, 1971.
6. Grzeschik KH, Allderdice PW, Grzeschik A, Opitz JM, Miller OJ, Siniscalco M: Cytologic mapping of human X-linked genes by use of somatic cell hybrids involving an X-autosome translocation. Proc Natl Acad Sci USA 69:69–73, 1972.
7. Ricciuti F, Ruddle FH: Assignment of three loci (PGK, HGPRT, G6PD) to the long arm of the human X chromosome by somatic cell genetics. Genetics 74:661–678, 1973.
8. Grzeschik K-H: Syrian hamster-human somatic cell hybrids: Isolation and characterization. Humangenetik 20:211–218, 1973.
9. Ricciuti F, Ruddle FH: Assignment of nucleoside phosphorylase to D-14 and localization of X-linked loci in man by somatic cell genetics. Nature [New Biol] 241:180–182, 1973.
10. McKusick VA, Ruddle FH: The status of the gene map of the human chromosomes. Science 196:390–405, 1977.

Aplasia of the Müllerian System: Evidence for Probable Sex-Limited Autosomal Dominant Inheritance

M. H. K. Shokeir, MD, PhD

INTRODUCTION

Absence of the vagina is observed in a number of disorders occurring in phenotypic females, sometimes described as male pseudohermaphrodites. In those individuals who suffer from aplasia or hypoplasia of the müllerian system it is the "müllerian" portion of the upper two-thirds to four-fifths of the total vaginal length that is affected. The lower one-fifth to one-third of the vagina, which is derived from the upward invagination of the urogenital sinus, is usually normal in isolated abnormalities of müllerian duct development. The external genitalia, which are usually also normally developed, are unmistakably female. The sex of rearing and the assigned gender are also always female.

At puberty secondary sex characteristics develop normally including breasts and pubic and axillary hair. Amenorrhea is, however, an invariable finding. Attempts at sexual intercourse may be both painful and difficult.

Familial examples of congenital absence of the vagina have been reported [1, 2]. Phaneuf [3] reported the occurrence of the disorder in 2 pairs of sisters whose mothers were sisters. Jones and Mermut [4], who described a similar disorder with familial distribution, cast doubt on the validity of the diagnosis in some earlier reports and expressed the view that many were examples of testicular feminization. Based on their observations, autosomal recessive transmission was felt to be the operative mode of inheritance.

In this communication, we wish to report 13 unrelated individuals who appear to present the complete form of aplasia of the müllerian duct derivatives. In 10 of them, other female relatives were similarly affected. In 8 such families the

Birth Defects: Original Article Series, Volume XIV, Number 6C, pages 147–165

pattern of distribution of affected individuals was most consistent with autosomal dominant transmission with sex limitation; in 2, autosomal dominant or recessive inheritance might have operated. In the remaining 3 families, no affected relatives were ascertained; therefore, no evidence could be adduced to support a genetic etiology, let alone a particular mode of inheritance. Nevertheless, even in these, autosomal dominant or autosomal recessive inheritance may have been operative and could not be ruled out.

Three further cases were observed where a rudimentary uterus or a muscular bud was ascertained at laparotomy. They likely represent examples of Rokitansky-Kuster-Hauser syndrome (RKHS). None of them had relatives known to have the disorder.

PRESENT INVESTIGATIONS

The patients reported here were seen for genetic assessment, karyotyping, and counseling. The presenting complaint was difficult or painful attempts at sexual intercourse in 9 of the 16 probands, amenorrhea in 6, and periodic pain with slight fever and general malaise in 1 patient. [The last patient was ultimately diagnosed, following surgical exploration, as an example of partial müllerian hypoplasia (Rokitansky-Kuster-Hauser syndrome) with mild endometriosis.] Among all those who complained of coital difficulty, history of primary amenorrhea could be readily elicited. Conversely, among those who complained of amenorrhea, history of sexual difficulty was revealed on further questioning in all but 2 patients, 15- and 18-year-old unmarried girls. Five of our patients were married, whereas the remaining 11 were single. The former ranged in age from 18 to 28 with a mean of 22 years whereas the latter ranged from 15 to 20 with a mean of 17 years.

The investigations carried out on all patients comprised clinical (including pelvic) examination, buccal smear and chromosome analysis, pituitary gonadotropin (FSH and LH) assay and repeated estrogen and progesterone assay. Furthermore, on 11 patients excretory urography was performed. All but 1 patient underwent laparoscopy and 5 were subjected to laparotomy. Ovarian biopsy was undertaken on 6 patients (3 times by laparoscopy and 3 times by laparotomy). In 1 patient (with the partial syndrome) the rudimentary uterine buds were removed and histologically examined (Fig. 1).

When the diagnosis of müllerian duct aplasia was entertained, other female members of the family were approached regarding history of amenorrhea, coital difficulty, or previous surgical construction of the vagina. Among them, some selected individuals were also clinically examined to elicit any pertinent findings. Initially in the study, those relatives with suggestive history and clinical examination underwent cytogenetic investigations and the relevant hormonal studies.

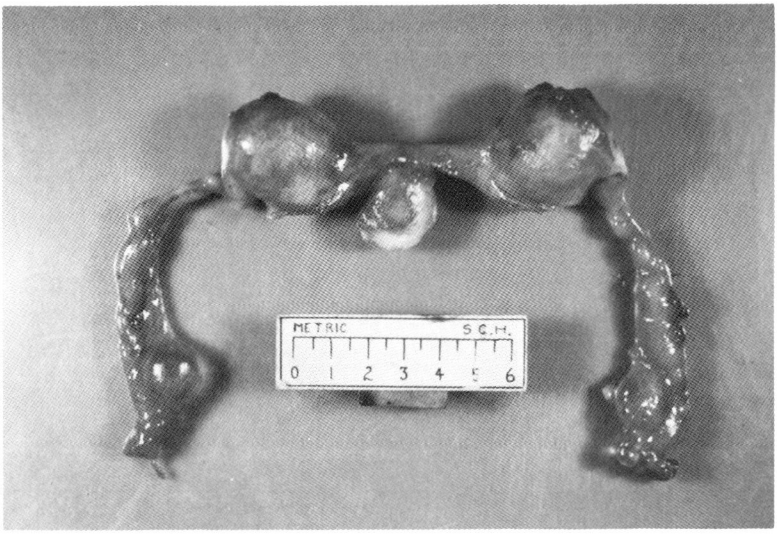

Fig. 1. Surgical specimen showing partial müllerian hypoplasia.

RESULTS

All 16 probands had absence of the upper two-thirds to four-fifths of the vagina, with only a shallow identation marking the position of the vagina at the introitus. The external genitalia were otherwise normal female. The patients were of normal stature (Fig. 2), feminine contour (Figs. 2, 3) with normally developed secondary sex characteristics. In particular, breast development was ample, and pubic and axillary hair abundant. Intellect, vision, and hearing were unimpaired. Libido was normally preserved and the affected individuals thought of themselves as normally responsive females. They felt the desire for and enjoyed sexual contact with males. Nevertheless, the absence of the vagina, particularly the sexual difficulty it engendered, generated anxiety, both considerable and understandable, among them.

In all but 3 patients, the cervix and uterine corpus were absent, and fallopian tubes were absent or rudimentary. In the remaining 3, small uterine rudiments in the form of muscular buds were found linked by fibrous bands to the site of the ovarian attachment. In one of them, small fragments of ectopic endometrial tissue were discernible, attached to the pelvic wall. The latter finding apparently provided the explanation for the periodic (every 5 weeks) abdominal and pelvic pain associated with slight fever (38.2°C) and general malaise which brought this patient to medical attention in the first place. Remarkably, this patient had been

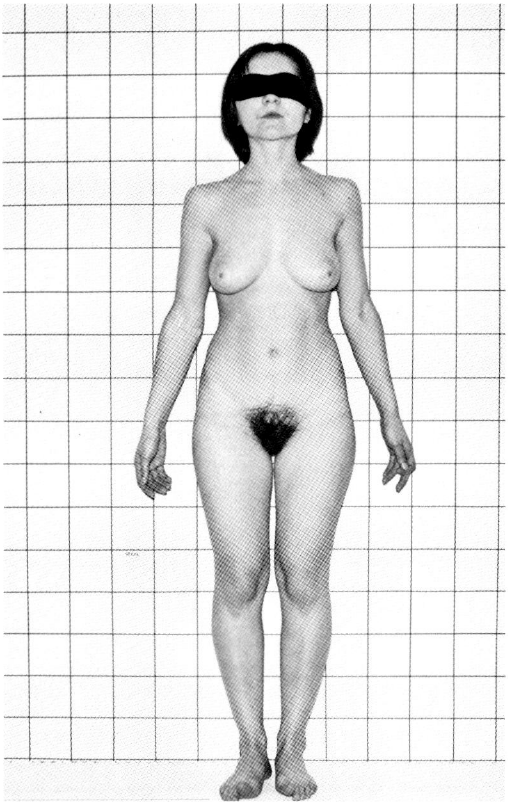

Fig. 2. Photo of patient showing normal stature and normal secondary sex characteristics.

married for over 3½ years and never had vaginal sexual intercourse, having herself experienced excruciating pain and her husband considerable difficulty in attempting it at the beginning of their marriage. They had what appeared to be a satisfying sexual relationship employing oral-genital contact with occasional anal intercourse.

Frequently, patients were referred who had a tentative clinical diagnosis of testicular feminization. The patients' initial complaint was amenorrhea or difficulty in consummating coitus. On 3 occasions it was the husband who brought his wife complaining about their difficulty in having satisfactory sexual relations. Surprisingly, on these 3 occasions the couple had been married for over 2 years and in 1 case for 5 years. The possibility of testicular feminization could readily be ruled out by the abundant axillary and pubic hair growth. Moreover, the appearance of the nipples and areolae were perfectly feminine.

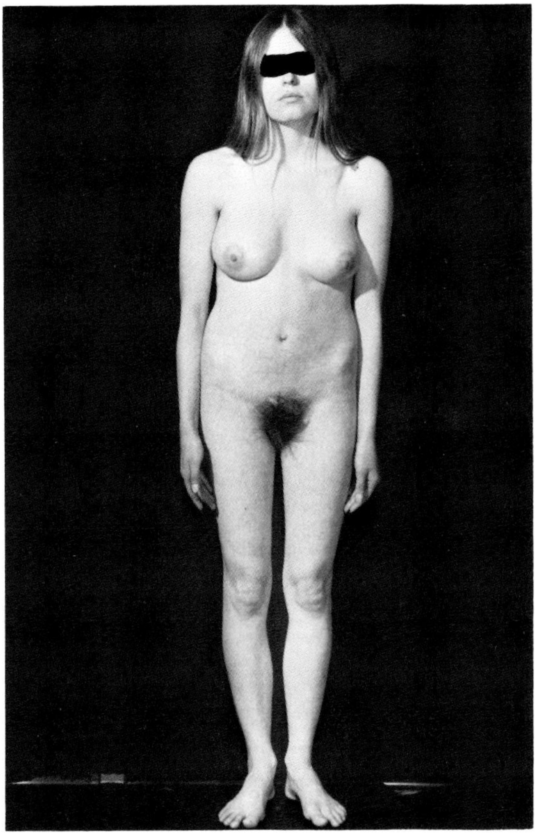

Fig. 3. Normal female contour in a case of müllerian aplasia.

The evidence available from both excretory urography and laparotomy point to an increased occurrence of urinary tract malformations in association with müllerian duct aplasia, especially in the partial form. The detected malformations include: double ureter, horseshoe kidney, absent or ectopic kidney, and hydronephrosis both partial and complete. Anomalies were detected in 2 of the 8 patients with complete müllerian aplasia. Both were sporadic cases. Furthermore, all 3 patients with partial aplasia (RKHS) were also found to have renal and/or urinary malformations.

The relevant findings on the reported patients with complete müllerian aplasia are summarized in Table 1. Table 2 summarizes the findings on the 3 reported patients with partial aplasia.

TABLE 1. Patients with Complete Müllerian Aplasia

Patient	Age of Presentation, Years	Marital Status	Presenting Complaint	Endocrine Investigations	IVP
Case 1	28	M	Coital difficulty	Normal cyclic pattern	Normal
Case 2	15	S	Primary amenorrhea	Normal	Normal
Case 3	19	S	Coital difficulty	Normal cyclic pattern	Not done
Case 4	20	S	Coital difficulty	Normal cyclic pattern	Normal renal shadows, normal excretory pattern
Case 5	15	S	Primary amenorrhea	Normal	Normal kidney shadows, normal function, unremarkable ureters and bladder
Case 6	16	S	Coital difficulty	Normal cyclic pattern	Normal
Case 7	15	S	Coital difficulty	Normal	Not done
Case 8	17	S	Primary amenorrhea	Normal cyclic pattern	Not done
Case 9	23	M	Coital difficulty	Normal cyclic pattern	Not done
Case 10	19	S	Coital difficulty	Normal	Not done
Case 11	18	S	Primary amenorrhea	Normal cyclic pattern	(R) kidney absent, (L) sided double ureter
Case 12	20	M	Coital difficulty	Normal cyclic pattern	(L) side uretero vesical reflux with (L) hydronephrosis, vesical verticulum
Case 13	18	S	Primary amenorrhea	Normal	Normal

[a] R and L refer to Right and Left here and in Table 2.

Laparoscopy	Laparotomy	Cytogenetic Studies	Female Relatives with Disorder
Normal ovaries, absent tubes and uterus	Not done	46,XX Normal banding pattern	2; sister, patrilineal first cousin
Normal ovaries, absent tubes and uterus	Not done	46,XX Normal banding pattern	3; 2 paternal half sisters, patrilineal first cousin
Normal ovaries, follicles seen, rudimentary tubes but no uterus seen	Not done	46,XX Normal banding pattern	5; 3 paternal aunts, 2 patrilineal second cousins
Normal ovaries, biopsy done, absent tubes and uterus	Not done	46,XX Normal banding pattern	2; patrilineal first cousins
Normal ovaries, absent tubes and uterus	Ovaries visualized, no tubes or uterus seen, Rt. ovarian biopsy done	46,XX Normal banding pattern	3; 2 patrilineal first cousins once removed, 1 patrilineal second cousin
Normal (R)[a] ovary, (L)[a] ovary not seen. Absent tubes and uterus	Not done	46,XX Normal banding pattern	2; 2 paternal great aunts
Normal ovaries, absent tubes and uterus	Not done	46,XX Normal banding pattern	2; 2 patrilineal first cousins
Normal (L) ovary, (R) ovary not seen, absent	Not done	46,XX Normal banding pattern	2; 2 patrilineal first cousins
Normal (L) ovary, (R) ovary not seen, absent tubes and uterus	Ovaries normally visualized, absent tubes and uterus, ovarian biopsy done	46,XX Normal banding pattern	2; 2 sisters
Normal ovaries, absent tubes and uterus	Not done	46,XX Normal banding pattern	1; sister
Normal ovaries, biopsy done, absent tubes and uterus	Ovaries visualized, single kidney with double ureters	46,XX Normal banding pattern	None
Normal (R) ovary, (L) ovary not seen, follicle seen, biopsy done, no tubes or uterus seen	Not done	46,XX Normal banding pattern	None
Not done	Not done	46,XX Normal banding pattern	None

TABLE 2. Patients with Partial Müllerian Hypoplasia (Rokitansky–Kuster–Hauser Syndrome)

Patient	Age of Presentation, Years	Marital Status	Presenting Complaint	Endocrine Investigations	IVP
Case 14	15	S	Primary amenorrhea	Normal	(L) kidney absent, double ureter on the right, partial hydronephrosis (R) kidney
Case 15	22	M	Periodic abdominal and pelvic pain	Normal cyclic pattern	(R) kidney absent, (L) kidney double ureter
Case 16	18	M	Coital difficulty	Normal cyclic pattern	Single horseshoe kidney, delayed, excretion, vesical diverticum

DISCUSSION

Inspection of the pedigrees of *Cases 1–8* (Figs. 4–11), provides evidence for autosomal dominant inheritance of aplasia of the müllerian ducts. Limitation of expression to females (individuals with XX sex chromosomal complement) is, of course, expected. The disorder appears to be transmitted to them through their unaffected carrier fathers. The distribution of the affected females in several successive generations, clustering of the affected individuals on the paternal side, their relationships to one another through male relatives (fathers, paternal uncles, paternal grandfathers, paternal great uncles and patrilineal male first cousins or first cousins once removed), and no transmission through the maternal line argue

Laparoscopy	Laparotomy	Cytogenetic Studies	Female Relatives with Disorder
Normal (R) ovary, follicle seen, (L) ovary not visualized, no tubes seen, small nubbins corresponding to rudimentary uterus	Not done	46,XX Normal banding pattern	None
Ovaries normal, follicle seen, muscular buds bilaterally visualized	Normal ovaries, corpus luteum on the right side, no tubes seen, 2 muscular buds identified connected by fibrous bands to the site of the cervix, endometriosis seen on the left side, single kidney with double ureters	46,XX Normal banding pattern	None
Ovaries normal, muscular bud and fibrous band unilaterally seen	Normal ovaries, biopsy done, no tubes seen, unilateral muscular bud connected by fibrous band to the lateral pelvic wall and to the location of the cervix was identified, horseshoe kidney	46,XX Normal banding pattern	None

for sex-limited autosomal dominant transmission. This is not surprising in a disorder that affects the female internal genitalia and precludes the reproduction of those who manifest it.

Autosomal dominant inheritance can be further explored when the numbers of those observed to be affected and those expected to be affected on the basis of this hypothesis are compared. When the families of all the cases where there is at least 1 affected individual beside the proposita are considered, ie *Cases 1–10,* 34 affected females are encountered among a total of 59 females distributed in 21 sibships. Utilizing the simple sib method of Weinberg [5], whereby the patients whose ascertainment led to the inclusion of their respective sibships are eliminated, 13 affected individuals in a total of 38 females at risk remain. On the

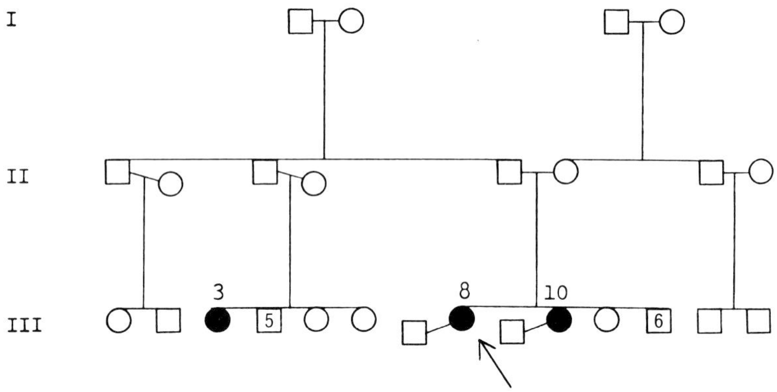

Fig. 4. *Case 1* family pedigree.

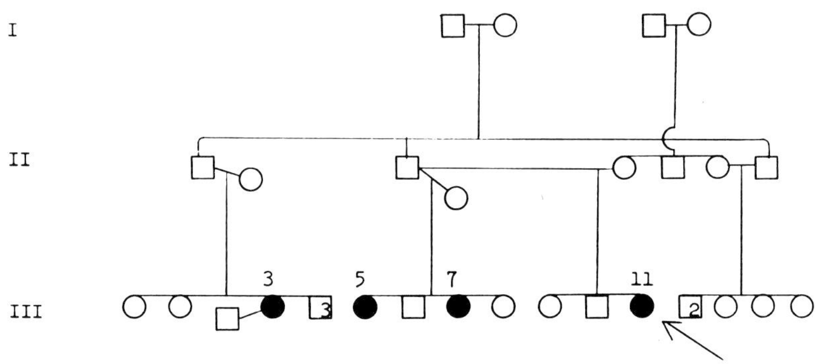

Fig. 5. *Case 2* family pedigree.

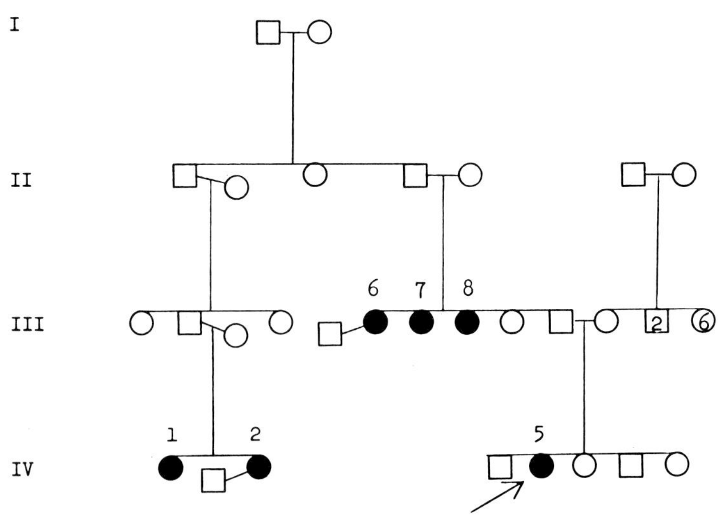

Fig. 6. *Case 3* family pedigree.

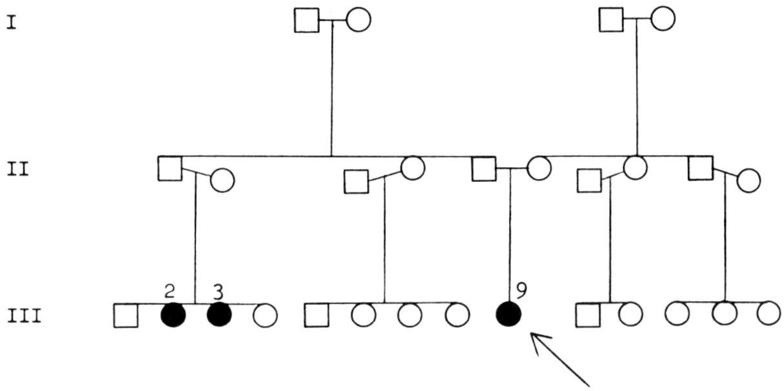

Fig. 7. *Case 4* family pedigree.

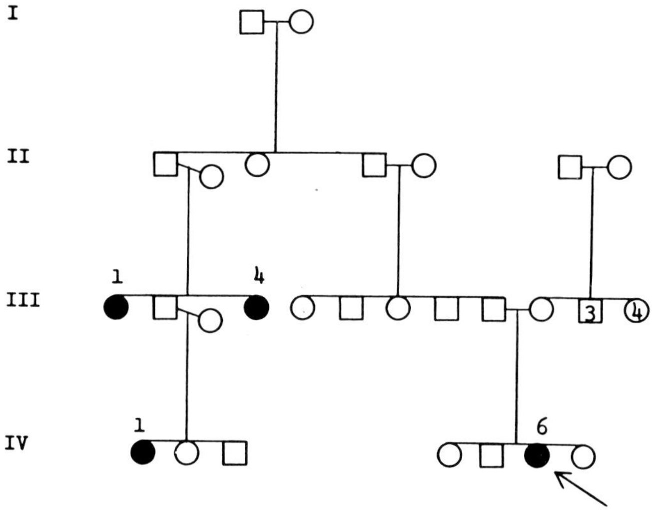

Fig. 8. *Case 5* family pedigree.

basis of sex-limited autosomal dominant inheritance 19 affected individuals are expected. Using Yates correction factor, the difference is not statistically significant ($^{Gk\,47}X^2_{(1)} = 3.18$, $0.05 < P < 0.10$). Without the use of Yates correction, the difference, though sizable, still falls short, though barely, from statistical significance at the 5% level ($^{Gk\,47}X^2_{(1)} = 3.79$, $0.05 < P < 0.10$).

Assuming multiple incomplete ascertainment and using the appropriate correction based on that assumption [6] the departure from expectation is even less

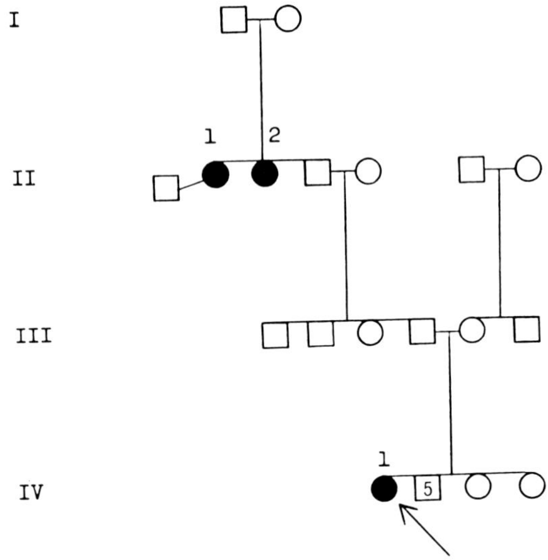

Fig. 9. *Case 6* family pedigree.

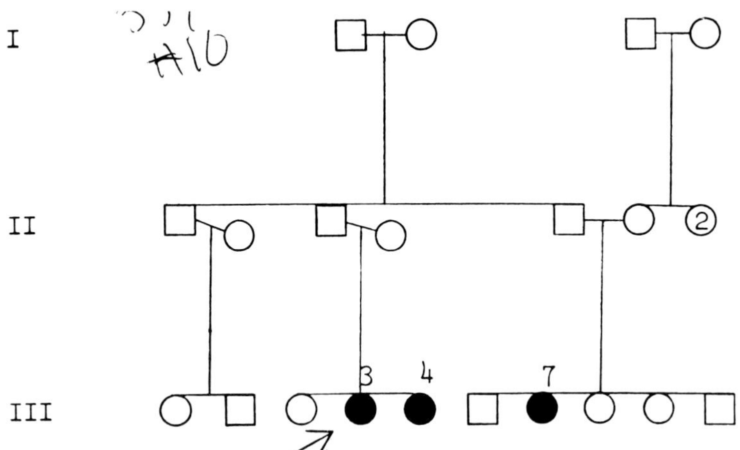

Fig 10. *Case 7* family pedigree.

conspicuous; 17 affected out of 46 are observed after correction. Using Yates correction factor ($^{Gk\ 47}X^2_{(1)} = 2.63, 0.10 < P < 0.25$). When Yates correction factor is not used the difference is still less than significant ($^{Gk\ 47}X^2_{(1)} = 3.13, 0.05 < P < 0.10$).

The observed distribution of affected females:unaffected females:males in the sibship in which patients were encountered did not depart significantly from the expected 1:1:2 ratios predicated on the premise of sex-limited autosomal dominant inheritance ($^{Gk\,47}X^2_{(2)} = 3.8399$, $0.10 < P < 0.25$).

The pedigrees of *Cases 9* and *10* (Figs. 12, 13), show 3 affected individuals in the former and 2 in the latter. In both instances the distribution of the affected individuals is restricted to but a single sibship. Therefore, the evidence presented

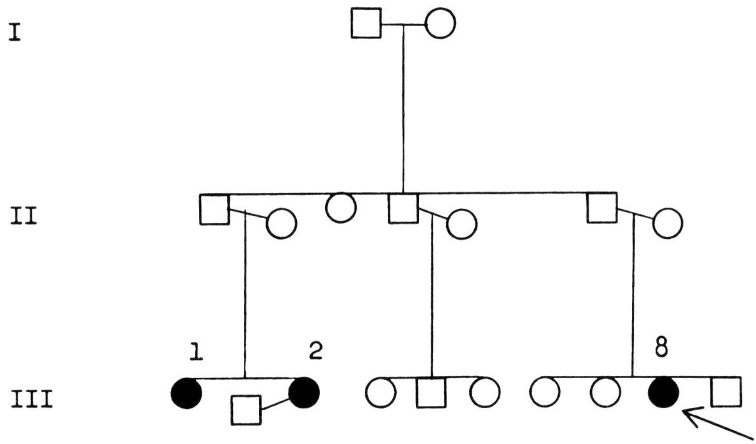

Fig. 11. *Case 8* family pedigree.

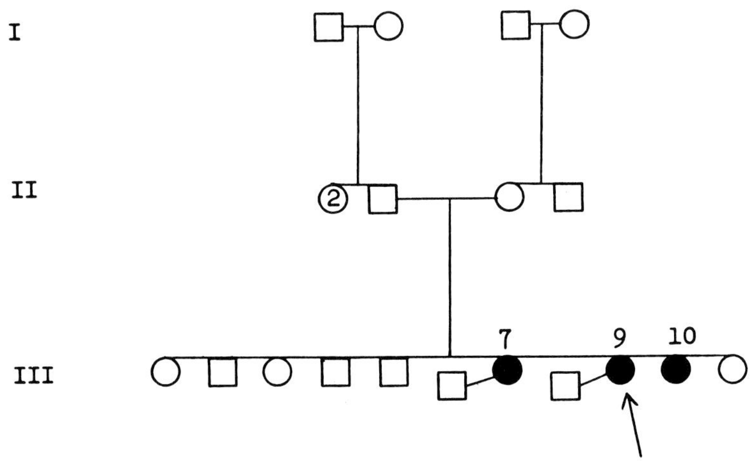

Fig. 12. *Case 9* family pedigree.

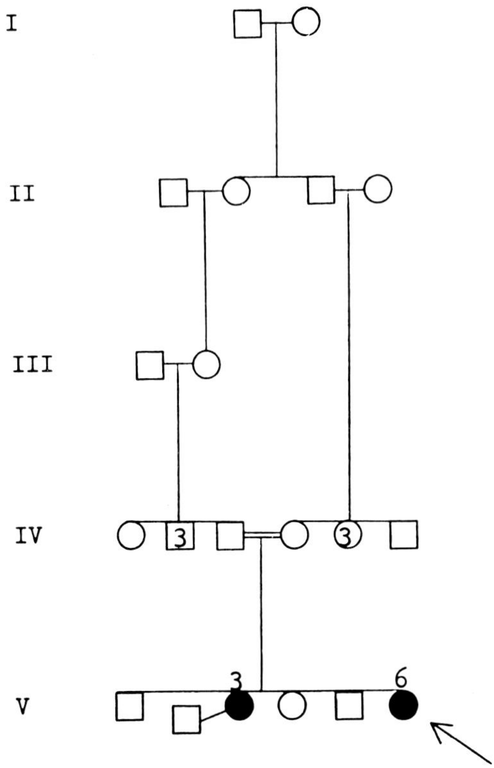

Fig. 13. *Case 10* family pedigree. Note consanguinity between the parents.

by these families, though in favor of genetic determination, is compatible with either sex-limited autosomal dominant or autosomal recessive transmission. Indeed, the latter mode of inheritance receives some support from the observed consanguinity in the mating producing the 2 affected individuals.

In each of the remaining 3 cases (*Cases 11, 12,* and *13;* Figs. 14–16), only 1 affected individual (the proband) was encountered in the family. The etiology and any possible role of genetic determination, therefore, remain speculative. In particular, *Case 11* is the only individual at risk and *Case 12* is not only the sole individual at risk but also her father is adopted with no history available of his biologic family. The evidence presented by *Case 13,* the youngest in a sibship of 11, may lend further, albeit weak, support to the hypothesis of sex-linked autosomal dominant inheritance. Her father was 52 years old at the time of her conception. In view of the well-recognized relationship between advanced paternal age and fresh autosomal dominant mutations [7, 8], it may be conjectured that the sperm involved in her conception bore the mutation for müllerian duct aplasia.

One may therefore surmise that in at least some cases, namely those with multigeneration familial distribution, the etiologic determination is probably genetic and the mode of inheritance is likely autosomal dominant with female

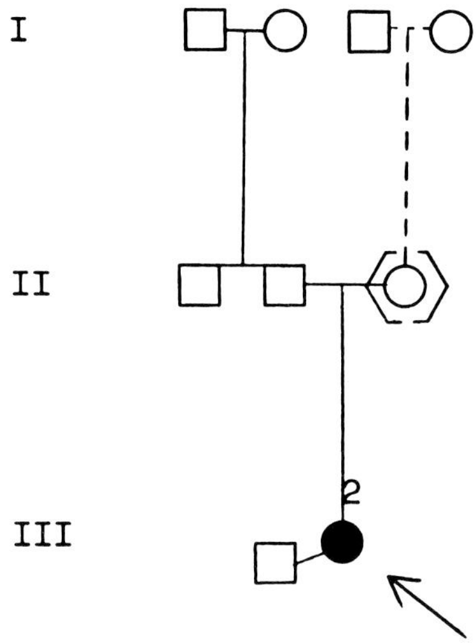

Fig. 14. *Case 11* family pedigree.

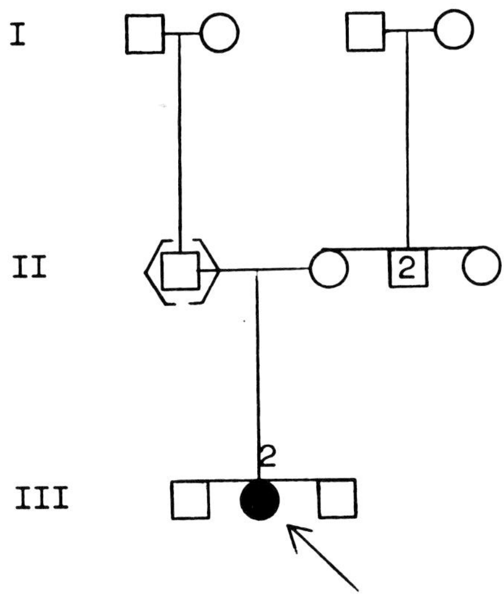

Fig. 15. *Case 12* family pedigree. Note father was adopted, no paternal family history available.

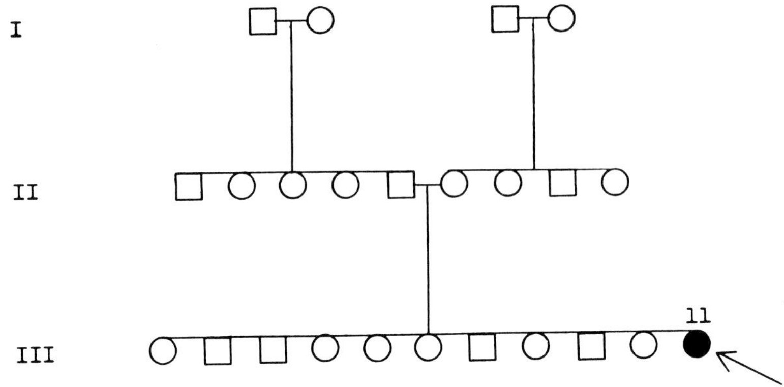

Fig. 16. *Case 12* family pedigree.

sex limitation. Other patterns of inheritance, such as sex-limited autosomal recessive and polygenic determination, as well as nongenetic etiology, have not been ruled out in the remaining cases. It would be premature and possibly incorrect to assign an exclusive pattern of genetic determination and mode of transmission to müllerian aplasia even in its complete expression. The lessons derived from the study of similar morphologic and functional disorders which all too often emphasize genetic heterogeneity in the causation of human malformations should not be overlooked.

It is apparent from the foregoing that the anomaly with which the reported women were affected implicated their müllerian duct development. Since it is not known whether the derivatives of the müllerian ducts failed to develop or were arrested in their development or had regressed following normal but partial development, it is perhaps wiser to use the term aplasia rather than agenesis to describe the observed abnormalities. It is clear from the results of urinary tract investigations that both the genital and the urinary systems were involved in the partial form of müllerian aplasia (Rokitansky-Kuster-Hauser syndrome). Patients with the latter disorder suffered renal, ureteral and vesicle malformations. The ovary, however, whose embryologic origin is independent of and distinct from that of the müllerian structures, developed normally in both the complete and partial varieties of müllerian aplasia. The appearance of normal, secondary sex characteristics, normal sex hormone assays, the cyclic nature of these hormone levels and the normal histologic findings on ovarian biopsy (Fig. 17) attest to the above distinction.

It is tempting to strike a further distinction between the complete and partial forms of müllerian aplasia. The latter appears to be a dysmorphogenetic event in which the internal genitalia, urinary system, and possibly the spine display evidence of malformations. Conversely, in the complete form (*Cases 1–12*) the lesions appear to be highly selective and confined to the internal genitalia – derivatives of the müllerian ducts. It is also in the complete variety (*Cases 1–9*) that evidence of sex-limited autosomal dominant transmission has been adduced and this mode of inheritance thus invoked. Is it that these female patients are

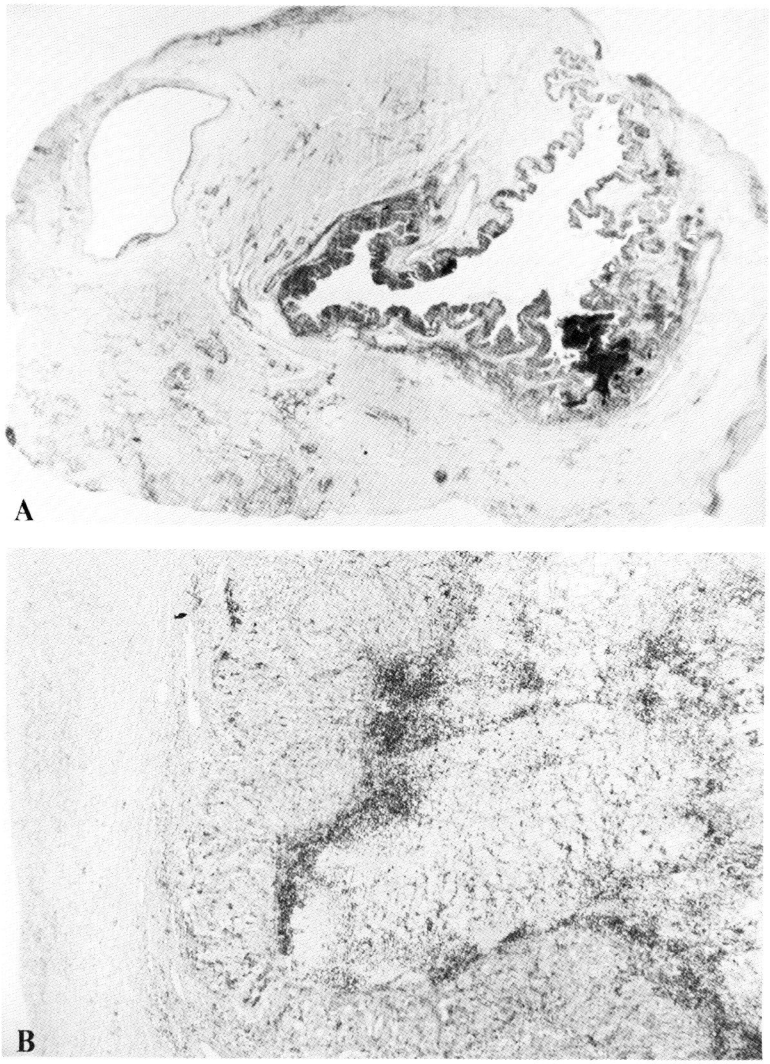

Fig. 17. A) Ovarian biopsy from a case of müllerian aplasia showing corpus hemorrhagicum. B) Microscopy from an ovarian biopsy of a case of müllerian aplasia showing normal cortex and luteal cells. Note organizing hemorrhage.

capable, during their embryogenesis, of elaborating and/or responding to a selective müllerian duct inhibitory factor, akin to that normally produced by males? The specific nature of the lesions as well as the proposed autosomal dominant inheritance lend some credence to this conjecture.

SUMMARY

Thirteen unrelated females ranging in age from 15 to 28 years old have recently been recognized to have aplasia of the müllerian duct derivatives. They presented one or both of two major complaints: amenorrhea and difficulty or pain on attempted sexual intercourse. Clinically all cases were characterized by absence of vagina and failure to palpate the uterus rectally; normal mature external genitalia; normal stature, intellect, hearing and vision; normal secondary sex characteristics including breast development, pubic and axillary hair. Libido was normally preserved and sexual self-identification was unambiguously feminine. Apart from the pelvic findings, the affected individuals were phenotypically unremarkable women.

Laparoscopic examination revealed absent uteri, absent or rudimentary tubes, but normally developed ovaries. Cytogenetic evaluation disclosed normal female karyotypes with normal banding patterns on all patients.

Endocrine investigations revealed female cyclic pattern consistent with normal ovarian endocrine function. Biopsy specimens on the ovaries of 5 patients were histologically unremarkable with evidence of normal follicular activity.

Investigation of the families of these patients revealed similarly affected individuals in 10 families. In 8 of these the pattern of transmission was consistent with autosomal dominant inheritance with sex limitation to XX individuals; in 2 the distribution of affected individuals was compatible with either sex-limited autosomal dominant or autosomal recessive inheritance. Though unlikely, polygenic inheritance remains a formal possibility. In 3 families, apart from the propositae, no other affected females were ascertained. In these, the etiology is either teratogenic or genetic with autosomal dominant or recessive inheritance, and sex limitation to females.

Three further cases with partial aplasia or the Rokitansky-Kuster-Hauser syndrome (RKHS) were ascertained. No other similarly affected individuals were encountered in their families.

REFERENCES

1. Anger D, Hemet J, Ensel J: Forme familial due syndrome Rokitansky-Kuster-Hauser. Bull Fed Soc Gynec Obstet Franc 18:229–234, 1966.

2. Bryan AL, Nigio JA, Counseller VS: One hundred cases of congenital absence of the vagina. Surg Gynecol Obstet 88:79–86, 1949.
3. Phaneuf LE: Discussion (congenital malformations of reproductive organs). Am J Obstet Gynecol 53:48, 1947.
4. Jones HW, Mermut S: Familial occurrence of congenital absence of the vagina. Am J Obstet Gynecol 114:1100–1101, 1972.
5. Weinberg W: Weitere Beitrage Zur Theorie der Vererbung 4. Uber Methoden und Fehlerquellen der Untersuchung anf Medelsche Zahlen bein Menschen. Arch Rass -u Ges Biol 9:165–174, 1912.
6. Nora JJ, Fraser FC: Incomplete multiple ascertainment. In "Medical Genetics: Principles and Practice." Philadelphia: Lea & Febiger, 1974, pp 87–88.
7. Penrose LS: Parental age and mutation. Lancet 2:312–313, 1955.
8. Harvey MA, Jones KL, Quan L, Hall BD, Smith DW: Additional data relative to paternal age and fresh gene mutation. Clin Res 22:219A, 1974.

EDITORIAL COMMENT

The author of this paper notes that the entity which he calls aplasia of the müllerian duct system is an entity distinct from the Rokitansky-Kuster-Hauser syndrome, the latter differing from the former by involving only *partial* aplasia of the müllerian duct system.

I believe it is worthwhile pointing out that others, including Simpson (Simpson JL: "Disorders of Sexual Differentiation." New York: Academic Press, 1977, p 342) consider the syndrome reported in this paper and the Rokitansky-Kuster-Hauser syndrome to be one-and-the-same entity.

Robert L. Summitt

Gonadal Dysgenesis in Association With Autosomal Chromosome Aberration*

Lawrence R. Shapiro, MD, Zoë R. Graves, BA, Dorothy Warburton, PhD, and Herbert A. Huss, MD

Gonadal dysgenesis is generally associated with a sex chromosome abnormality, and several forms appear to be inherited in an autosomal recessive or X-linked recessive fashion [1–4]. A few cases of autosomal chromosome aberrations in association with gonadal dysgenesis have been reported to involve translocation of the B and C groups [5], asymmetric or enlarged number 1 chromosomes [6, 7], and abnormal D-group chromosomes [8–12]. The majority of these studies were done without the benefit of chromosome banding and are considered by many to be inconclusive; however, the most frequent association is with D-group chromosome abnormalities and gonadal dysgenesis. While the bulk of these cases were reported prior to the capability of banding, several recent reports utilizing banding techniques have been published [11, 12].

In view of the association of autosomal chromosome abnormalities and gonadal dysgenesis, autosomal gene effects may be involved in female sex differentiation in addition to the genes on the X chromosome.

CASE REPORT

A 17-year, 9-month-old female was evaluated because of primary amenorrhea and the lack of normal breast development. This patient was the second born to a 29-year-old mother and a 35-year-old father. Her past history was negative, with normal growth and development and satisfactory performance in school.

*This work was supported in part by The New York State Department of Health Birth Defects Institute and The National Foundation–March of Dimes

Birth Defects: Original Article Series, Volume XIV, Number 6C, pages 167–170

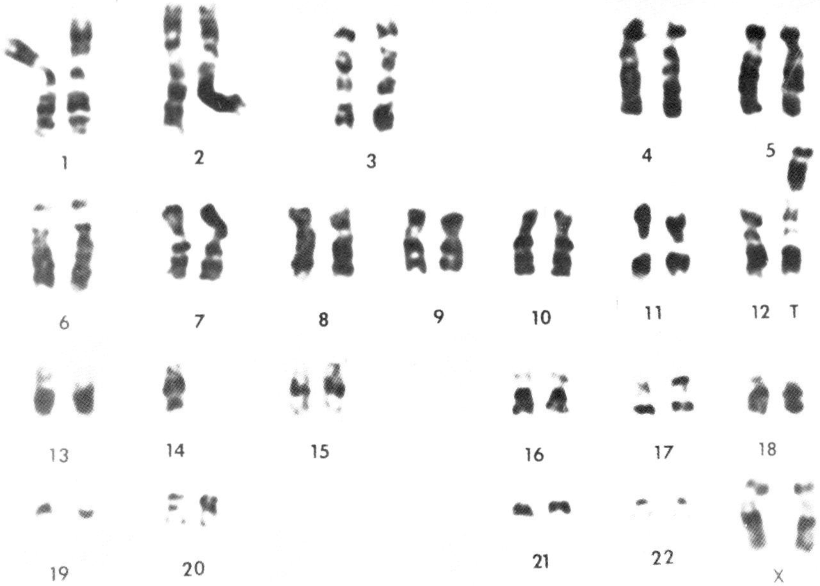

Fig. 1. Karyotype of trypsin-Giemsa banded chromosomes showing 12/14 translocation and two normal X chromosomes.

Physical examination revealed a pleasant and alert adolescent who appeared her stated age. Her height was 157cm; her weight was 40 kg. No breast development was noted, and there was only early axillary and pubic hair development with normal external genitalia. Palmar dermatoglyphic analysis was not unusual, and the total finger ridge count was 109. Vaginal examination under anesthesia revealed a normal vagina and cervix. Exploratory laparotomy revealed a uterus measuring 4 cm from cervix to fundus. The fallopian tubes and fimbriae were intact, but the gonads appeared to be fibrous streaks underlying the fallopian tubes bilaterally. Histologic section of the gonads revealed fibrous stroma with no primordial follicles or ovarian elements.

Karyotype analysis was first carried out in this patient before banding techniques were available. All cells from a peripheral blood culture and fibroblast cultures from skin and streak gonad had 45 chromosomes with a missing D- and C-group chromosome and an abnormal, large submetacentric chromosome with a prominent secondary constriction.

When trypsin-Giemsa banding was used, the translocation was found to involve chromosomes 12 and 14 (Fig. 1). There was no visible loss of chromosome 12 material. Attached to the short arm of chromosome 12 was an additional dark band followed by a large nonstaining area and then apparently all of chromosome 14. Both X chromosomes appeared normal, and DNA replication studies using tritiated thymidine and autoradiography showed one of the X chromosomes to be late-replicating.

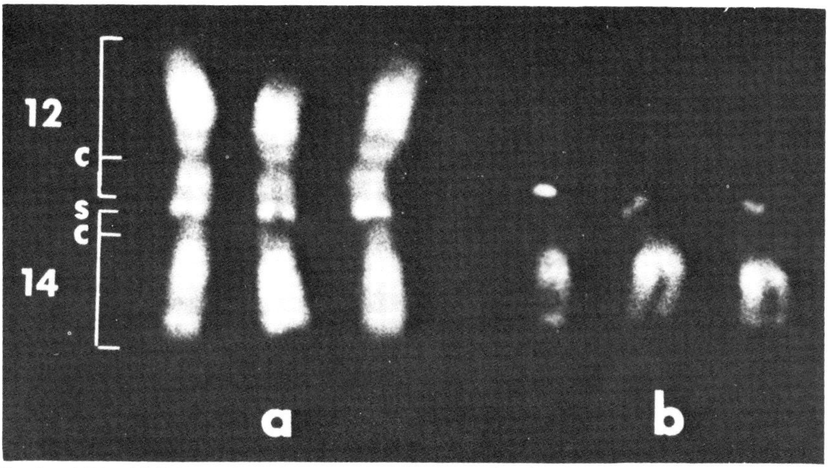

Fig. 2. a) The 12/14 translocation chromosome from 3 cells with quinacrine fluorescent staining: c, centromere positions; s, position of satellite region of chromosome 14. b) Quinacrine fluorescent staining of chromosome 14 with brilliant satellites from 3 cells of the father.

Quinacrine fluorescent staining gave the added information that the extra band proximal to the secondary constriction of the translocation chromosome was an intercalary satellite which was brilliantly fluorescent and originated from one paternal chromosome 14 (Fig. 2).

Several possible explanations can be offered for the autosomal chromosome abnormality described above and the associated gonadal dysgenesis:

1) It may be fortuitous – the autosomal aberration associated with an XX sex chromosome constitution might be detected only because of the individual's presenting complaints relating to the gonadal dysgenesis.

2) Hemizygosity for a mutant gene producing gonadal dysgenesis in the presence of two X chromosomes is another possibility. The loss of genetic material from chromosome 12 and/or 14 was not demonstrated but cannot be excluded. If it were shown that the same autosome was consistently involved, the possibility of hemizygosity would deserve strong consideration.

3) Position effect, considered as the inactivation of genes that have been translocated adjacent to heterochromatin [13], is a theoretic consideration, but this cannot be proven in this particular case.

4) An undetected and unrelated X chromosome abnormality could conceivably be present in other tissues to account for the patient's clinical features, or a complex translocation involving the X chromosome cannot be excluded with the current techniques, although the evidence does not imply this.

It is intriguing to note that the current case involves a number 14 chromosome, and the cases in the literature have a preponderance of D-group chromosome involvement without further specific identification.

Although it is possible that the relationship of the patient's gonadal dysgenesis and autosome aberration may be coincidental, it seems possible that the loss or rearrangement of genetic material from chromosome 12 and/or 14 is directly related to the gonadal dysgenesis.

REFERENCES

1. Jones HW Jr, Ferguson-Smith MA, Heller RH: The pathology and cytogenetics of gonadal agenesis. Am J Obstet Gynecol 87:578–600, 1963.
2. Ferguson-Smith MA: Karyotype-phenotype correlations in gonadal dysgenesis and their bearing on the pathogenesis of malformations. J Med Genet 2:142–155, 1965.
3. Simpson JL, Christakos AC, Horwith M, Silverman FS: Gonadal dysgenesis in individuals with apparently normal chromosomal complements: Tabulation of cases and compilation of genetic data. In Bergsma D (ed): Part X. "The Endocrine System." Baltimore: Williams & Wilkins for The National Foundation–March of Dimes, BD:OAS VII(6):215–228, 1971.
4. Simpson JL: Gonadal dysgenesis and abnormalities of the human sex chromosome: Current status of phenotypic-karyotypic correlations. In Bergsma D (ed): "Genetic Forms of Hypogonadism." Miami: Symposia Specialists for The National Foundation–March of Dimes BD:OAS XI(4):23–59, 1975.
5. Mann JD, Valdmanis A, Capps SC, Puite RH: A case of primary amenorrhea with a translocation involving chromosomes of groups B and C. Am J Hum Genet 17:377–383, 1965.
6. Cooper HL, Hernits R: A familial chromosome variant in a subject with anomalous sex differentiation. Am J Hum Genet 15:465–475, 1963.
7. Philip J, Frydenberg O, Sele V: Enlarged chromosome No. 1 in a patient with primary amenorrhea. Cytogenetics 4:329–339, 1965.
8. Walker S, Harris R: Investigation of family showing transmission of a 13–15 chromosomal translocation (Denver classification). Br Med J 2:25–26, 1962.
9. Cohen MM, Capraro VJ, Takagi N: Pericentric inversion in a group D chromosome (13–15) associated with amenorrhea and gonadal dysgenesis. Ann Hum Genet 30:313–323, 1967.
10. Nielsen J, Friedrich U: Pure gonadal dysgenesis. Clin Genet 3:52–58, 1971.
11. Khudr G, Naftolin F, Benirschke K, Zarate A, Guzman-Toledano R: Unusual translocations and reproductive failure. Obstet Gynecol 41:542–546, 1973.
12. Laszlo J, Gaal M, Bosze P: Chromosome studies in ovarian hypoplasia. Clin Genet 9:61–70, 1976.
13. Biederman B, Bowen P: Balanced translocations involving chromosome 12: Report of a case and possible evidence for position effect. Ann Genet 19:257–260, 1976.

Pregnancy in Five Women With 45,X /46,XX and 45,X/47,XXX Gonadal Dysgenesis

M.H.K. Shokeir, MD, PhD

The syndrome of 45,X gonadal dysgenesis was probably first described by Funke [1] who recognized the constellation of abnormalities in a 16-year-old girl who was short in stature (only 135 cm in height), had webbed neck, low-set ears, micrognathia and high-arched palate. She had primary amenorrhea, no breast development and small nipples, and virtually no pubic or axillary hair. As a newborn infant, she had swollen hands and feet. The definitive description, however, had to await the report by Turner [2], who recognized the features of sexual infantilism, congenital webbed neck, cubitus valgus, short stature, and retarded osseous development in several females. The eponym Turner syndrome, has largely replaced earlier appellations for the disorder. The finding by Varney et al [3] of elevated urinary gonadotropins supported an earlier observation by Meyer [4] of follicle-stimulating hormone in the urine of a woman who ostensibly manifested this syndrome. It was not until 1944 that Wilkins and Fleischmann [5] carried out biopsy of the gonads to discover tissue resembling ovarian stroma but deficient in the parenchyma and bereft of follicles. They, therefore, coined the term ovarian agenesis. A decade later Wilkins et al [6] and Polani et al [7] applied the newly developed technique of sex chromatin [8] and discovered that patients with Turner syndrome are X-chromatin negative.

The exact chromosomal basis for this condition was established by Ford et al [9] who recognized the presence of 45 chromosomes with X chromosome monosomy in the nuclei (from bone marrow) of a patient with gonadal dysgenesis (Turner syndrome).

There appears to exist a rather rigorous natural selection against fetuses with Turner syndrome which reduces the frequency of liveborn Turner females with a 45,X karyotype to 1 in 10,000 [10]. This is reflected in the substantial frequency (5.3%), among abortuses, of 45,X Turner syndrome [11]. It was further estimated that the probability of an 45,X conceptus surviving to be born alive at term is around 2.5% [11]. This high frequency of fetal wastage afforded the

Birth Defects: Original Article Series, Volume XIV, Number 6C, pages 171–184

opportunity to examine the gonads of 45,X fetuses which were found to be relatively normal with demonstrable germ cells in normal number [12, 13]. This finding indicates that there is no agenesis of germ cells in the ovary of 45,X Turner female fetuses but there is a relatively late damage or dysplasia to the fetal ovary in this disorder. This is usually attended with increase in connective tissue stroma of the affected ovaries. The very presence of normal ovarian germ cells at some stage of development of the fetal ovary in 45,X Turner syndrome may provide a possible basis for the occasional persistence of viable follicles with potentially fertile ova at a later point in life. In this communication we wish to report on pregnancy in 5 women who displayed features of Turner syndrome. In 2 of them the chromosome complement was uniformly 45,X with no evidence of mosaicism. In another 2, however, 45,X/47,XXX mosaicism was acertained although the majority of cells were of the 45,X variety. In the fifth patient 45,X/46,XX mosaicism was detected, with the proportion of cells in the 2 lines being approximately equal.

PRESENT INVESTIGATIONS

Case 1

At the time of her first presentation, the patient was a 23-year-old unmarried female who had a somewhat late menarche at 14 years of age. Her periods were scanty, irregular, and only occasionally painful. She was married at 16 years of age and became pregnant a year later. The pregnancy, which was initially largely unremarkable, ended in a spontaneous abortion at 14 weeks gestation. General fatigability and secondary amenorrhea brought her to medical attention at age 23.

Following her miscarriage she never menstruated. Her secondary amenorrhea had lasted until the time of her recent presentation at 42 years of age. There were no definite menopausal symptoms following her confinement. She began, however, having bone pains, particularly in her back and thighs. She was given repeated courses of estrogen therapy and underwent withdrawal bleeding following the discontinuation of each. In general, she had prematurely aged, appearing at the time of her last presentation to be considerably older than her stated age.

She measured 137.5 cm in both height and span (Fig. 1). She weighed 38 kg. She had slight antimongoloid slant of the palpebral fissures, high-arched palate, mild retrognathia, short neck with mild webbing, and low nuchal hair line. She had bilateral palmar simian creases and both her 4th metacarpals were somewhat short. Cubitus valgus was noted bilaterally. Pigmented nevi were widespread.

No axillary and only scant pubic hair was observed. Her breasts were developed in bulk but were mostly fat with very little lobulated breast tissue. The areolae

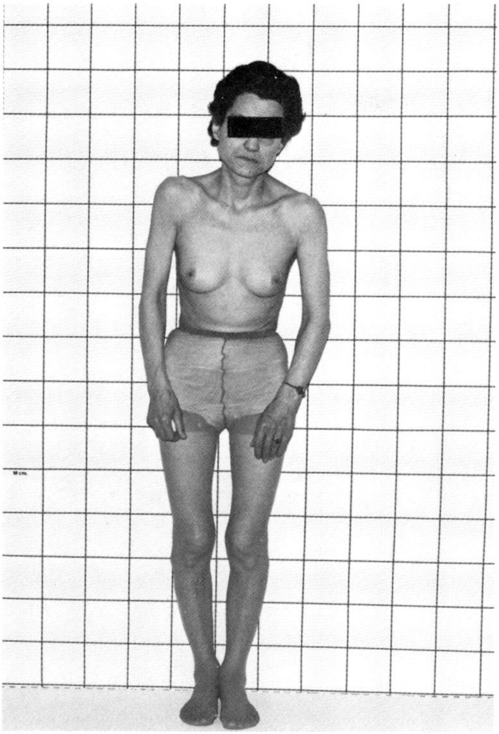

Fig. 1. Photo showing short stature of *Case 1.*

and nipples were poorly developed. Interestingly, the patient manifested severe expression, similar to that seen in hemizygous males, of the X-linked recessive disorder, amelogenesis imperfecta (Fig. 2). There were no longitudinal striations characteristic of involvement in heterozygous females.

Buccal smear revealed X-chromatin negative nuclei. Chromosome analysis of lymphocytes showed 45 chromosomes with a single X chromosome. This was confirmed in fibroblasts from skin biopsy specimens.

Endocrine studies were consistent with a postmenopausal pattern with no measurable plasma levels of estradiol and elevated follicle-stimulating hormone (FSH) and luteinizing hormone (LH) levels.

Case 2

The patient presented for diagnosis and management of a recent episode of dyspnea, chest pain, back pain associated with fever, and general malaise. She

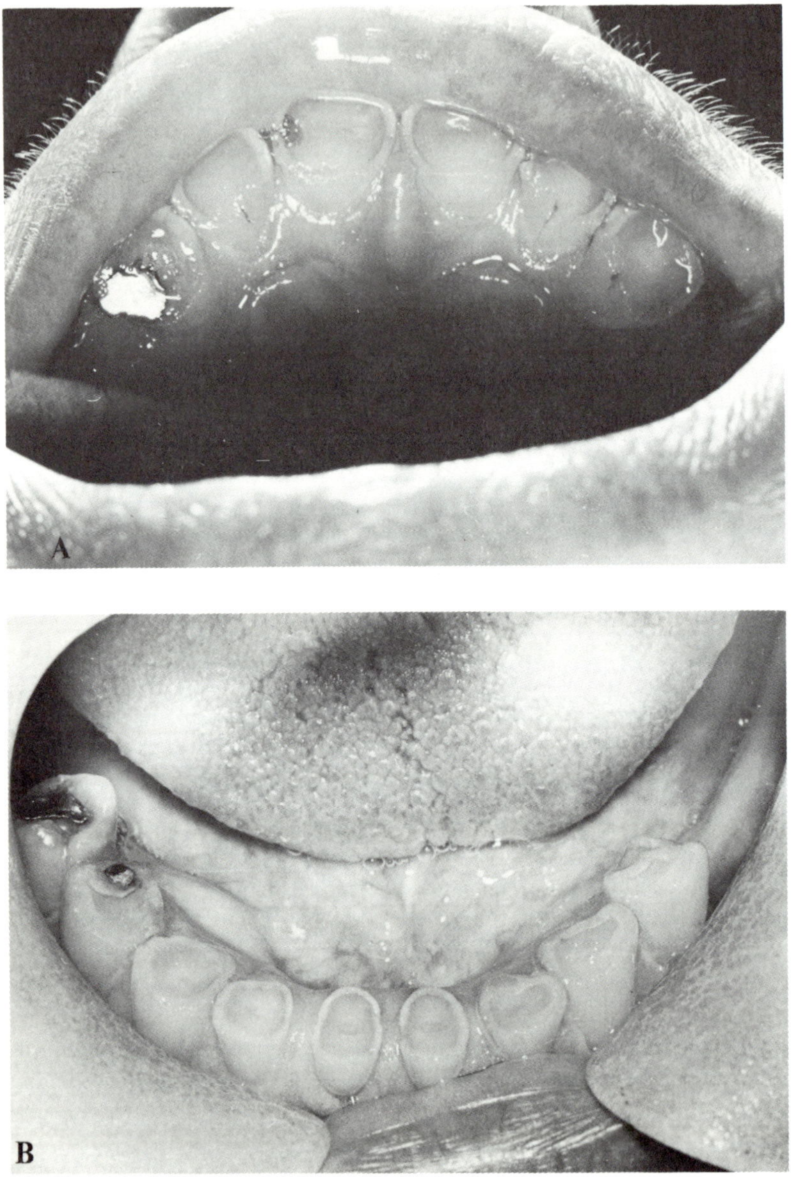

Fig. 2. Amelogenesis imperfecta in teeth of A) the upper jaw, B) lower jaw of *Case 1.*

had been married since age 19 and had first been evaluated at age 20 for infertility and secondary amenorrhea. She gave a history of menarche at 13 years of age and regular periods (every 30 days) which lasted, on average, 4 days and were frequently painful. The patient had become pregnant at 15 years of age. The pregnancy was unremarkable and ended uneventfully in the birth of an apparently normal male infant. He was adopted at a few days of age.

What brought the patient to medical attention on her recent presentation at age 38 was pleurisy which had been complicated by empyema. Although it was somewhat difficult to determine the sequence of events, it appeared that a deformity of the thoracic spine (kyphoscoliosis) led to atelectasis of a pulmonary lobe which became infected. The ensuing pneumonitis was complicated by pleurisy and empyema. The origin of the spinal deformity, which the patient denied having as a younger woman and which was not recorded in the initial clinical encounter, might have been a skeletal lesion due to prolonged estrogen deficit.

On examination, the patient measured 140 cm in height and weighed 40 kg (the patient stated that she had lost 5 cm in height in the preceding few years). She had downward slanting palpebral fissures, small chin, low-set auricles, and marked neck webbing. She had a shield-like chest with widely spaced nipples and pectus excavatum. Furthermore, she had a pronounced chest deformity and kyphoscoliosis with rotational deformity of the spine. Her lower limbs showed genu varum and her hips coxa vara. Her breasts were small, flabby and atrophic; axillary hair was absent and pubic hair very sparse.

Plasma FSH and LH were markedly elevated with unmeasurable serum estradiol.

Buccal smear was X-chromatin negative.

Chromosome analysis of blood and skin, the latter from 2 locations, revealed 45 chromosomes with a 45,X karyotype.

Case 3*

This 33-year-old, gravida 1, para 1, married female presented because of secondary amenorrhea. The patient had experienced menarche at 10 years of age. Her periods were regular with occasional dysmenorrhea, usually lasted 5 days, and were 28 days apart. Following her marriage, she conceived when she was 21 years of age. Her pregnancy was uneventful, delivery and puerperium unremarkable. She was delivered of a healthy female infant who was alive and well at the time of presentation. Specifically, there was no history of antepartum or post-

*This case is now in print [14]. The present author was the clinical geneticist who evaluated both mother and daughter and carried out the relevant genetic studies.

partum bleeding which might have evoked suspicion of Sheehan syndrome. Following her confinement her periods became irregular, shorter in duration, and less frequent. Presumably to regulate her menses, she was given oral contraceptives for 1 year (at age 25) and expectedly had regular withdrawal bleeding while receiving this treatment. However, on cessation of oral contraceptives, oligomenorrhea returned to be replaced at age 31 years by amenorrhea. Concomitant with the latter she had menopausal symptoms of hot flushes, anxiety, depression, and excessive nocturnal perspiration. These symptoms had largely subsided at the time of presentation.

On examination, this patient presented features highly reminiscent of Turner syndrome: she was 138 cm in height, had antimongoloid slant of the palpebral fissures, high-arched palate, low-set auricles (Fig. 3), short neck, low nuchal hairline, slight neck webbing, simian creases on both palms, short 4th metacarpal bones bilaterally, and somewhat increased carrying angles at the elbows, though not full-fledged cubitus valgus. Her breasts, though formed, were thin and pendulous. She had axillary and pubic hair.

Skull x ray was unremarkable with normal pituitary fossa; short 4th metacarpals were confirmed radiologically.

Endocrine investigations revealed extremely low levels of estradiol and elevated (postmenopausal) gonadotropins (FSH and LH).

Chromosome analysis was undertaken on blood and skin (the latter from 2 sites). It showed evidence of mosaicism. There were no normal 46,XX cells; all the metaphases examined were either 45,X or 47,XXX with the former cell line being by far the preponderant. The ratios averaged 9:1 in the blood and 5:1 in the skin.

Chromosome studies on the daughter revealed a 46,XX complement in blood and skin.

Case 4

This 35-year-old female presented for treatment of hypothyroidism. On more thorough clinical assessment the suspicion of gonadal dysgenesis was raised. She had been amenorrheic for some 14 years prior to presentation. Her periods had, however, been regular between menarche at age 15 years and a pregnancy at age 20. Following the birth of her only child (a boy), she never regained her menses and suffered symptoms which retrospectively could be attributed to premature menopause. Her baby who was found to have a ventricular septal defect was successfully operated on. Chromosome studies on his lymphocytes revealed a 46,XY complement.

On examination (Fig. 4) a short (height 145 cm), moderately obese (weight 54 kg), intelligent, cooperative female was encountered. Her face showed antimongoloid slant of the palpebral fissures, low-set rotated auricles, long philtrum,

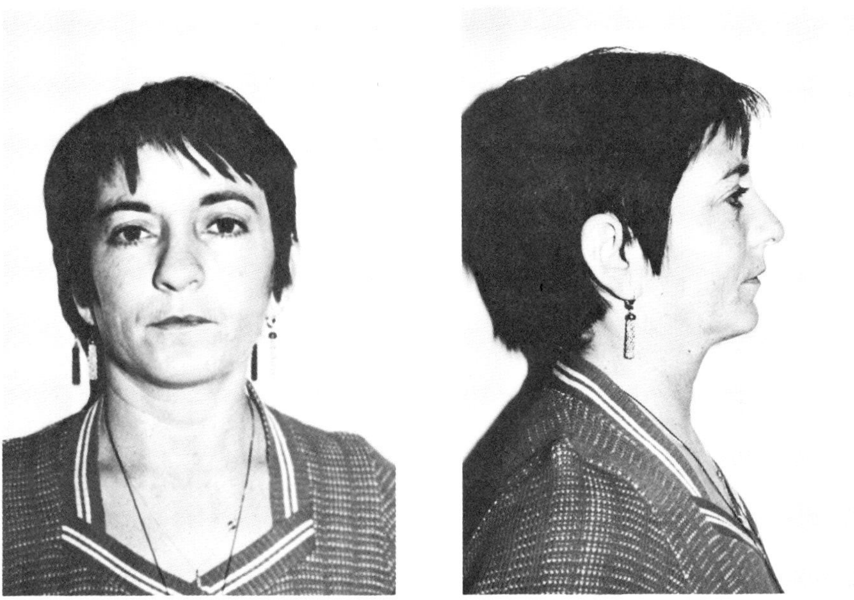

Fig. 3. *Case 3.* Note low-set auricles.

and small chin (Fig. 5). Her neck was short and showed mild webbing (Fig. 5). Cubitus valgus (Fig. 6) and short 4th metacarpals (Fig. 7) were found. Her breasts were atrophic and pendulous (Fig. 6) and she had no axillary and virtually no pubic hair. She had moderate coxa vara and mild genu valgum (Fig. 4).

Pelvic examination showed somewhat atrophic vulva with a narrow vaginal orifice and an erythematous but atrophic lining. The vagina bled easily on attempted bimanual examination, which revealed a small uterus with palpable adnexa on the left but not the right side. The skin was dry and scaly.

On endocrine evaluation results indicated a postmenopausal pattern, with elevated plasma FSH and LH and virtually unmeasurable estradiol levels.

Laparoscopic examination revealed a streak gonad on the right side and a small hypoplastic ovary on the left. When biopsied the latter showed essentially stromal tissue with some corpora albicantia with virtually no detectable follicular activity.

Cytogenetic studies showed X-chromatin negative nuclei on buccal smear. Chromosome analysis of lymphocytes revealed uniformly 45,X nuclei. Chromosome analysis on skin from the abdomen showed mosaicism with 45,X and 47,XXX cell lines. The proportion of 45,X cells was 92%, the remaining 8% being the triple X line. On a second skin biopsy specimen from the thigh the

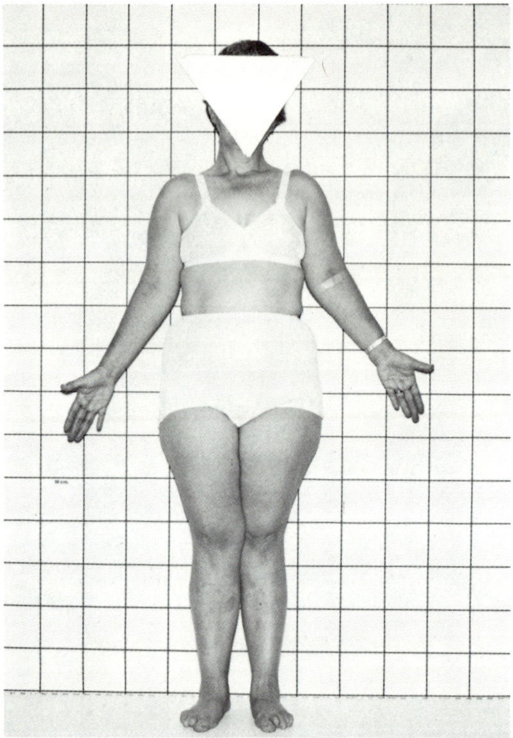

Fig. 4. Photo of *Case 4* showing short stature and mild genu valgum.

ratio of 45,X/47,XXX cells was 83:17, respectively. Cultured cells from the left ovarian stroma grew poorly and only few metaphases were available for analysis. These were all 45,X cells.

Case 5

This 19-year-old unmarried female presented with a history of secondary amenorrhea, fatigue, anorexia, weight loss, vaginal discharge, and pain during intercourse. The patient had experienced menarche at 12 years of age and continued to have regular menstrual periods, which lasted approximately 4 days, every 30–32 days, until she conceived when she was 15 years old. Since the pregnancy resulted from a casual affair with a relative and in view of her youthful age, she felt emotionally incapable of carrying on with the pregnancy or of mothering the baby. Therapeutic abortion was carried out at 14 weeks of gestation. A female fetus was identified. The patient continued to bleed and to have

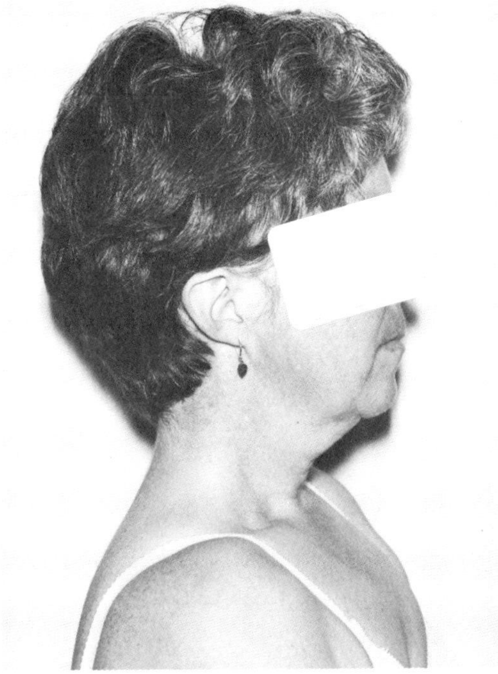

Fig. 5. *Case 4,* showing low-set auricles and long philtrum.

vaginal discharge following the termination, and dilation and curettage was performed 7 weeks later. For the next 3 months the patient had apparently regular, albeit scant, menstrual bleeding. She then became amenorrheic and remained so until her recent presentation. Chromosome studies were carried out on the product of conception, and a 46,XX chromosomal complement was found.

On examination, the patient measured 156 cm in height and weighed 45 kg. Her skin was pale and sallow. Her breasts, which appeared fairly well formed, were thin and pendulous and were deficient, on palpation, in lobulated glandular tissue. Vaginal examination showed a tight introitus with a somewhat erythematous and atrophic mucous membrane suggesting a poorly estrogenized vagina. The impression was supported by plasma estrogen levels which were low and FSH and LH levels which were elevated in the menopausal range.

The patient lacked most of the physical stigmata of Turner syndrome except for a short neck, short 4th metacarpals and metatarsals, and increased carrying angles at the elbows.

Cytogenetic studies carried out on peripheral lymphocytes showed 2 populations of cells: one 45,X and the other 46,XX. They were present in a ratio of

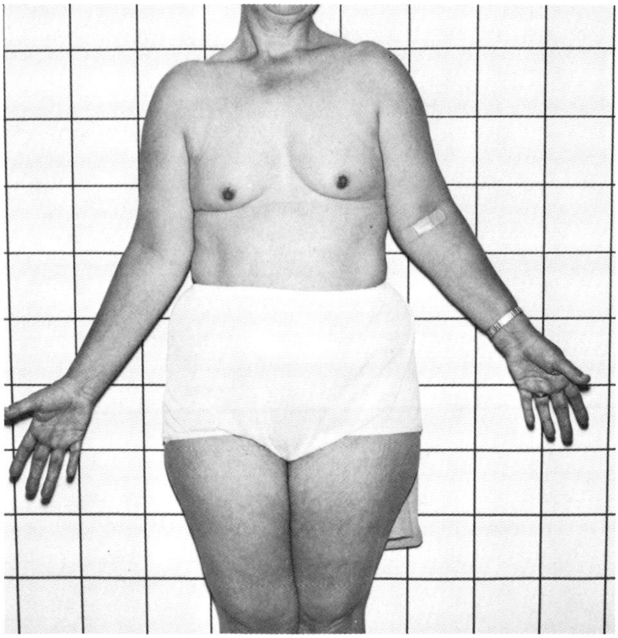

Fig. 6. *Case 4,* showing webbed neck, cubitus valgus, and atrophic pendulous breasts.

52:48, respectively. Skin biopsy specimens were obtained from 2 sites (forearm and abdomen). Fibroblast culture from these showed a similar distribution, although the proportion of 45,X cells was lower than that found in the blood, 38:62.

Laparoscopic examination showed normal uterus and tubes but fibrous tissue replacing the ovaries. Attempts to biopsy the ovarian streaks were unsuccessful.

DISCUSSION

One's initial reaction to pregnancies in women with 45,X gonadal dysgenesis is to doubt the diagnosis of Turner syndrome. It is well known that females with 45,X Turner syndrome are not only hypogonadic but also infertile; their ovaries are almost always replaced by streaks of connective tissue in which germ cells are but very rarely found [15]. Nevertheless, on review of the literature several documented examples of pregnancy among women with 45,X Turner syndrome and 45,X/46,XX or 45,X/47,XXX mosaicism have apparently taken place. In particular, 5 pregnancies have been ascribed to 4 women with a 45,X karyotype, at least in peripheral lymphocytes. Four of the pregnancies produced liveborn

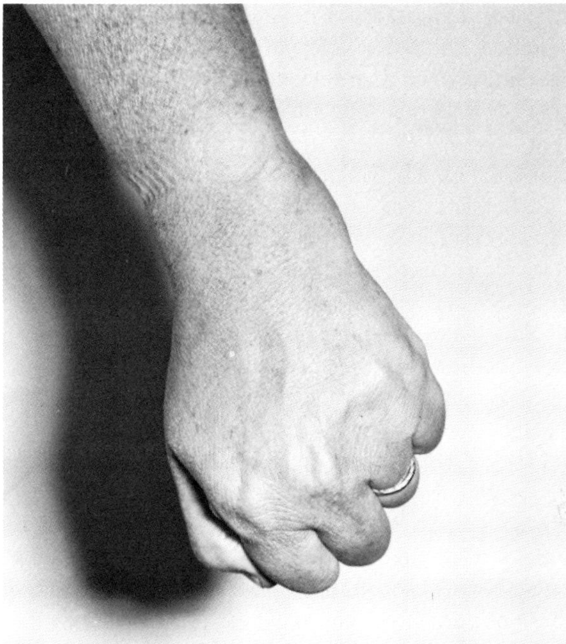

Fig. 7. Short 4th metacarpal bone, *Case 4.*

babies while the fifth resulted in a stillbirth [16–19]. Many more pregnancies have now been reported in females mosaic for X chromosome monosomy in which 2 cell lines, one with a single X and the other with the normal complement of 2 X chromosomes, coexist [20–22]. In the 45,X/47,XXX mosaics, 6 pregnancies in 4 females are on record, all of which ended in live births [14, 23, 24]. More complicated forms of mosaicism, in which 3 or more cell lines occur, have been encountered in women who were fertile [25–27].

Three explanations come to mind. First, patients with 45,X Turner gonadal dysgenesis may have viable ova in their ovaries, imparting to them potential or actual fertility during a part of their life in which reproduction may occur. Among 45,X females normal appearing ovaries with a normal number of germ cells have been observed not only in embryos and fetuses [13] but also in affected live newborn females [12]. Moreover, a biopsy specimen from the ovary of a 45,X girl revealed some normal follicles [28]. Baker [29] reported that among normal females the number of germ cells sharply fell in the fetal ovary at 5 months gestation only to continue a slower decline after birth. In the case of 45,X Turner gonadal dysgenesis such decline may be hastened leading to the nearly uniform failure of ovarian function among affected patients. Such acceleration of germ cell disappearance in this disorder may itself display a broad spectrum of variation, with a few affected individuals entering reproductive age with some preservation of ovarian follicular activity. Menstrual function has also been reported in 45,X Turner individuals [30, 31] which is compatible with ovarian function.

Second, even when blood, skin, and buccal mucosal cells are found to be uniformly 45,X, the possibility of an ovary or a segment thereof harboring cells which have the normal female chromosome complement cannot be entirely ruled out. However, it becomes an increasingly improbable situation as more metaphases from cells derived from various body tissues are found to be consistently 45,X. Nonetheless the possibility cannot be eliminated with complete surety. Even when biopsy of the ovary is undertaken, either by laparoscopy or laparotomy, there is no guarantee that the segment containing cells with normal chromosomal make-up will not be missed. This is obviously the most feasible explanation for *Case 5* wherein a normal population of cells exists in both blood and skin. It is probably also the explanation in *Cases 3* and *4* where 1 of 2 cell lines detected was 47,XXX. When these cells populate the ovary or a portion thereof, potential fertility may exist, since 47,XXX women are fertile, albeit less so than normal 46,XX females [32, 33]. The ascertained offspring of triple-X women have been of normal chromosomal make-up suggesting that either fertilization preferentially involves the single-X bearing eggs or that the disomic chromosome complement is regularly included in one of the polar bodies, leaving the ovum with but a single X [34]. This has been the experience in *Cases 3* and *4* where a 46,XX and 46,XY offspring were observed respectively.

Third, it is possible that fertility in 45,X Turner syndrome and in 45,X mosaic individuals is grossly underestimated. Considering the phenotypic variability of the disorder(s), many such patients, particularly those who actually reproduced or menstruated, might never have been studied cytogenetically and, therefore, never diagnosed. It is noteworthy that it was for the most part the cessation of ovarian function with the attendant secondary amenorrhea and other complications that impelled our patients to seek attention. The suspension of ovarian activity was tantamount to a premature menopause brought about by the disappearance of the residual follicles in the case of 45,X individuals (*Cases 1* and *2*) or the exhaustion of the functional follicles with 47,XXX or 46,XX chromosome make-up of mosaic individuals (*Cases 3, 4* and *5*). Implicit in the above argument, therefore, is an ascertainment bias in favor of menstruating fertile women with gonadal dysgenesis who may become pregnant but who then quickly run out of viable ova, experience secondary amenorrhea, and suffer symptoms of premature menopause.

Finally, the demonstrated fertility of the above women with 45,X chromosome make-up might serve as a reminder of the commonality of mammalian species. It may be of some evolutionary significance that XO mice have normally preserved ovaries containing viable gametes. Such female mice are, therefore, sexually normal and fertile [35].

REFERENCES

1. Funke: Pterygium Colli. Dtsch Z Chir 63:162, 1902.
2. Turner HH: A syndrome of infantilism, congenital webbed neck and cubitus valgus. Endocrinology 23:566–574, 1938.
3. Varney RF, Kenyon AT, Koch FC: Association of short stature, retarded sexual development and high urinary gonadotropin titers in women. J Clin Endocrinol 2:137–145, 1942.
4. Meyer R: Uber Gemebliche Anomalien und ihre Beziehung zu anigen Geschwusten der ovarien. Arch Gynaekol 145:2, 1931.
5. Wilkins L, Fleischmann W: Ovarian agenesis. Pathology, associated clinical symptoms and bearing on the theories of sex differentiation. J Clin Endocrinol 4:357–375, 1944.
6. Wilkins L, Gumbach MM, Van Wyk JJ: Chromosomal sex in "ovarian agenesis." J Clin Endocrinol 14:1270–1271, 1954.
7. Polani PE, Hunter WF, Lennox BC: Chromosomal sex in Turner's syndrome with coarctation of the aorta. Lancet 2:120–121, 1954.
8. Barr ML, Bertram EG: A morphological distinction between neurones of the male and female, and the behaviour of the nuclear satellite during accelerated nucleoprotein synthesis. Nature 163:676–677, 1949.
9. Ford CE, Jones KW, Polani PE, de Almeida LC, Briggs: A sex-chromosome anomaly in a case of gonadal dysgenesis (Turner's syndrome). Lancet 1:711–713, 1959.
10. Hook EB, Hamerton JL: The frequency of chromosome abnormalities in consecutive newborn studies – Results by sex and by severity of phenotypic involvement. In Hook EB, Porter IH (eds): "Population Cytogenetics. Studies in Humans." New York: Academic Press, 1977, pp 63–79.
11. Carr DH: Chromosome anomalies as a cause of spontaneous abortion. Am J Obstet Gynecol 26:308–326, 1967.
12. Carr DH, Haggar RA, Hart AG: Germ cells in the ovaries of XO female infants. Am J Clin Pathol 49:521–526, 1968.
13. Singh RP, Carr DH: The anatomy and histology of XO human embryos and fetuses. Anat Rec 155:369–383, 1966.
14. Reyes FI, Koh KS, Faiman C: Fertility in women with gonadal dysgenesis. Am J Obstet Gynecol 126:668–670, 1976.
15. Jones HW, Ferguson-Smith MA, Heller RH: The pathology and cytogenetics of gonadal agenesis. Am J Obstet Gynecol 87:578–600, 1963.
16. Grace JH, Quinlan DK, Edge WEB: 45,X Lymphocyte karyotype in a fertile woman. Am J Obstet Gynecol 115:279–282, 1973.
17. Nakasima J, Robinson A: Fertility in a 45,X female. Pediatrics 47:770–772, 1971.
18. Bahner F, Schwartz G, Harnden DG, Jacobs PA, Heinz WK: A fertile female with XO sex chromosome constitution. Lancet 2: 100–101, 1960.
19. Groll M, Cooper M: Menstrual function in Turner syndrome. Obstet Gynecol 47:225–226, 1976.
20. Lieber E, Berger J: Fertility in a 45,X/46,XX patient. Lancet 1:199, 1973.

21. Predescu V, Christodorescu D, Tantn C, Ciovirnache M, Constantinescu E: Repeated abortions in a woman with XO/XX mosaicism. Lancet 2: 217, 1969.
22. Bishun NP, Rashad MN, Morton WRM, Mannion PL, Neely MR, Burke G: Chromosomal mosaicism in a case of repeated abortion. Lancet 1: 936, 1964.
23. McLean N, Court-Brown WM, Jacobs PA, Mantle DJ, Strong JA: A survey of sex chromatin abnormalities in mental hospitals. J Med Genet 5:165–172, 1968.
24. Lewis FJW, Poulding RH, Eastham RDK: Acute leukemia in an XO/XXX mosaic. Lancet 2: 306, 1963.
25. Kim JH, Hsu LYF, Paciuk S, Christian S, Quintana A, Hirschhorn K: Cytogenetics of fetal wastage. N Engl J Med 293:844–847, 1975.
26. Girand F, Hartung M, Brusquet Y, Mattei JF, Sebahous A, Serment H, Stahl A, Bernard R: La fertilite des femmes 45,X/46,XX et 45,X/46,XX/47,XXX. Ann Genet 13:255, 1970.
27. Hsu LYF, Palo-Garcia F, Grossman D, Kutinsky E, Hirschhorn K: Fetal wastage and maternal mosaicism. Obstet Gynecol 40:98–103, 1972.
28. Lamy M, Josso N, de Grouchy J, Bitan A: Anomalies des gonosomes 20e Congres des Pediatres de Lange Francaise. L'Expansion Scientifique Francaise, Paris, 1965, p 41.
29. Baker TG: A quantitative and cytological study of germ cells in human ovaries. Proc Soc Lond [Biol] 158:417–433, 1963.
30. Kaufman S: Vaginal bleeding in gonadal dysgenesis. Pediatrics 37:26–33, 1966.
31. McDonough PG, Byrd JR, Freedman MA: Gonadal dysgenesis with ovarian function. Obstet Gynecol 37:868–872, 1971.
32. Stewart JSS, Sanderson AR: Fertility and oligophrenia in an apparent triplo-X female. Lancet 2:21–23, 1960.
33. Barr ML: The sex chromatin. In Overzier C (ed): "Intersexuality." New York: Academic Press, 1963, pp 48–71.
34. Hamerton JL: X-chromosome polysomy in the human female. In "Human Cytogenetics." New York: Academic Press, 1971, vol II, p 106.
35. Welshons WS, Russell LB: The Y chromosome as the bearer of male-determining factors in the mouse. Proc Natl Acad Sci USA 45:560–566, 1959.

Selected Abstracts

INTERSEX SYNDROMES: CYTOGENETIC AND CLINICO-PATHOLOGIC EVALUATION OF 24 PATIENTS WITH AMBIGUOUS GENITALIA

John D. Blair

Department of Pathology, Cardinal Glennon Memorial Hospital for Children and St. Louis University School of Medicine, St. Louis, MO

The clinical and cytogenetic features of 24 individuals with ambiguous external genitalia are reviewed and correlated with the histopathology of the gonads available from 16 patients who underwent exploratory laparotomy or postmortem examination. Nine patients had mixed gonadal dysgenesis: 7 with 45,X/46,XY karyotypes, 1 with 45,X/46,XY/47,XYY chromosome complement; and a 45,X/46,XYq– patient with hernia uteri inguinalis. Four patients in this group had gonadoblastoma in dysplastic testis and/or "streak" gonads. Thirteen patients were classified as "male pseudohermaphrodites" with normal 46,XY karyotypes. Included in this group are 6 patients from the Dominican Republic referred to the author for chromosome analysis, who subsequently were found to have 5 alpha-reductase deficiency by Imperato-McGinley et al, [Science 186:1213, 1974; also 1974 Birth Defects Conference]. Two cases of genital ambiguity were found in association with unusual karyotypes: apparent 69,XXX in a newborn who had multiple anomalies, ambiguous genitalia, and intraabdominal testes with marked interstitial cell hyperplasia; the other patient was a sex-chromatin positive intersexual individual with 46,XX karyotype in blood cells; an ovary was present on one side and a testis on the other; temporarily this patient is classified as a "true hermaphrodite." These two cases will be discussed in detail.

Birth Defects: Original Article Series, Volume XIV, Number 6C, pages 185–187

This review, which includes cases studied at different institutions (Cleveland Metropolitan General Hospital and the University of Kentucky Medical Center) in the last 15 years, brings out the nosologic difficulties and inadequate understanding of basic pathobiologic processes of sexual differentiation and development. It emphasizes the different pathogenetic mechanisms which might be involved in the production of genital malformations and points out the necessity for further investigation of fetal endocrine factors which probably account for most cases currently classified as "male pseudohermaphrodites."

MALE PSEUDOHERMAPHRODITISM: XY GONADAL DYSGENESIS

Virginia P. Johnson, Hubert Theissen, and John Elston

Department of Obstetrics and Gynecology and Center for the Developmentally Disabled, University of South Dakota Medical School, Vermillion, SD

Two sibs, 17 and 14 years, have identical phenotype with XY chromosomes, streak gonads, fallopian tubes, rudimentary uterus, normal vagina and female external genitalia. At puberty there was amenorrhea, absence of sex hair, and no breast development. Hormone studies showed high FSH, low testosterone and low estrogen levels. Microscopic examination showed gonadal dysgenesis. X-linkage is suggested with four maternal "aunts" affected, some of whom had clitoridectomy at puberty. The paper is reported to show "grades" of gonadal dysgenesis with similarity within sibship and differences between sibships. The importance of continued counseling is stressed, as is follow-up to allow for early diagnosis in the future generation.

ABNORMAL DEVELOPMENT OF EXTERNAL GENITALIA ASSOCIATED WITH MATERNAL HORMONAL INTAKE

Samia A. Temtany, Adel H. Loutfy, Shahira Boulos, Moushira Abdel-Salam, Isis Ghaly, Mostafa Raafat, and H.H. de Hondt

Department of Medical Research and Cell Biology Unit, National Research Center, The Children Hospital and The Cancer Institute, Cairo University, Cairo, Egypt

We studied 5 cases of intersex whose mothers gave definite history of intake of hormones during pregnancy. Four cases proved to be female pseudohermaphrodites (see Table 1). *Case 5* was a male with the rare anomaly of transposition of external genitalia. His phallus was bifid and very hypoplastic. He died of uremia at age 2 months. The mother of *Case 1* received long-acting synthetic progesterone for 2 months as a treatment for threatened abortion. Mother of *Case 2* received combined progesterone and estrogen for induction of abortion. Mother of *Case 3* received combined progesterone and estrogen as a pregnancy test and also received progesterone monthly throughout pregnancy as a treatment for threatened abortion. Mother of *Case 4* received combined progesterone and estrogen as a pregnancy test. The mother of *Case 5* took contraceptive pills throughout pregnancy. The parents of *Case 4* were first cousins. In all cases in addition to clinical examination, sex-chromatin, chromosomal studies from peripheral blood cultures, urinary 17-ketosteroids, laparotomy and gonadal biopsies were performed. Our observations emphasize the danger of maternal intake of hormones during pregnancy and suggest that the severity of malformations is related to the dose and duration of hormonal intake.

Table 1. Female Pseudohermaphrodites Associated With Maternal Hormonal Intake.

Case no.	Sex of rearing	Age (years)	Single urogenital orifice	Hyper-trophied clitoris	Fused labia minora	Fused labia majora
1	F	13	+	+	–	–
2	F	3	+	+	–	–
3	M	½	+	+	+	+
4	F	1½	+	+	+	–

SECTION II: CHROMOSOME ABNORMALITIES

Clinical Syndromes Associated With Partial Duplications of Chromosomes 2 and 3: dup(2p),dup(2q),dup(3p),dup(3q)

Uta Francke, MD

INTRODUCTION

The precision with which chromosomal aberrations are being characterized is increasing, and duplications and deletions have been reported for many autosomal regions. For several of these, recognizable clinical syndromes have been delineated or proposed [1]. This review covers syndromes associated with duplications ("partial trisomies") involving chromosomes 2 and 3: duplication 2p and 2q, 3p and 3q. Since chromosomes 2 and 3 can be identified with standard staining, some cases reported in the prebanding era have been included in addition to those documented with banding techniques. In cases of dup(3) the involvement of long or short arm could not previously be distinguished. After defining the syndromes associated with dup(3p) and dup(3q) through the patients studied with chromosome banding, it has become possible on the basis of clinical findings to assign some early cases to the appropriate category.

First, a few comments about nomenclature: The term "partial trisomy" is enjoying a widespread use in the human cytogenetic literature, although it is, strictly speaking, incorrect. The term "trisomy" means "having a single extra chromosome in the cell in addition to the normal diploid number" [2]. Thus, the presence of an extra chromosome consisting of part of a normal chromosome, should be called "partial trisomy," as in the case of 47,+14q–, ie partial trisomy 14. In contrast, if the extra material is incorporated into another chromosome, resulting in a karyotype with a diploid chromosome number, such as in unbalanced offspring of balanced translocation heterozygotes, the term "partial trisomy" is not appropriate. Based on nomenclature used in cytogenetics of other organisms, the term "duplication" has been proposed to describe the situation.

Birth Defects: Original Article Series, Volume XIV, Number 6C, pages 191–217

The Paris Conference Nomenclature booklets [3, 4] provide abbreviated formulas for the description of karyotypic findings, but a verbal terminology is not given. For the moment, I prefer to use the terms "duplication," "deletion," and "duplication/deficiency."

An alternative nomenclature has been proposed [5] which is attractively brief but not consistent with the Paris Committee recommendations. In analogy to the naming of deletion syndromes (eg cri-du-chat syndrome – 5p– syndrome) the symbol "+" has been used for a duplication; thus duplication of 5p would be "5p+ syndrome." However, "+" after p or q stands for elongation of the respective chromosome arm. The term "5p+ syndrome" may accurately describe the abnormal chromosome in the rare cases of inverted ("mirror") or tandem duplications. But in the more common cases of unbalanced translocations, the extra 5p material would be located somewhere else and the patient would be expected to have two normal chromosomes 5. To call this "5p+ syndrome" may become very confusing to parents and nongenetic health professionals. An international agreement of the naming of unbalanced chromosome constitutions would be desirable.

DUPLICATION (2p) SYNDROME

The extent of the duplicated regions of 2p in 9 documented cases is indicated in Figure 1. All of them are offspring of balanced translocation carriers from 5 unrelated families [6–10]. The translocations involve the terminal regions of 14p (*Cases 1* and *2*), 7q (*Cases 3* and *4*), 18p (*Case 5*), 3p (*Cases 6–8*), and 22q (*Case 9*). Although the translocations are not obviously reciprocal, submicroscopic deletions of the terminal regions of the recipient chromosomes are assumed. A patient reported in 1969 [11] has been included as *Case 10*; the paternal translocation has recently been interpreted as t(2;6)(p21 or p23;p23 or p25) by banding methods (Bender, personal communication, 1977).

Cytogenetic, historic, and physical examination data on the 10 documented cases are listed in Table 1.

The similarity of facial features can be appreciated in Figure 2. High forehead and wide depressed nasal bridge are consistent. A more severe degree of hypertelorism is apparently associated with duplication of a more proximal region of 2p (*Cases 9* and *10*). Maxillary hypoplasia leads to downslanting palpebral fissures. Puffiness of the eyelids is present at an early age but disappears later. The nasal tip appears pulled downward, giving rise to triangular-shaped nostrils. The philtrum may be wide. The mandible is often narrow with a pointed tip.

Figure 3 illustrates the slender body configuration with long narrow trunk, scoliosis, dolichostenomelia, long fingers and toes, hypoplastic musculature, and

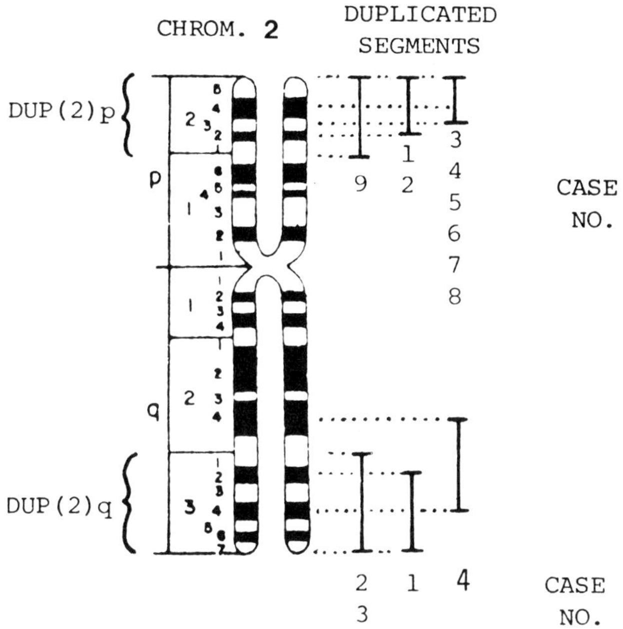

Fig 1. Standard ideogram [3] of chromosome 2, indicating extent of the duplicated regions in 9 cases with duplication (2p) syndrome and 4 cases with duplication (2q) syndrome.

a small penis buried in an overriding scrotum. Dermatoglyphics are not consistently abnormal. A peculiar finding in *Cases 4* and *5* is pyramid-shaped soft tissue pads in the interdigital areas of the palm. Whorl patterns on top of them indicate that they were present in early embryonic life. Such pads were noted independently in *Case 3*. In addition, there are an increased number of fine creases, hyperextensible fingers, and subluxed proximal interphalangeal (PIP) joints (Fig. 4).

Occasional abnormalities not listed in Table 1 include hemangiomata, variably deformed external ears, retinal detachment, persistent hyaloid canal, cataracts, sparse or absent eyelashes, congenital heart defects, dextrocardia, hip dislocation, extra or missing ribs, abnormal segmentation of the sternum, supernumerary lateral maxillary incisors, splaying of distal phalanges, clinodactyly, and seizures. The total lack of intelligible speech in all of these patients cannot entirely be explained on the basis of mental retardation. Hearing tests in some patients revealed normal or slight loss of hearing.

TABLE 1. Duplication (2p) Syndrome

Case	*1*	*2*	*3*	*4*	*5*
Reference	[6] (Obs. 1)	(Obs. 2)	[7][a]	[8] (Case 1)	(Case 2)
Banding method	R		Q	G	
Parental translocation	t(2;14)(p22;p11)mat		t(2;18) (p23;p11)pat	t(2;7)(p23;q36)mat	
Sex	M	M	F	M	F
Gestation	term	term	41 wk	40 wk	42 wk
Birthweight (gm)	3,200	3,450	3,700	2,350	2,260
Birth length (cm)	50	52	53	46	53
Age at report	7 yr	3 mo	9 mo (4 yr)	8 yr	12 yr
Growth retardation	< 3rd %	...	< 3rd %	< 3rd %	< 3rd %
Psychomotor retardation	profound	...	profound	profound	profound
Craniofacies					
High prominent forehead	+	+	+	+	+
Frontal upsweep of hair	+	+	+	+	+
Hypertelorism	+	+	–	+	+
Epicanthal folds	+	–	+	+	–
Maxillary hypoplasia	+	+	+	+	+
Short nose with prominent tip ("pug nose")	+	+	+	+	+
Broad alveolar ridges/ narrow palate	+	+	+	+	+
Small mandible/pointed chin	+	+	+	+	+
Ptosis	+	–	–	+	+
Strabismus	+	–	+	+	–
Myopia	+	–	+	+	+
Dacryostenosis	–	–	+	+	+
Skeleton					
Pectus excavatum	–	–	–	–	+
Long, narrow trunk	+	+	+	+	+
Dolichostenomelia	–	–	+	+	+
Long fingers and toes	+	+	+	+	+
Hyperextensible fingers	+	+	+	+	+
Flat feet	+	+	+	+	+
Wide-spaced toes	+	+	+	+	+
Genitalia					
Overriding ("shawl") scrotum	–	–	...	+	...
Hypoplastic external genitalia	–	–	...	+	+
Cryptorchidism	...	...	...	+	...
Figure	2e,m	2h,i	3a	2j,o, 3c	2n, 3c

Symbols: + present; – not present; ... not applicable or not known.
[a]Also Magenis and Prescott, personal communication, 1977.
[b]Also Bender, personal communication, 1977.
†Proband died at 16 months.

6	*7*	*8*	*9*	*10*
	[9]		[10]	[11][b]
(Case 1)	(Case 2)	(Case 3)		
	G		G	S, G
t(2;3)(p23;p27)mat			t(2;7)(p21;q36.3)pat	t(2;6)(p21 or 23;p23 or 25) pat
M	M	M	F	F
term	term	term	term	term
3,525	3,465	3,125	1,060	3,100
...	...	...	...	47
5 yr	1 yr	18 yr	11 mo	16 mo†
< 5th %	< 5th %	< 5th %	marked	severe
marked	delayed	profound	severe	profound
+	+	+	+	+
–	–	+	...	...
+	+	+	+	+
+	+	–	–	–
+	+	+	+	+
+	+	+	+	+
+	+	+	...	+
+	+	+	–	+
+	+	–	–	–
+	+	+	–	+
...	...	...	...	...
–	+	–	...	...
–	+	+	+	...
–	+	+	+	...
–	–	+	...	–
...	...	+	+	+
...	...	...	–	...
...	...	+	...	...
...	+	+	...	–
+	+	+	...	...
+	+	+	+	...
...	+	...	...	...
2f,g, 3b	2a,b, 3b	2k, 3d	2c,d	2e

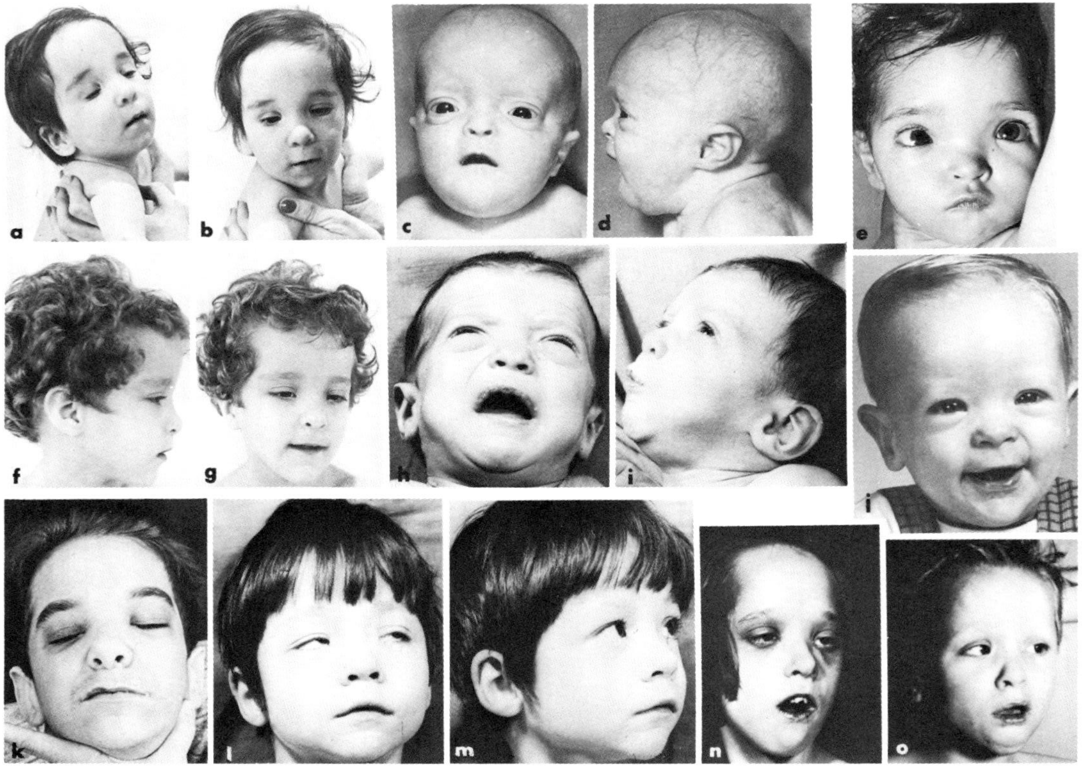

Fig 2. Dup (2p) syndrome. a, b) *Case 7,* f, g) *Case 6*, k) *Case 8*, from Cassidy et al [9]; c,d) *Case 9*, from Yunis [10]; e) *Case 10*, from Bender et al [11]; h, i) *Case 2*, l, m) *Case 1*, from Stoll et al [6]; j, o) *Case 4*, n) *Case 5*, from Francke and Jones [8].

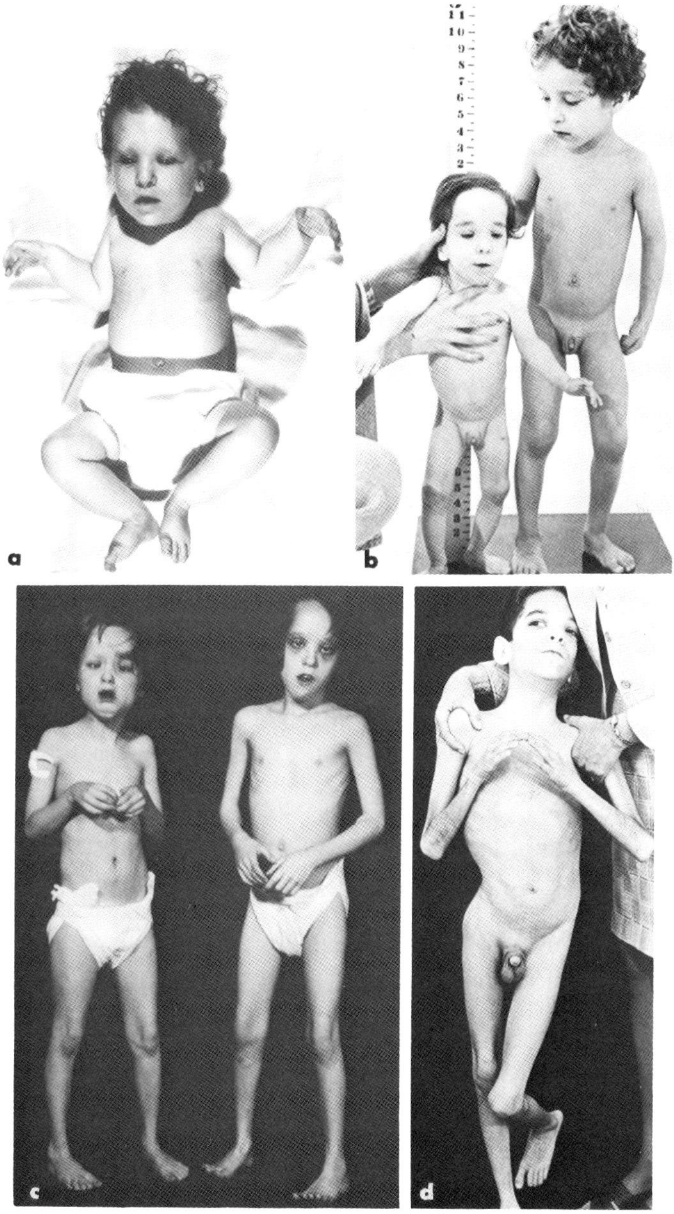

Fig 3. Dup(2p) syndrome. a) *Case 3*, Magenis et al [7]; b) *Cases 6* and *7*, d) *Case 8*, from Cassidy et al [9]; c) *Cases 4* and *5*, from Francke and Jones [8].

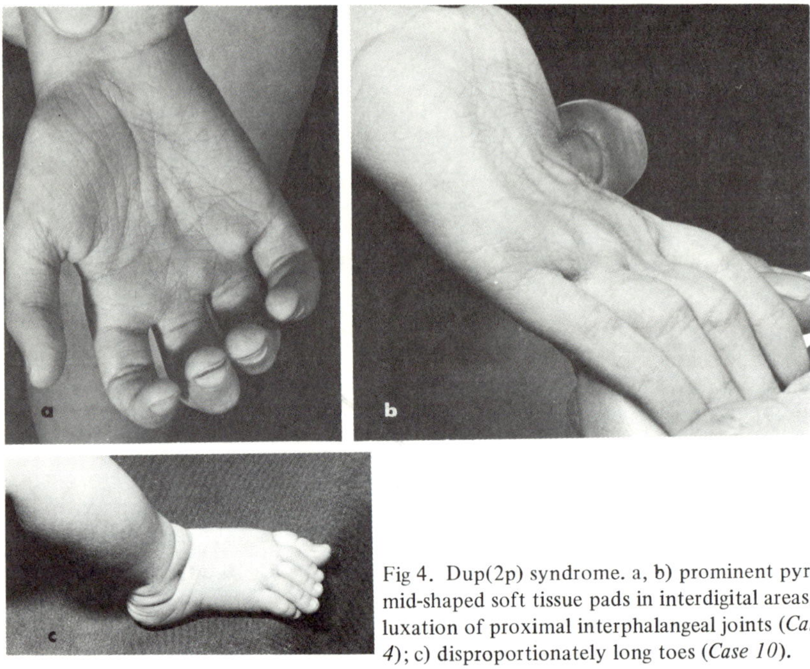

Fig 4. Dup(2p) syndrome. a, b) prominent pyramid-shaped soft tissue pads in interdigital areas, subluxation of proximal interphalangeal joints (*Case 4*); c) disproportionately long toes (*Case 10*).

Certain features of the patients with dup(2p) syndrome are reminiscent of the Aarskog syndrome [12], such as short stature, hypertelorism, maxillary hypoplasia, ptosis, pectus excavatum, scoliosis, shawl scrotum, and cryptorchidism. However, the Aarskog syndrome, caused by mutation of an X-linked gene is associated with normal intelligence or mild retardation, different facial appearance, and brachydactyly. Despite the overlap in certain physical findings, the overall clinical picture is sufficiently different to allow distinction of the two syndromes.

Two early reports of duplication (2p) due to parental t(2p–;3p?+) translocations are now interpreted as dup(3q)/del(3p) due to pericentric inversion of chromosome 3 and will be discussed with the dup(3q) syndrome [13, 14]. A large kindred with t(2p–;10q+) containing several individuals with dup(2p) is under study. The duplication involves a smaller region of 2p and the clinical effects are milder, with retardation in the severe rather than the profound range ([15]; and Wasdahl, personal communication, 1977).

DUPLICATION (2q) SYNDROME

The phenotype associated with duplication of the distal one-third to one-half of the long arm of chromosome 2 is less well defined due to the paucity of reported cases. The extent of the 2q duplications is indicated in Figure 1 for the 4 cases documented with chromosome banding [16–19]. Cytologic and the most consistent clinical findings are listed in Table 2. A distinctive facies is suggested by published photographs of *Cases, 1, 3,* and *4* (Fig. 5) with hypertelorism, prominent glabella, depressed nasal bridge, and hypoplastic nasal bones. The nostrils are completely anteverted in *Cases 1* and *3* but not in *Case 4*, who has duplication of a more proximal region of 2q. The nasolabial distance is increased and the philtrum is well developed. A prominent Cupid's bow gives the mouth a characteristic shape resembling a "chapeau de gendarme" as described in the French genetic literature [16, 19].

In addition to those listed in Table 2, the following abnormalities have been reported: congenital heart defects, kyphosis and scoliosis, sacrococcygeal fistula, clubfoot, absent or hypoplastic nails, and abnormal renal calix or malrotation of a kidney. One male (*Case 3*) had undescended testes, and inguinal and hiatal hernias. Dermatoglyphics were variably abnormal, distal axial palmar triradii (t′ or t″) being the most consistent feature. Two cases had an increased number of digital arches (*Cases 6* and *9*) and had 5 radial loops.

Chromosome 2 can be identified by morphology in standard stained preparations. Of the patients reported in the prebanding era, the one studied by Ricci and colleagues in 1968 [20] most likely had the same syndrome. The female infant was small for gestational age and died at 75 days. She was microcephalic with hypertelorism, flat nasal bridge, micrognathia, low-set ears, abnormal dematoglyphics, and coarctation of the descending aorta. She was the unbalanced offspring of a phenotypically normal t(2q−;Dq+) translocation carrier. Her karyotype (46,XX,Dq+) signifies duplication of a region of a 2q and a q terminal deletion from an unidentified D group chromosome. In contrast, the patient reported by Bijlsma et al [21] who was studied with standard staining and autoradiography, does not appear to fit the dup(2q) syndrome clinically. It is likely that the rearrangement involves a different region or that it is more complex than a simple reciprocal translocation. A repeat study with banding should clarify the diagnosis.

Rosenthal and colleagues reported repeatedly in abstracts [22, 23] on a kindred with a t(2;9)(q33;p24) translocation containing 5 retarded males with duplication 2q33→2qter/deletion 9p24→9pter. The presence of abnormal nasal bones, hypertelorism, and optic atrophy are mentioned. An unrelated female patient with dup(2)(q31→qter) had mental retardation, hypertelorism, flat occiput, frontal bossing, and urinary tract anomalies [23].

Two sibs, offspring of a balanced nonreciprocal ins(12;2) insertion carrier mother, whose chromosomes were published in 1972 [24], are alive with moderate mental retardation but no major congenital malformations. Their facies (Fig. 5f,g) show striking resemblance to reported patients with the dup(2q) syndrome (L. Shapiro, personal communication).

In summary, the existing information on 13 cases with variable duplication of distal 2q suggests that this syndrome is characterized by moderate to severe

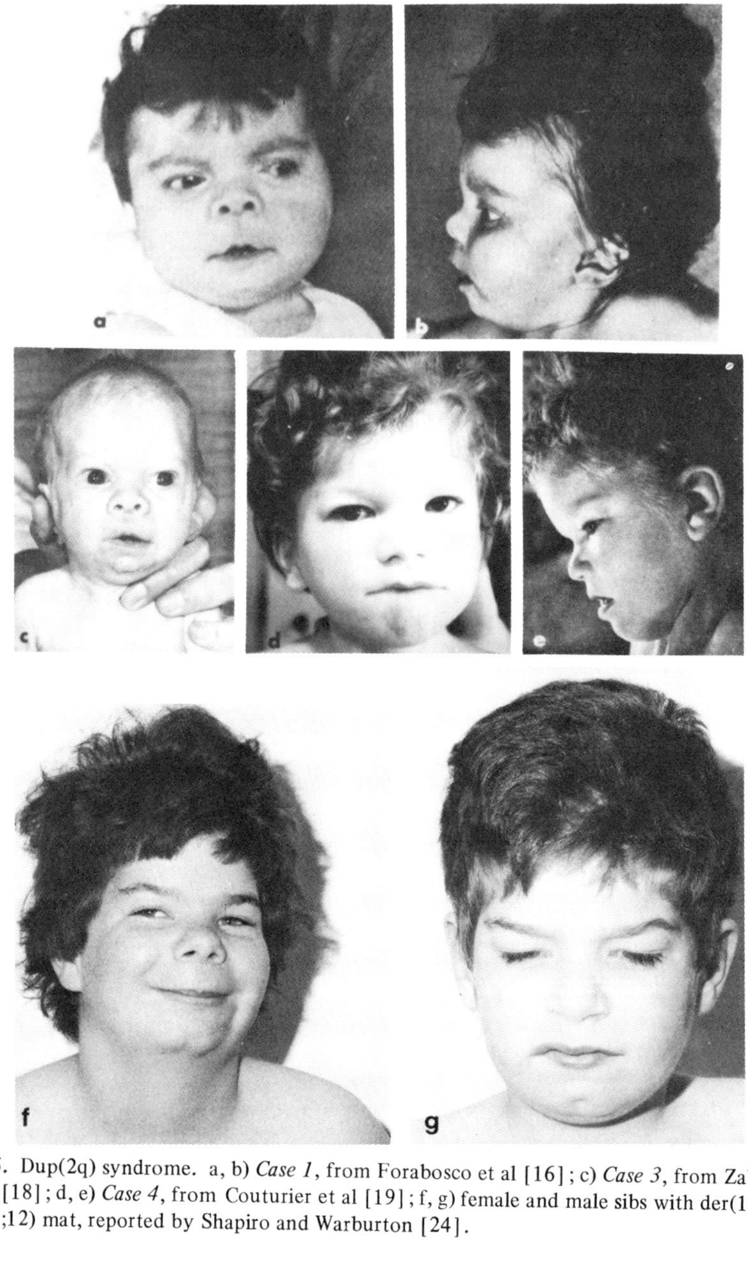

Fig 5. Dup(2q) syndrome. a, b) *Case 1*, from Forabosco et al [16]; c) *Case 3*, from Zabel et al [18]; d, e) *Case 4*, from Couturier et al [19]; f, g) female and male sibs with der(12), ins(2;12) mat, reported by Shapiro and Warburton [24].

TABLE 2. Dup(2q) Syndrome

Case	*1*	*2*	*3*	*4*
Reference	[16]	[17]	[18]	[19]
Banding method	R, T	Q	G	R, T
Parental translocation	t(2;13)(q32;q33)mat	t(1;2)(p36;q31)pat	t(2;12)(q31;q24)mat	ins(6;2)(p22;q24q34)
Duplicated segment	2q32→2qter	2q31→2qter	2q31→2qter	2q24→2q34
Deleted segment	13q33→13qter	1p36→1pter	12q24→12qter	none
Sex	F	F	M	M
Gestation	term	term	term	8 mo
Birthweight (gm)	2,700	(3rd %)	2,400	3,000
Birth length (cm)	...	(25th %)	46	...
Age at report	14 mo	22 mo	3 wk	6 yr
Growth	low normal	25th %	...	retarded
Psychomotor retardation	severe (DQ 30)	severe (DQ 28)	...	profound (IQ < 25)
Prominent forehead	+	+	+	+
Depressed nasal bridge	+	+	+	+
Hypertelorism	+	+	+	+
Strabismus	+	+	–	–
Nystagmus	+	–	+	–
Very short nose	+	–	+	+
Long philtrum	+	–	+	+
Distinct cupid's bow ("chapeau de gendarme")	+	...	+	+
Low-set posteriorly rotated ears	+	+	+	–
Small mandible	+	+	+	–
Short neck	+	+	+	–
Short stubby hands	+	+	+	+
Clinodactyly	+	+	+	–
Dry, wrinkled skin	+	–	+	...
Hypotrophic muscles	+	+	+	+
Persistence of archaic muscles	+	+	+	...
Figure 5	a, b	–	c	d, e

Symbols: +present; –not present; ...not known.

(occasionally profound) mental retardation, characteristic facies, abnormal dermatoglyphics, and lack of major congenital anomalies, and that it is consistent with survival into adulthood.

DUPLICATION (3p) SYNDROME

An 18-month-old girl with an inherited dup(3)(p22→pter) chromosomal imbalance is presented and compared to 7 well-documented cases from the literature. A chromosome 3 distal short arm duplication is associated with a clinically recognizable syndrome consisting of a characteristic facies, congenital heart defect, incomplete development of male genitalia, and typical dermatoglyphics. Although the duplication may involve one-third to one-half of 3p, comparable in size to a chromosome 21, pre- and postnatal growth is usually normal and psychomotor development is in the moderately retarded range.

Case Report

The proband was born after a 40-week pregnancy during which the mother took Bendectin daily and Librax during the fifth month only. Weight gain during pregnancy was 12 lbs. Fetal activity started in the fourth month and was vigorous throughout the remainder of pregnancy. Delivery was spontaneous vaginal from a vertex presentation. Apgar was 9. Birthweight was 3,200 gm, length 50 cm, and head circumference 33.6 cm. As a newborn the infant had noisy respirations attributed to laryngotracheomalacia. A systolic cardiac murmur was noted at 1 month of age. She was diagnosed to have acyanotic heart disease and was treated with digoxin. Cardiac catherization at 6 months revealed a patent ductus arteriosus (PDA) with a large left-to-right shunt and a small atrial septal defect (ostium secundum). The PDA was ligated at 11 months of age. Her condition has been stable postoperatively without medication.

Developmental milestones were delayed. She smiled at 4–6 months, sat with support at 7 months, and alone at 15 months. She pulled herself to standing position at 15 months and walked with assistance at 18 months. She does not crawl. She started to feed herself and drink from a cup at 15 months. She vocalizes "Baba" and "Dada."

Family history. There are 2 normal older brothers, 11 and 7 years old. The parents are Mexican-American, both were 29 years old at the patient's birth. The father's only full-sister died at birth of an unknown cause; he has 6 half-sibs through his parents' second marriages, but no reliable clinical information is available.

Physical examination. At 8 months of age, proband's height was 69.5 cm (50th %), head circumference was 41.5 cm (3rd %), weight was 6.8 kg (below

10th %). The head was symmetric with a flat occiput; transillumination was negative. The face appeared broad and short. There was hypertelorism with an inner canthal distance of 2.8 cm and an outer canthal distance of 8.5 cm. Palpebral fissures were horizontal. Eye examination revealed no abnormalities. The external ears had elongated lobes and measured 4 cm on the right, and 3.8 cm on the left. The left helix was folded over. The philtrum was short and prominent. The angles of the mouth turned downward (Fig. 6a and 6b). The palate was wide, and no teeth had erupted yet. There was mild retrognathia. The neck was short but had full range of motion. There was redundant skin at the nape of the neck. The thorax showed mild pectus excavatum and flaring of the lower ribs. The nipples were hypoplastic and appeared low-set. Internipple distance was 22.8% of the chest circumference (1 SD < mean). A loud systolic murmur was heard over the entire precordium and a thrill was palpable at the lower left sternal border in the 4th to 5th intercostal space. Liver and spleen were not enlarged. The infant had no hernias and normal external female genitalia. The 4th toes overlapped the 5th bilaterally. The infant had good muscle tone and head control, and would bear weight when placed upright. She reached for objects and smiled responsively. The range of supination and pronation of the forearms appeared increased, radiographs of the elbows were normal, serum thyroxine level was 11.4 μg/ml.

Dermatoglyphics. The patient had strikingly prominent pyramid-shaped soft-tissue pads over the distal phalanges of fingers and toes. The digital dermatoglyphic patterns consisted of 9 whorls and 1 radial loop, on the right 5th finger. This represents a significantly higher number of whorls than was found in the parents (mother: 6 ulnar loops, 2 whorls, 2 double loops on the thumbs, Father: 6 ulnar loops, 4 whorls).

When reexamined at 18 months, the patient's height was 81 cm (50th %) weight 8.3 kg (< 3rd %), and head circumference 44 cm (< 2nd %). Although she was well-nourished, active, alert, and responsive, her musculature appeared underdeveloped. She had 7 maxillary and 8 mandibular normal teeth. The unusual features noted previously were still present. Her face now appeared more square due to prominent cheeks and short mandible (Fig. 6h). A cardiac murmur was no longer present. The skin was strikingly mottled at room temperature. There were no focal neurologic signs. The child was functioning at approximately a 9-month level, the highest scores being in the adaptive and personal-social areas.

Chromosome studies. The patient's chromosomes prepared from peripheral blood cultures were analyzed with trypsin-Giemsa (GTG) banding. Her karyotype was 46,XX,13p+; the abnormality consisted of a moderate amount of additional chromosomal material translocated onto the short arm of a chromsome 13. While the mother's karyotype was normal, the father carried a balanced translocation between the short arms of a chromosome 3 and a chromosome 13 (Fig. 7). The breakpoint in the short arm of chromosome 3 was interpreted to be

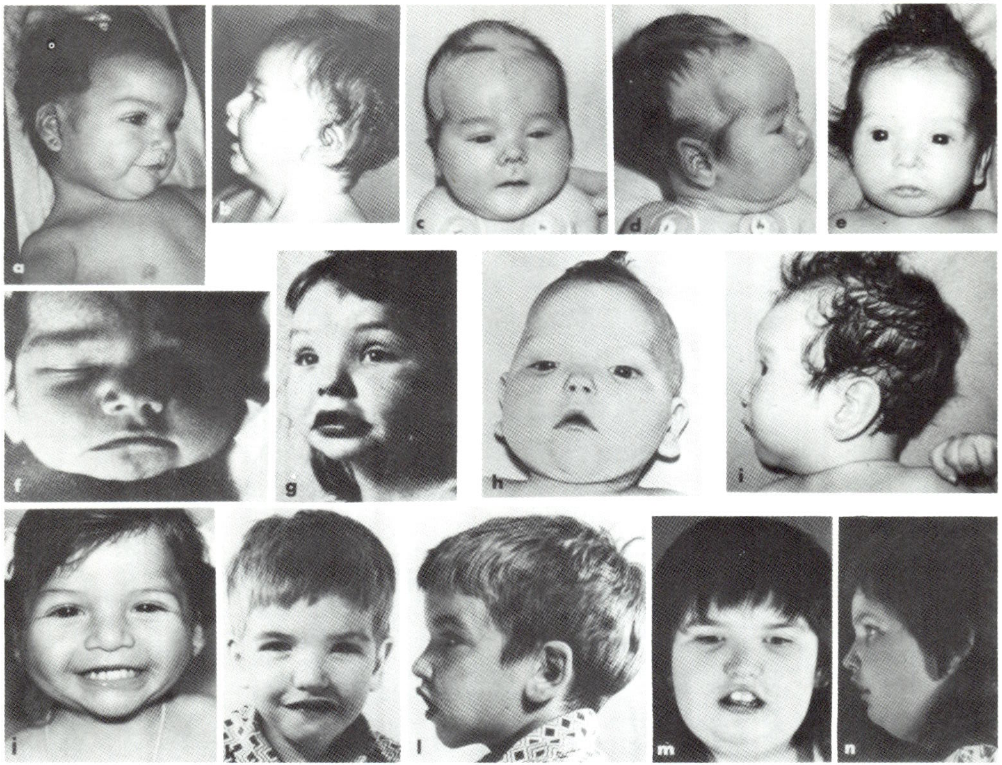

Fig 6. Dup(3p) syndrome. a, b, j) *Case 1,* this report; c, d) *Case 8,* from Yunis [28]; e, i) *Case 5,* from Ballesta and Vehi [26]; f) *Case 2,* g) *Case 3,* from Rethoré et al [25]; h) case with dup (3p)/del(18)q from Aarskog [30]; l) *Case 7,* m, n) *Case 6,* from Say et al [27].

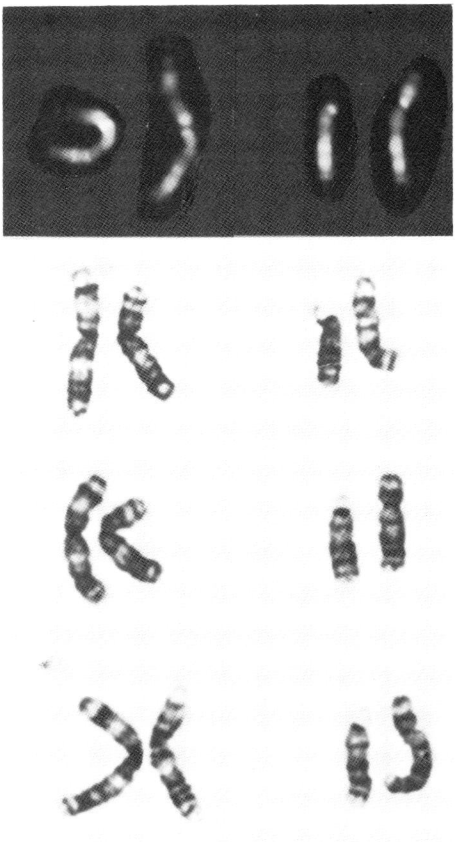

Fig 7. Partial QFQ and GTG-banded karyotypes from father of *Case 1*, showing reciprocal translocation t(3;13)(p22;p12). The rearranged chromosomes der(3) and der(13) are placed to the right of each pair.

located in band p22 using the Paris nomenclature ideogram. The break in chromosome 13 had occurred at the border of the short arm and the satellite stalk. Fluorescent quinacrine (QFQ) indicated that the Q-positive short arm of chromosome 13 was retained in the translocation chromosome. The translocation was clearly reciprocal, since the der(3) chromosome carried satellites at the p terminal end and participated in satellite association with acrocentrics. The father's karyotype is designated 46,XY,t(3;13)(p22;p12). The patient, who has inherited the 13p+ or der(13) chromosome together with the normal chromosome 3 from her father, has a duplication of region 3p22→3pter and a deficiency of the satellite region of chromosome 13. Her karyotype designation is 46,XX,-13, +der(13),t(3;13)(p22;p12)pat. Her abnormal phenotype is thought to be due to the distal 3p duplication, since deletion of satellites from a chromosome 13 (as, for example, in t(13q14q) heterozygotes) is not associated with clinical abnormalities. The two older brothers had normal karyotypes. Additional members of the father's family were not available for study.

Delineation of dup(3p) Syndrome

When the clinical findings in this patient are compared with seven cases reported in the literature [25–28], a distinct pattern of malformation emerges. All these patients have duplication of the distal part of the short arm of chromosome 3 due to a balanced chromosomal rearrangement in a parent. The duplicated regions are indicated diagramatically in Figure 8. Assuming that the translocations in the parents are reciprocal, the patients should also have deletions of terminal parts of the reciprocal chromosomes. Deletions of possible clinical significance may be present in *Case 5* and *Case 8* (Table 3). The patient reported by Sachdeva et al [29] as having dup(3p) syndrome is more likely to have duplication of the long arm of chromosome 3, as suggested by the published karyotype. Therefore, this case has been omitted from the following discussion of the dup(3p) syndrome.

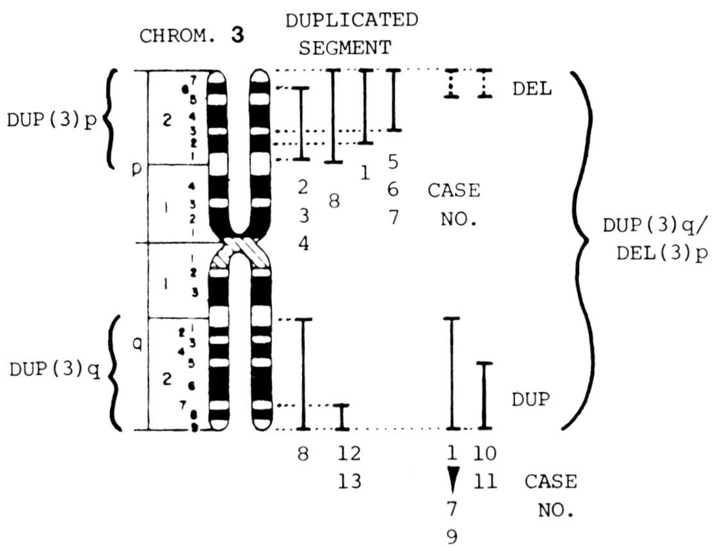

Fig 8. Standard ideogram [3] of chromosome 3 illustrating the duplicated regions in 8 cases with dup(3p) syndrome and in 3 cases with dup(3q). For 10 cases with dup(3q)/del(3p) due to der inv(3) the presumed deleted as well as the duplicated regions are indicated.

Cytogenetic and clinical data on the above case and on seven other well-documented patients are listed in Table 3. *Case 3* had ambiguous genitalia at birth and was reared as a female. The facies characteristic of this syndrome consists of a prominent but narrow forehead with peculiar circumscribed temporal indentions (Fig. 6), square-shaped face due to prominent zygomatic bones, full cheeks and short receding mandible, hypertelorism, short and prominent philtrum, and downturned angles of the mouth. The neck is short, with redundant skin surrounding it in the newborn period. The external ears are normally set. Various abnormalities of the pinna have been described that are apparently not consistent. Congenital heart defects are common and have led to cyanosis in a few cases and congestive heart failure at birth in one. The ductus arteriosus failed to close spontaneously and required ligation in several cases. Incomplete development of male genitalia was present in 5 of the 6 males and consisted of small penis, hypospadias, and undescended or dysgenetic testes. Dermatoglyphic analysis consistently revealed an increased number of digital whorls as compared to first-degree relatives.

Additional, infrequently reported abnormalities include cleft lip and palate and cleft soft palate, esophageal atresia, atresia of the transverse colon, common mesentery, megacolon, clubfeet, supernumerary nipple, and wide space or deep plantar crease between 1st and 2nd toes. *Case 8* had bilateral syndactyl between the 3rd and 4th fingers and camptodactyly [28]. These findings were attributed to the associated deletion of the 22q terminal segment. Ocular abnormalities are rare: Eccentric pupils and exotropia were seen once each. Radiographic examination of the skeleton was performed on five patients. The only abnormality found was coxa valga in one case. Intravenous pyelograms were done on 4 patients. A dilated ureter and a unilateral duplicated ureter were found once each.

This syndrome should be clinically diagnosable on the basis of the characteristic short square facies, congenital heart defect, hypoplastic male genitalia, and increased number of digital whorls. Aarskog [30] in 1969 reported a male with duplication (3?)/deletion (18q) resulting from a reciprocal translocation. He had the typical facies of the dup(3p) syndrome, small penis with glandular hypospadias and chordee, hypoplastic scrotum, and cryptorchidism. Growth was along the 25th percentile and psychomotor development was markedly retarded. He died at 16 months of bronchopneumonia. At autopsy, atrial septal defect and persistent left superior vena cava were found. Partially atretic ear canals may represent a manifestation of the associated 18q− deletion. An older brother, who was not karyotyped, had died at 7 weeks of age of complex cardiopulmonary malformations. Recently a female sib was born with a similar clinical picture. Chromosome analysis with G-banding confirmed the involvement of the *short* arm of chromosome 3: dup(3p)/del(18q) (D. Aarskog, personal communication, 1977).

E. Yunis et al [31] have recently reported a female patient with a de novo 3p+ chromosome which they interpreted as "trisomy 1q32→1qter" on the basis

TABLE 3. Dup(3p) Syndrome

Case	*1*	*2*	*3*	*4*	*5*	*6*	*7*	*8*
Reference	This report		[25]		[26]	[27]		[28]
		(III-7)	(III-8)	(III-9)		(Case 1)	(Case 2)	
Banding method	G, Q	R	R, Q	R	G	G, Q	G, Q	G
Parental translocation	t(3;13)(p22;p12)pat		ins(7;3)(q31;p21p26)mat		t(3;4)(p23;p16)mat	t(3;15)(p23;p12)mat		t(3;22)(p21.2→q13.3)pat
Duplicated segment	3p22→3pter		3p21→3p26		3p23→3pter	3p23→3pter		3p21.2→3p27
Deleted segment	13p12→13pter		–		4p16→4pter	15p12→15pter		22q13.3→22q13.5
Sex	F	M	(F(XY)	M	M	F	M	M
Gestation (wk)	40	term	term	38	...	43	term	term
Birthweight (gm)	3,200	3,700	3,830	2,190	2,360	3,300	4,080	3,345
Birth length (cm)	50	47	49	47	47	...	...	...
Age at report	18 mo	4 mo†	4 yr	3 d†	4 mo	9 yr	6 yr	6 wk
Growth	50th %	< 3rd %	50th %	...	3rd %	50th %	...	10th %
Psychomotor development	IQ ~ 50	...	IQ 47	...	...	IQ 37	IQ 27	...
Craniofacial								
Brachycephaly	+	+	–	+	+	–	–	–
Frontal bossing	+	+	+	+	+	+	+	+
Temporal indentation	+	–	+	+	+	–	–	+
Square face	+	+	+	+	+	+	+	+
Prominent cheeks	+	+	+	+	+	+	+	+

Hypertelorism	+	+	+	+	+	+	+	–
Epicanthal folds	–	+	+	+	+	–	+	–
Short philtrum	+	+	+	...	+	+	+	–
Prominent philtrum/ upper lip	+	+	+	...	+	+	+	+
Downturned angles of mouth	+	+	+	+	+	+	+	–
Retrognathia	+	+	–	+	+	+	+	+
Short neck	+	+	+	+	+	–	+	+
Cardiac defect	PDA, ASD	PDA, ASD VSD	–	PDA, VSD	VSD, PS	–	PS(?ASD)	PDA
Genitalia								
Micropenis	...	+	+	+	–	...	–	+
Hypospadias	...	–	–	–	–	...	+	–
Dermatoglyphics								
Excess whorls	+	+	+	+	+	–	+	+
Figure 6	a, b, j	f	g	–	e, i	m, n	k, l	c, d

Symbols: + present; – not present; ... not applicable or not known; Abbreviations: PDA: patent ductus arteriosus; ASD: atrial septal defect; VSD: ventricular septal defect; PS: pulmonary stenosis.

†Deceased.

of G-banding patterns. The child's clinical description and photographs are in perfect agreement with the dup(3p) syndrome as outlined above. The abnormal chromosome may be interpreted as a "mirror" or tandem duplication of the distal 3p.

On the basis of the 10 known cases one can conclude that the prognosis for growth and survival is good if the child survives infancy and is treated appropriately for congenital heart defects. The dup(3p) patients can be expected to function in the trainably mentally retarded range and should benefit from special education. Chromosome analysis with banding methods is required to confirm the diagnosis and to identify family members carrying a balanced chromosomal rearrangement.

DUPLICATION (3q) SYNDROME

Duplication of the distal portion of the long arm of chromosome 3 has been found in unbalanced offspring of balanced translocation carriers in several families, but most of the known cases are unbalanced offspring of heterozygotes for a pericentric inversion of chromosome 3 (Fig. 8). In case of a pericentric inversion a loop is formed ("inversion loop") for a synapsis of homologous regions during meiosis. An uneven number of crossover events within the inversion loop results in unbalanced gametes. In the Canadian kindred reported by Allderdice et al [32] the inversion was due to breaks in 3p25 and 3q21. Offspring with del(3q)/dup(3p) have not been identified. However, numerous offspring with dup(3q)/del(3p) have been produced. Those with clinical documentation are included as *Cases 2–7* in Figure 8 and Table 4. Additional families with inversions involving the same or similar breakpoints have been reported from other parts of the world (*Case 1* [33] and *Case 9* [34]).

A pericentric inversion of chromosome 3 looks morphologically like a t(2p−;3?+) translocation. Three patients reported before banding [13, 14] as dup(2p) offspring of such "translocation" carriers most likely had the dup(3q)/del(3p) syndrome as suggested by clinical criteria. In one family this was confirmed by recent banding studies (Summitt, personal communication, 1977). These patients are included in Table 4 but not in Figure 8; two of them are shown in Figure 9 (c, h, d, and i).

Dup(3q) due to parental translocation rather than inversion has been documented in 3 cases (*Case 8* [34], *Case 12* and *13* [35]). In addition, 4 unbalanced members of a t(3?−;Gp+) kindred reported by Falek et al [36] as "Familial de Lange syndrome with chromosome abnormalities" appear to have the dup(3q) syndrome and have been included in Table 4 but not in Figure 8. Figure 8 compares the chromosome findings in only those cases that are well-documented clinically as well as with banding methods. *Case 8* [34] and *Cases 12* and *13* [35] are due to translocations, with the associated terminal deletions involving

TABLE 4. Dup(3q) Syndrome: Most Consistent Clinical Features

	Dup(3q)/del(3p) der inv(3)		Dup(3q) der,t(3;?)	
Cytologic technique	B	S	B	S
No. of cases (total = 20)	10	3	3	4
Sex M/F	6/4	2/1	1/2	1/3
Average birthweight (gm)	2,931	2,838	3,443	3,625
Average birth length (cm)	46.8	48.2	48.5	49.3
Postnatal growth	3%	3%	normal[a]	3%
Psychomotor (IQ estimates)	20	20	50–60[a]	26
Hirsutism	2	1	–	4
Low hairline	2	3	–	4
Bushy eyebrows/long lashes	2	3	–	4
Characteristic facies	9	3	3	4
Cleft palate (bifid uvula)	1(2)	2	1	(2)
Short (webbed) neck	(5)	1	(1)	2
Pectus excavatum	3	1	1	–
Kyphosis/scoliosis	3	–	–	1
Short limb	4	–	–	–
Congenital heart defect	5	1	–	1
Duplicated ureter	2	1	–	1
Hydronephrosis/hydroureter	2	–	1	–
Cryptorchidism	5	2	–	–

B – banding method; S – standard stain.

[a]Cases 12 and 13, the only survivors in this group, had duplication of only a small region (3q27→3qter), which accounts for milder clinical manifestations.

2pter (*Case 8*) and 2qter (*Cases 12* and *13*). As might be expected from the difference in the extent of the duplication, *Case 8* had severe malformations and died at 32 days of age, while *Cases 12* and *13* are mild-to-moderately retarded teenage sibs of normal height. *Cases 1–11* are offspring of inv(3) carriers with the dup(3q)/del(3p) syndrome. Most individuals with this chromosome constitution have died early in infancy and little is known about them. From the report by Allderdice et al [32] only those cases that were studied in some detail have been included (as *Cases 2–7*). *Cases 9–11* are from 2 unrelated families reported by Fineman et al [35].

It appears from comparison of the clinical findings in all the cases that the majority of abnormalities in the dup(3q)/del(3p) patients are due to the dup(3q) and not to the del(3p) chromosome imbalance. Characteristic facial features, as illustrated in Figure 9, are low hairline, bushy eyebrows, long eyelashes, occasional hypertelorism and epicanthal folds, upslanting palpebral fissures, short nose with upturned nostrils, prominent maxilla, receding mandible,

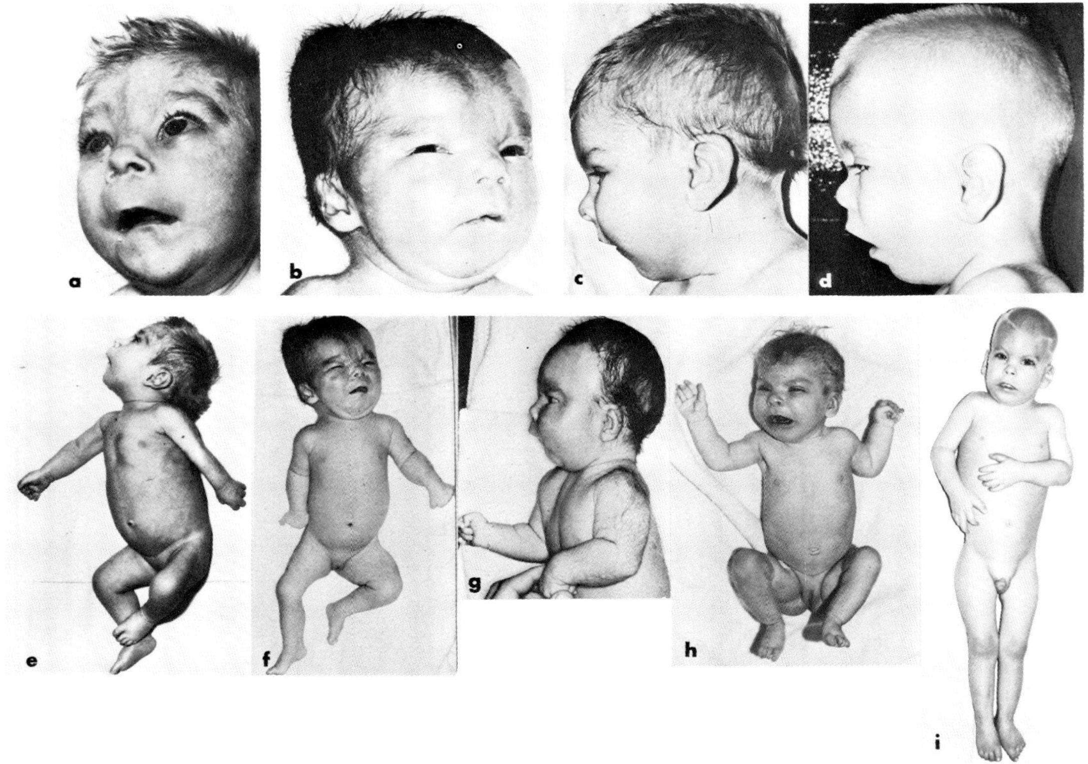

Fig 9. Dup(3q) syndrome. a, e) *Case 1* with der inv(3), from Boué et al [33]; b, f, g) *Case 8* with der, t(2;3), from Chiyo et al [34]; c, h) and d, i) sister and brother with der, inv(3), from Summitt [14].

long philtrum, thin upper lip, downturned angles of the mouth, and short neck. Other frequent findings are irregular shortening of the limbs, pectus excavatum, clinodactyly, camptodactyly, and an undescended testicle.

In an attempt to determine the relative frequencies of manifestations we have listed the most consistent clinical abnormalities separately for 4 different groups of patients, dup(3q)/del(3p) derived from a pericentric inv(3) parent, confirmed (B) or not confirmed (S) by chromosome banding studies, and der(3),t(3;?) derived from a balanced reciprocal translocation carrier, confirmed (B) or not confirmed (S) by banding (Table 4). Most of the abnormalities have occurred in all 4 groups, although the cases are quite heterogeneous with respect to size of the duplication and kind of associated deletion. In addition to those listed, the following abnormalities have been observed occasionally or at autopsy: congenital glaucoma, strabismus, distended scalp veins, dilated cerebral ventricles, hypoplastic cerebellum, omphalocele, dilated or malrotated colon, Hirschsprung disease, generalized seizures, double vagina and bicornuate uterus in a female, rudimentary uterine horns in a male, horseshoe kidney, cystic kidney and small glomerular cysts, small penis and shawl scrotum, dislocated elbows, hypoplastic ribs, and postaxial polydactyly.

Certain features of the patients with duplication of at least 3q25→3qter overlap with the Cornelia de Lange syndrome (CDL) [37] (Table 5). It is easy to see that by comparing isolated features one could arrive at that diagnosis. One major distinction between the two syndromes is the lack of significant intrauterine growth retardation in the dup(3q) patients. The facies are also quite different if one considers the overall "Gestalt" rather than the list of isolated findings. Although arms and legs in the dup(3q) cases have occasionally been reported as short, there were no severe malformations of the upper limbs. The dup(3q) patients rarely had 2/3 syndactyly of the toes, and for some of them this was also present in otherwise unaffected family members. However, since dup(3q) individuals have been reported as representing familial cases of CDL [36], two points should be emphasized: 1) Patients with a questionable diagnosis of (nonclassic) CDL and/or with a family history of CDL should have their chromosomes examined; 2) the recurrence risk of CDL in sibs that is usually quoted as 1–3% may be partially based on the admixture of dup(3q) patients. Therefore, CDL with normal chromosomes might have a lower or no significant recurrence risk.

DISCUSSION

The pathogenetic mechanisms are obscure by which chromosomal duplications or deletions produce phenotypic abnormalities. The considerable overlap in symptomatology, ie the observation that certain anomalies are found frequently in disorders involving different chromosomes, suggests that the

Table 5. Comparison of Abnormalities in Cornelia de Lange Syndrome With Dup(3q) Syndrome (q21 to 25→qter)

	CDL	Dup(3q)
Mental retardation	++/+++	+++
Microbrachycephaly	+	+/–
Hirsutism	+	+
Bushy eyebrows	+	+
Synophris	+	–
Hypertelorism	–	+
Long, curly eyelashes	+	+
Small nose with anteverted nares	+	+
Thin upper lip	+	+
Downturned angles of mouth	+	+
High arched palate	+	+
Micrognathia	+	+
Short neck	+	+
Micromelia	+	+
Flexion contracture of elbows	+	–
Proximally implanted thumbs	+	+
Clinodactyly 5th finger	+	+
Syndactyly 2/3 toes	+	+
Cryptorchidism	+	+

Less frequently observed in both syndromes: strabismus, low-set ears, cleft palate, congenital heart defects, malrotation of gut.

specific genes located in the duplicated or deleted chromosome regions may not play the major role. One exception would be the hemizygous expression of a specific recessive gene in case of deletion of the normal allele.

Most manifestations that are observed in patients with chromosomal imbalance could be explained on the basis of two mechanisms: incoordinate ("dysharmonic") growth during embryonic development, and secondary effects of brain dysfunction. The brain abnormality, in turn, is most likely due to a defect in brain cell growth and migration. Development can be interpreted as the interaction of three processes: cell division, cell migration, and cell differentiation. A chromosomal imbalance might interfere with each one of them. In regard to differentiation, histologic abnormalities in organ structure have not often been reported in chromosomal imbalance states, with the exception of defects of testes and ovaries which are often immature and abnormal. As to cell function, abnormal levels of metabolic intermediates, hormones or enzymes have rarely been detected in these patients. The Down syndrome is a notable exception because activities of various intracellular enzymes are elevated, at least in erythrocytes. Although a gene dose effect has recently been demonstrated for some gene products, and has been exploited for regional mapping of the respective gene loci, these observations have not thrown light on the pathogenesis of malformations. For example, it is not clear how the abnormal phenotypes associated with the deletion or the duplication of the distal short arm of chromosome 8 could be related to the decreased, or respectively increased levels of the enzyme glutathione reductase in the cells of the patients [38, 39].

In contrast, most of the malformations observed could be explained on the basis of disturbed regulation of growth. For instance, if the palatal shelves do not grow sufficiently to meet by the time they are supposed to fuse, cleft palate will result. Similarly, congenital heart defects may be produced by lack of coordinate growth of certain structures within the developing heart at a defined time. Many examples come to mind: maxillary or midface hypoplasia, renal cysts, malrotation of the gut, micropenis, short digits, etc.

Chromosomal imbalance does not necessarily cause lack or delay of growth. Several chromosomal syndromes have been described that are associated with normal intrauterine and even postnatal growth, eg trisomy 8 mosaicism, dup(3p) and dup(20p) [40]. If there is delay of growth, it is certainly not due to the extra chromosome prolonging the cell cycle. The extra Y does not inhibit growth of XYY individuals. In fibroblast cultures from individuals with trisomy 8 mosaicism, we have found that the percentage of trisomic cells does not decrease, but may even increase, over extended periods of cell growth in culture. In mosaic trisomy for chromosome 9 the situation is quite different. Here the trisomic cells appear to be at a disadvantage under standard culture conditions and disappear rapidly, owing either to overgrowth by cells with a normal karyotype or to a decreased life span of the trisomic cells. These in vitro observations are consistent with clinical observations that trisomy 9 mosaic individuals are more severely affected than those with mosaic trisomy 8, not only with respect to body growth but also to brain development and function.

It appears that quantitative disturbances of genetic material exert their effect by interfering with programmed coordinate growth of tissues during morphogenesis and produce malformations that are by themselves not specific for the particular chromosome regions involved. However, the pattern of malformations may be quite characteristic. That part of the body where dysharmonic growth produces the most obvious changes is the face. Even subtle differences in the growth of facial bones and soft tissue will produce a strikingly different appearance. Therefore, the facies, the overall "Gestalt" of the head and face, may be the single most useful characteristic in the clinical recognition of the many new (and old) chromosomal syndromes. Fortunately, the diagnosis does not have to rely on photographic memory and intuition alone; confirmation (or correction) by the cytogenetic laboratory is required.

ACKNOWLEDGMENTS

Drs. R.M. Peterson, D. Chadwick, and S. Pappelbaum contributed to the evaluation of dup(3p) *Case 1*. Drs. D. Aarskog, K. Bender, S. Cassidy, H.-A. Chiyo, R. Fineman, E. Magenis, G. Prescott, B. Say, L. Shapiro, R. Summitt, W. Wasdahl, and J. Yunis have helped greatly by supplying unpublished information and photographs.

N. Oliver and M.G. Brown provided technical assitance and the San Diego Regional Center for the Developmentally Disabled provided financial assistance.

REFERENCES

1. Yunis JJ (ed): "New Chromosomal Syndromes." New York: Academic Press, 1977.
2. "Webster's New World Dictionary of the American Language." 2nd college Ed. New York and Cleveland: World, 1970.
3. Bergsma D (ed): "Paris Conference (1971): Standardization in Human Cytogenetics." White Plains: The National Foundation–March of Dimes, BD:OAS VIII(7), 1972.
4. Bergsma D (ed): "Paris Conference (1971), Supplement (1975): Standardization in Human Cytogenetics." White Plains: The National Foundation–March of Dimes, BD:OAS XI(9), 1975.
5. Lewandowski RC, Yunis JJ: New chromosomal syndromes. Am J Dis Child 129: 515, 1975.
6. Stoll C, Messer J, Vors J: Translocation t(2;14) equilibrée chez une mère et trisomie partielle d'une partie du bras court d'un chromsome No. 2 chez deux de ses enfants. Ann Genet (Paris) 17:193, 1974.
7. Magenis RE, Koler RD, Lovrien E, Bigley RH, DuVal MC, Overton KM: Gene dosage: Evidence for assignment of erythrocyte acid phosphatase locus to chromosome 2. Proc Natl Acad Sci USA 72:4526, 1975.
8. Francke U, Jones KL: The 2p partial trisomy syndrome. Am J Dis Child 130:1244,1976.
9. Cassidy SB, Heller RM, Chazen EM, Engel E: The chromosome 2 distal short arm trisomy syndrome. J Pediatr 91:934, 1977.
10. Yunis JJ: Duplication 2p syndrome. Am J Dis Child. (In press)
11. Bender K, Reinwein H, Gorman LZ, Wolf U: Familiäre 2/C-Translokation: 46,XY, t(2p–;Cp+) und 46,XXCp+. Humangenetik 8:94, 1969.
12. Aarskog D: A familial syndrome of short stature associated with facial dysplasia and genital anomalies. J Pediatr 77:856, 1970.
13. Lee CSN, Bowen P, Rosenblum H, Linsao L: Familial chromosome 2, 3 translocation ascertained through an infant with multiple malformations. N Engl J Med 271:12, 1964.
14. Summitt RL: Familial 2/3 translocation. Am J Hum Genet 18:172, 1966.
15. Wasdahl WA, Larson LM, Coleman M: New familial translocation involving chromosome 10 – preliminary studies. Mamm Chrom Newsl 17:7, 1976.
16. Forabosco A, Dutrillaux B, Toni G, Tamborino G, Cavazzuti G: Translocation equilibrée t(2;13)(q32;q33) familiale et trisomie 2q partielle. Ann Genet (Paris) 16:255, 1973.
17. Warren RJ, Panizales EG, Cantwell RJ: Inherited paritial trisomy 2: 46,XX,1p+;t(1;2)(p36;q31). In Bergsma D (ed): "New Chromosomal and Malformation Syndromes" Miami: Symposia Specialists for The National Foundation–March of Dimes, BD:OAS XI(5):177, 1975.
18. Zabel B, Hansen S, Hartmann W: Partial trisomy 2q and familial translocation t(2;12)(q31;q24). Hum Genet 32:101, 1976.

19. Couturier J, Aurias A, Prieur M, Barois A: Partial trisomy for the long arm of chromosome 2 due to malsegregation of a maternal insertion: ins(6;2) (p22;q24;q34). Ann Genet (Paris) 20:52, 1977.
20. Ricci N, Dallapiccola B, Cotti G: Translocation 2-D familiale. Ann Genet (Paris) 11:111, 1968.
21. Bijlsma JB, de France H, Bleeker-Wagemakers, Wijffels JCHM: Duplication deficiency syndrome in familial translocation (2q−;5p+). Humangenetik 12:110, 1971.
22. Rosenthal IM, Beligere N, Thompson F, Pruzansky S, Reinglass H: Trisomy of the distal portion of the long arm of chromosome 2, a new familial syndrome associated with mental retardation and a characteristic facies. Am J Hum Genet 26:72A, 1974.
23. Rosenthal IM, Kaye CI, Beligere N, Pruzansky S: Trisomy of distal portion of long arm of chromosome 2, an abnormality found in 2 unrelated families associated with mental retardation. In Armendares S, Lisker R (eds): "V International Congress on Human Genetics." Amsterdam: Excerpta Medica, 1976, p 149.
24. Shapiro LR, Warburton D: Interstitial translocation in man. Lancet 2:712, 1972.
25. Rethoré MO, Lejeune J, Carpentier S, Prieur M, Dutrillaux B, Seringe Ph, Rossier A, Job J-C: Trisomie pour la partie distale du bras court du chromosome 3 chez trois germains. Premier exemple d'insertion chromosomique ins(7;3) (q31;p21;p26). Ann Genet (Paris) 15:159, 1972.
26. Ballesta F, Vehi L: Trisomie partielle pour la partie distale du bras court du chromosome 3. Ann Genet (Paris) 17:287, 1974.
27. Say B, Barber N, Bobrow M, Jones K, Coldwell JG: Familial translocation (3p15p) with partial trisomy for the upper arm of chromosome 3 in two sibs. J Pediatr 88:447, 1976.
28. Yunis JJ: Trisomy for the distal end of the short arm of chromosome 3. Am J Dis Child (In press)
29. Sachdeva D, Smith GF, Justice P: An unusual chromosomal segregation in a family with a translocation between chromosomes 3 and 12. J Med Genet 11:303, 1974.
30. Aarskog D: A familial 3/18 reciprocal translocation resulting in chromosome duplication-deficiency (3?+−18q−). Acta Paediatr Scand 58:397, 1969.
31. Yunis E, Egel H, Zuniga R, Ramirez E, Torres de Caballero OM, Leibovici M: "de novo" trisomy 1q32→1qter and monosomy 3p25→3pter. Hum Genet 36:113, 1977.
32. Allderdice PW, Browne N, Murphy DP: Chromosome 3 duplication q21→qter deletion p25→pter syndrome in children of carriers of a pericentric inversion (3) (p25q21). Am J Hum Genet 27:699, 1975.
33. Boué J, Hirschhorn K, Lucas M, Gautier M, Moszer M, Bach Ch: Aneusomies de recombinaison. Conséquences d'une inversion péricentrique d'un chromosome 3 paternel. Ann Pédiatr 21:567, 1974.
34. Chiyo H-A, Kuroki Y, Matsui I, Niitsu N, Nakagome Y: A case of partial trisomy 3q. J Med Genet 13:525, 1976.
35. Fineman RM, Hecht F, Ablow RC, Howard RO, Breg WR: Chromosome 3 duplication-deficiency (3q+/3p−) syndrome. Pediatrics (In press)
36. Falek A, Schmidt R, Jervis GA: Familial de Lange syndrome with chromosome abnormalities. Pediatrics 37:92, 1966.
37. Motl ML, Opitz JM: Studies on malformation syndromes XXV A. Phenotypic and genetic studies of the Brachman-de Lange syndrome. Hum Hered 21:1, 1971.
38. George DL, Francke U: Gene dose effect: Regional mapping of human glutathione reductase on chromosome 8. Cytogenet Cell Genet 17:282, 1976.
39. De la Chapelle A, Icen A, Aula P, Leisti J, Turleau C, de Grouchy J: Mapping of the gene for glutathione reductase on chromosome 8. Ann Genet (Paris) 19:253, 1976.
40. Francke U: Abnormalities of chromosomes 11 and 20. In Yunis J (ed): "New Chromosomal Syndromes." New York: Academic Press, 1977, pp 245–272.

X-Autosome Translocations: A Review*

Robert L. Summitt, MD, Robert E. Tipton, MD, R.S. Wilroy, Jr., MD, Paula R. Martens, BS, and Jeffrey P. Phelan, LCDR, USNR

The essential property of the Lyon hypothesis [1–4] is the resulting presence in all mammalian somatic cells of one, and only one, genetically active X chromosome. Ample evidence is available in support of this, in normal mammals [3, 4] and in X-polysomy states [5, 6]. Further support is provided by the selective survival, in an XX organism with an intrinsic structural abnormality of one X chromosome, of only those cells in which the structurally altered X is inactivated (or by the *preferential* inactivation of the structurally altered X) [7–16]. However, the discovery in mammals of balanced and unbalanced X-autosome translocations has introduced problems to the single-active-X hypothesis. This paper will review the knowledge gained from the investigation of these most interesting chromosome rearrangements, and while enigmas remain, the major portion of that knowledge supports or is in keeping with the single-active-X hypothesis. This paper will, in addition, review the information gained from studies of the phenotypes of humans bearing balanced and unbalanced X-autosome translocations and their relationships to the specific chromosome rearrangements involved.

The evidence to be presented supports the existence of a single site or center in the X chromosome that governs and initiates X-inactivation, which then spreads in both directions along the length of the X chromosome. It further supports the proposal of Russell [4], Therman and co-workers [15, 17], Cohen and associates [18], and Summitt and associates [19] that this center is located in the proximal portion of the long arm of the X, but not in the *extreme* proximal portion. This paper will not deal with the current knowledge relative to mapping

*This study was supported in part by Special Project No. 900, Division of Health Services, MCHS, HSMHA, DHEW, by a grant from The National Foundation–March of Dimes, and by the U.S. Navy Bureau of Medicine and Surgery, Clinical Investigation Program No. CR-08-6-12.

Birth Defects: Original Article Series, Volume XIV, Number 6C, pages 219–247

of X-borne or autosomal gene loci, gained through studies utilizing X-autosome translocations.

A total of 40 bona fide or claimed X-autosome translocations have been reported [18–51]. Some have occurred as balanced and unbalanced translocations in different members of the same family. These different forms of the translocations are counted separately in reaching the total of 40. A consideration of X-autosome translocations should divide them into those that are balanced and those that are unbalanced. Among patients with balanced translocations are a group with normal phenotypes (Table 1), a group with gonadal dysfunction but with normal somatic phenotypes (Table 2) and a group with abnormal somatic phenotypes (Table 3). Among reported patients with unbalanced translocations is a group with features of the Turner syndrome (Table 4) and a group with abnormal somatic phenotypes (Table 5). These groups will be considered separately.

BALANCED X-AUTOSOME TRANSLOCATIONS

Seven balanced X-autosome translocations have been described in the presence of normal phenotypes (Table 1) [19–26]. Some of the carriers of these balanced translocations have produced offspring with unbalanced states of the translocations. In each of the balanced translocation carriers the *normal X* is the late replicating X. Examples include the cases of Summitt [19] and of Tipton [25]. In the former (Fig. 1), the karyotype includes a balanced translocation in which virtually the entire long arm of X (estimated breakpoint Xq11) is translocated to the short arm of 21 (estimated breakpoint 21p11), and the reciprocal made up of the short-arm centromere of X and presumably the short-arm satellite portion of 21 (although in no cell were satellites demonstrable). While no direct evidence exists, reciprocity is presumed. Figure 2 shows the pattern of late replication in this translocation. In every informative cell the normal X was the late replicator. Figure 3 is the karyotype of the patient reported by Tipton [25], showing a balanced X-8 translocation in which the long arm of an X chromosome (estimated breakpoint q26-q28) is elongated by the translocation to it of virtually the entire long arm of chromosome 8 (estimated breakpoint q12), along with the small reciprocal made up of the short-arm centromere of 8 and the distal portion of Xq. Figure 4 demonstrates the late replication of the normal X.

If X-inactivation is initiated, according to the hypothesis mentioned earlier, in a site in the proximal long arm of X, inactivation initiated in a site in the proximal long arm of the translocated X (Xt) of Summitt's case, for example, could not extend to the short arm, because the short arm is separated from the inactivation site. If, then, the Xt were the late replicator (and thus inactivated), the net result would be the presence in such cells of 1 normal active X and an

TABLE 1. Balanced X-Autosome Translocations With Normal Phenotypes Including Gonadal Function

Source	Karyotype	X-Breakpoint	Late Replicating X	Phenotype
Allderdice [20]	46,X,t(Xq−;14q+) *banded*	Xq13	Normal X	Normal
Buckton [22]	46,X,t(Xp−;14q+) *not banded*	Xp11	Normal X	Normal
Summitt [19]	46,X,rcp(X;21) (q11;p11) *banded*	Xq11	Normal X	Normal
de la Chapelle [23]	46,X,rcp(X;3) (q28;q13) *banded*	Xq28	Normal X	Few minor congenital anomalies
Davis [24]	46,X,t(Xp11;16q24) *banded*	Xp11		Normal
Tipton [25]	46,X,rcp(X;8) (q26;q12) *banded*	Xq26-28	Normal X	Normal
Leisti [26]	46,X,rcp(X;9) (q11;q32) *banded*	Xq11	Normal X	Normal

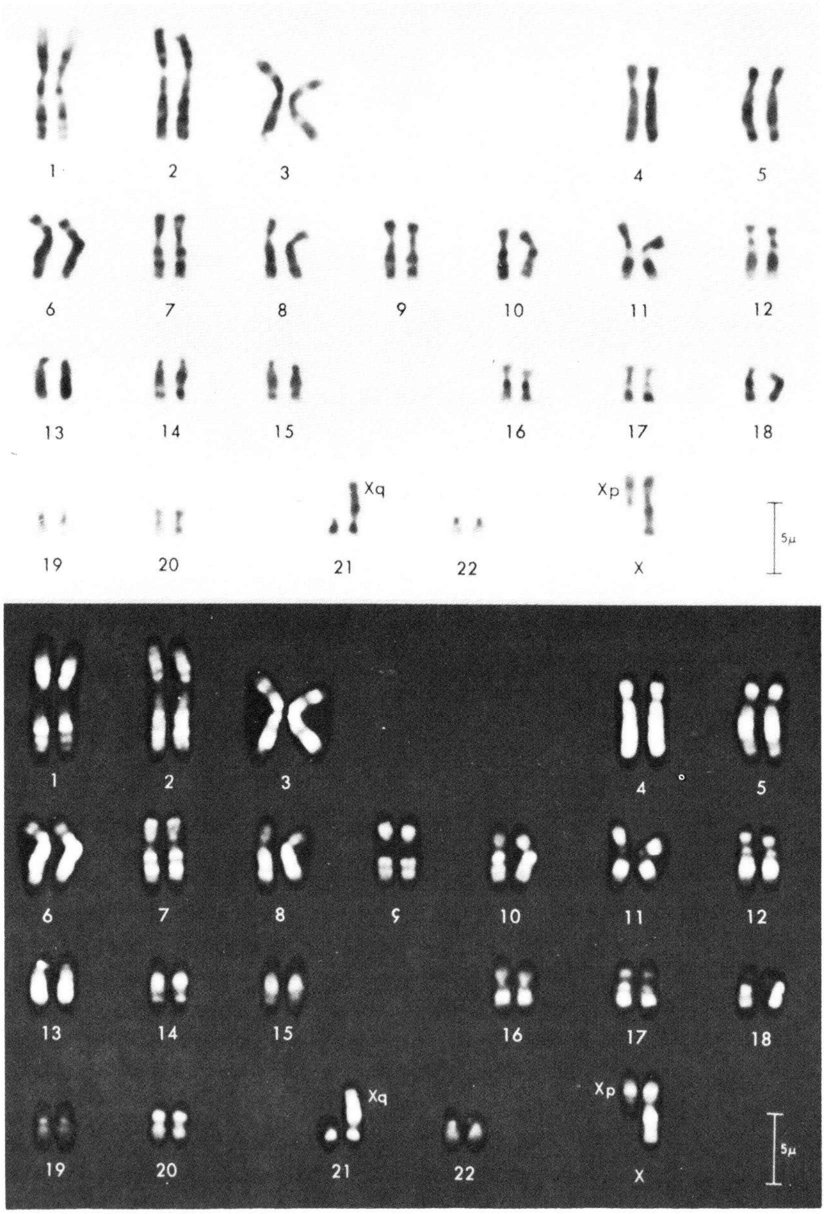

Fig. 1. Karyotype 46,X,rcp(X;21) (q11;p11), showing major portion of Xq translocated to 21p. (Figs. 1, 2, 6–8 reprinted from Summitt et al: X-autosome translocation in normal mother and effectively 21-monosomic daughter. J Pediatr 84:539–546, 1974, with permission.)

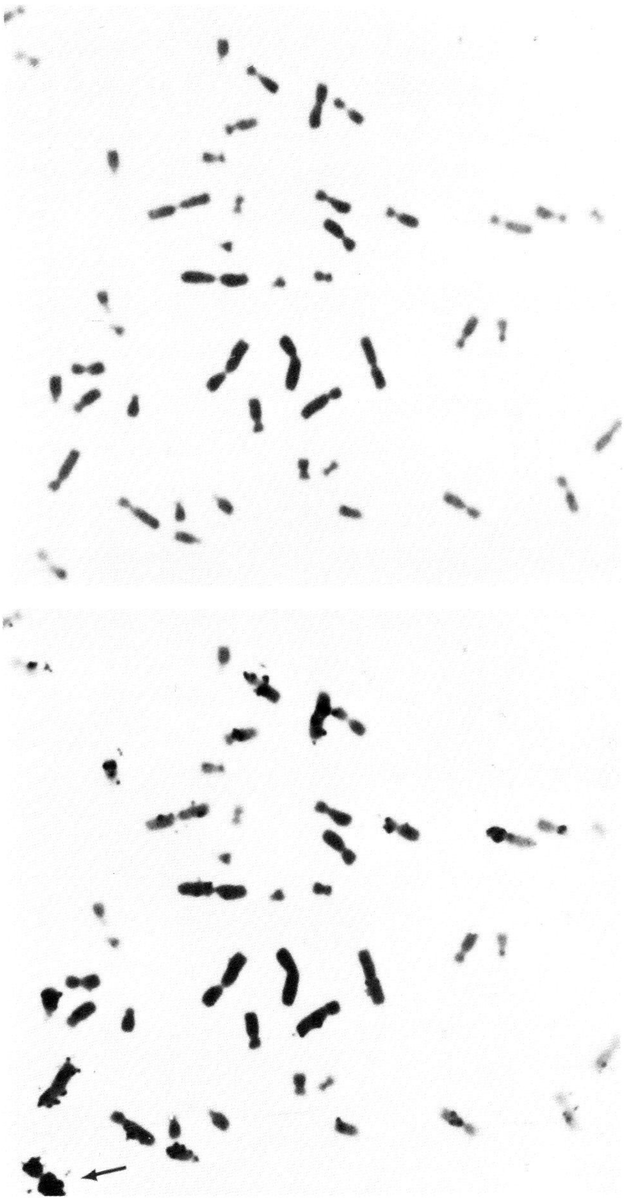

Fig. 2. Autoradiography of cell from patient whose karyotype is shown in Fig. 1. Note that the late replicating X is the normal X.

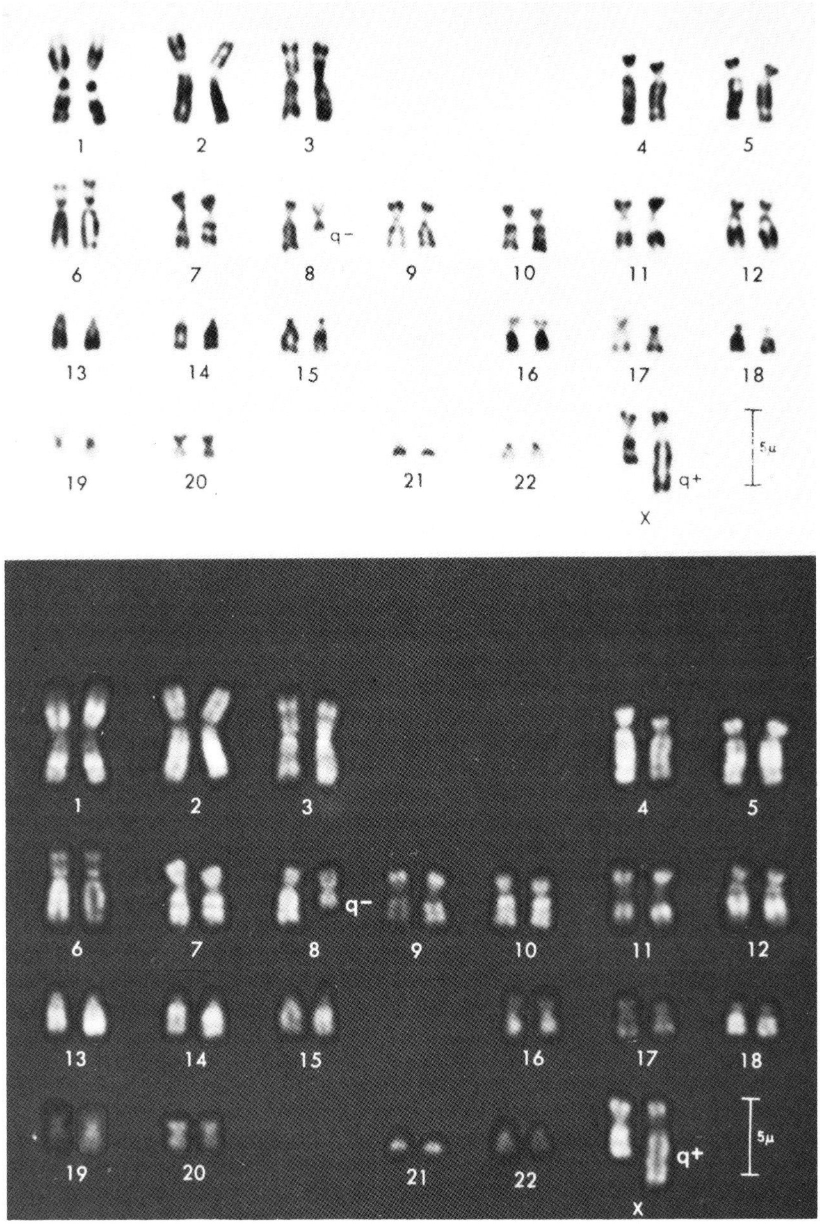

Fig.3. Karyotype 46,X,rcp (X;8) (q26;q12), showing major portion of 8q translocated to Xq. (Figs. 3, 4, 9–11 reprinted from Tipton RE: "Cytogenetic Studies in Subjects With Mental Retardation and Congenital Anomalies Using Banding Techniques." Master of Science Thesis, University of Tennessee, 1974, with permission.)

additional active Xp. The hypothesis, in this context, suggests that the resulting genetic imbalance exerts a negative selective force such that cells containing more than one active X do not survive, and that in all surviving cells the normal X is inactivated. In such cells, even though the active X is separated into 2 segments, its function is not impaired. An exception to this which depends on the location of the X-breakpoint will be presented later. While no supportive evidence exists, added to the negative selective force described above might be extension of inactivation from the center in proximal Xp to the attached long arm of 21.

The generalization possible on the basis of examples such as this is that if a single inactivation center exists in the long arm of X, separation of a portion of that X by translocation to an autosome would interrupt the spread of inactivation from the initiating center. The result, if the Xt chromosome is inactivated, would be the presence in such cells of more than 1 active X. Such a genetically imbalanced cell would be the object of negative selection. The alternative to the hypothesis of selective survival of cells in which the normal X is inactivated is the *preferential inactivation*, via some unknown mechanism, of the normal X. These are the same alternatives which apply to the inactivation of the structurally altered X chromosome in bearers of an intrinsic structural X abnormality such as a ring X, an isochromosome Xq, or a deleted X.

Thirteen balanced X-autosome translocations have been reported (Table 2) in the presence of normal *somatic* phenotypes but with evidence of hypogonadism [27–39] and, in some, anatomically demonstrated gonadal dysgenesis [27–31, 34–36, 38]. As is the case in phenotypically normal carriers of balanced X-autosome translocations, the normal X is the late replicating X in all cases thus studied. A possible exception is the case of Mann [29, 30] in which, in addition to the normal X, the X-derived portion of the Xt may be late replicating. Available photographs are difficult to interpret, and at most the labeling over the X-portion of the Xt is quantitatively much less than that over the normal X. Mann's discussion indicates that he believes that only the normal X has the autoradiographic labeling pattern indicative of the type of late replication expected of the inactive X.

The dilemma regarding these 13 cases is why, if in all carriers of a balanced X-autosome translocation the normal X is inactive and a complete X, although divided, remains active, do some have normal gonadal function while others do not. Sarto and associates [28] have suggested the presence of a "critical region" in the long arm of the X chromosome. If the X-breakpoint in an X-autosome translocation occurs within this critical region, then either by virtue of some type of *position effect* or because of effective hemizygosity for a mutant recessive allele producing gonadal dysgenesis, the result is hypogonadism. We have favored the former [35] possibility even though no direct evidence exists for position effect in man. Our inclination toward position effect is lent some sup-

Fig. 4. Autoradiography of cell from patient whose karyotype is shown in Fig. 3. Note that the late replicating X is the normal X.

4

TABLE 2. Balanced X-Autosome Translocations With Gonadal Failure

Source	Karyotype	X-Breakpoint	Late Replicating X	Phenotype
Thorburn [27]	46,X,t(Xq+;Cq−) *not banded*	Xq2	Normal X	Primary amenorrhea, hypogonadism, streak gonads
Sarto [28]	46,X,t(Xq−;12q+) *banded*	Xq23	Normal X	Primary amenorrhea, hypogonadism, underdeveloped ovaries
Mann [29, 30]	46,X,t(Xq−;5q+) *banded*	Xq13	Normal X *and X-portion of Xt*	Primary amenorrhea, streak gonads
Kallio [31]	46,X,t(Xq−;21q+) *banded*	Xq23	Normal X	Primary amenorrhea, streak gonads
Lucas [32]	46,X,Xq−,−15,+t(X;15) (q13;p13) *banded*	Xq13	Normal X	Primary amenorrhea, sexual infantilism
Gerald [33]	46,X,t(Xq−;19q+) *? banded*	Mid Xq	Normal X	Primary amenorrhea at 15, mental retardation
Gilgenkrantz [34]	46,X,t(X;2) (q21;p25) *banded*	Xq21	Normal X	Primary amenorrhea, streak gonads
Phelan [35]	46,X,rcp(X;4) (q26;q21) *banded*	Xq26	Normal X	Secondary amenorrhea, hypogonadism, streak gonads
Sauer [36]	46,X,t(X;7) (q21;p22) *banded*	Xq21		Secondary amenorrhea, hypogonadism, streak gonads
Sujansky [37]	46,X,t(X;15) (q13;p12) *banded*	Xq13	Normal X	Primary amenorrhea
Dutrillaux [38]	46,X,rcp(X;2) (q21;p25) *banded*	Xq21	Normal X	Primary amenorrhea, streak gonads
Dutrillaux [38]	46,X,t(X;8) (q21;p23) *banded*	Xq21	Normal X	Oligomenorrhea, ovarian hypoplasia
Palmer [39]	46,X,t(X;22) (q22;q13) *banded*	Xq22	Normal X	Primary amenorrhea

port from the existence of phenotypically abnormal carriers of apparently balanced reciprocal *autosomal* translocations [52, 53]. Sarto [28] originally proposed minimal limits for the critical region based upon the analysis of cases available at that time. Figure 5 schematically depicts the X chromosomes involved in the 21 reported balanced X-autosome translocations in adults, with the G-banding map on the extreme left. The region in which the X-breakpoint is estimated is shaded in each X. The number beneath each X is the appropriate reference number. The region of Xq extending from point A to point B represents Sarto's original minimal estimate of the critical region. Sarto and associates commented, however, that the critical region *might* encompass virtually the entire Xq. In reviewing subsequently reported cases, we have proposed [35] extension of the minimal limits of the critical region to include the region of Xq between points C and D. Accordingly, each patient carrying a balanced X-autosome translocation in which the X-breakpoint falls between points C and D has gonadal dysfunction; all those in which the X-breakpoint is elsewhere are normal. Furthermore, we would suggest that the available evidence also indicates that, relative to *proximal* extension of the critical region, point C establishes its *maximal* limit. This is supported by normal gonadal function in the case of Allderdice with a q13 X-breakpoint [20, 21], and the cases of Summitt [19] and of Leisti [26] with q11 X-breakpoints. In addition, Xq28 must be outside the maximal distal extension of the critical region, as exemplified by the cases of de la Chapelle [23] and Tipton [25]. The mechanism whereby such a position effect would lead to gonadal dysgenesis remains to be elucidated.

Five balanced X-autosome translocations have been reported in patients with mental retardation-somatic anomaly syndromes (Table 3). The case of Cohen [18] involves X and 9. The normal X replicates late in all cells examined, but in addition the X-derived portion of the Xt replicates late in 32% of cells. The somatic abnormalities might be attributable to inapparent extension of inactivation from the X-portion of the Xt to the adjacent 9-derived segment. Still, if the late replication of the Xt in 32% of cells is indicative of its genetic inactivation, then a portion of cells are effectively *nullisomic* for virtually all of Xq. This remains an extremely bothersome enigma.

The case of Thelen [40] involves X and 18. In this case, X inactivation may be random, a phenomenon which requires explanation. One possibility is that the segment of distal Xp separated from the inactivation center is so small that for it to remain active exerts no negative selective effect. A second possibility is that the portion of distal Xp separated from the inactivation center and which remains active is a portion of the X which is *not normally inactivated* even in the "inactive" X. The patient's phenotype is that the 18q− deletion syndrome, the result of extension of inactivation from the X-derived portion to the 18-derived segment of the inactivated Xt. Certainly, here inactivation has spread to involve the segment of autosome translocated to X.

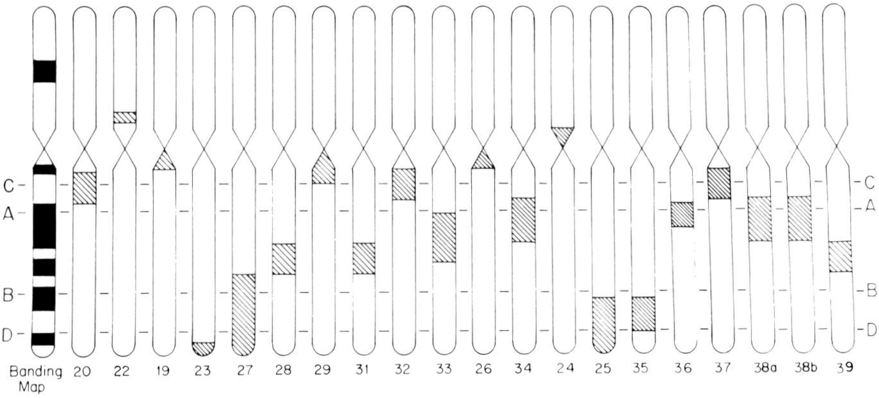

Fig. 5. Diagram of estimated breakpoints (hatched segments) in balanced X-autosome translocations in adult females, with indication of minimal limits of Sarto's "critical region" (A–B) and of extension of these minimal limits (C–D) based upon additional cases. Numbers below chromosomes are reference numbers. Cases with normal gonadal function include 19, 20, 22, 23, 24, 25 and 26; those with hypogonadism 27, 28, 29, 31, 32, 33, 34, 35, 36, 37, 38a, 38b, 39. On extreme left is G-banded X chromosome. (Modified from Sarto et al: X inactivation in man: A woman with t(Xq–;12p+). Am J Hum Genet 25:262–270, 1973; and Phelan et al: Balanced reciprocal X-4 translocation in a female with early secondary amenorrhea. Am J Obstet Gynecol 129:607, 1977.)

The case of Jenkins [41] involves an X-22 translocation, and while the translocation is reported as balanced, the short-arm centromere-extreme proximal long arm of 22 is absent. So the translocation is not completely balanced. Autoradiography was not mentioned in Jenkins' report, so we have no information on the late replicating X. The only reported abnormalities in the patient were mental retardation and neurologic dysfunction. An alternative explanation is that the phenotype and abnormal karyotype are coincidental and not causally related. In the case of Punnett [42], the Wiedemann-Beckwith syndrome may well be coincidental with the X-autosome translocation.

The case of Dutrillaux [38] is an apparently balanced reciprocal X-15 translocation, in which the patient has severe mental retardation and multiple congenital anomalies. The most acceptable explanation relative to this severely altered phenotype in the presence of an apparently balanced X-15 translocation in which the normal X is the late replicator is, again, a form of position effect similar to that mentioned in reference to balanced reciprocal autosomal translocations [52, 53].

The primary enigmas presented in these last 5 cases are the apparent effective Xq-nullisomy in 32% of cells from the case reported by Cohen, and the MR-MCA syndrome in the case of Dutrillaux.

TABLE 3. Balanced X-Autosome Translocations With Abnormal Phenotypes

Source	Karyotype	X-Breakpoint	Late Replicating X	Phenotype
Cohen [18]	46,X,t(Xq−;9p+) *banded*	Xq1	Normal X 100% X-portion of Xt 32%	Developmental delay and MCA
Thelen [40]	46,X,t(Xp+;18q−) *not banded*	Distal Xp	Normal X 21.5% Entire Xt 78.5%	18q deletion syndrome
Jenkins [41]	45,X,−22,t(X;22) (q28;q11) *banded* *may be unbalanced*	Xq28		MR
Punnett [42]	46,X,t(X;1) (q26;q12) *banded*	Xq26		Wiedemann-Beckwith syndrome *coincidental?*
Dutrillaux [38]	46,X,t(X;15) (Xp11;15q11 or Xq11;15p11) *banded*	Xp11 or Xq11	Normal X	MR, MCA, normal sex development

UNBALANCED X-AUTOSOME TRANSLOCATIONS

Among the 16 reported unbalanced X-autosome translocations, 3 [43–45] have produced the Turner syndrome (Table 4). On closer examination, it must be pointed out that the karyotypes of 2 of these 3 patients are mosaic and include a 45,X cell line [44, 45]. In the report by Gouw and associates [43], one X chromosome is morphologically normal while the other is elongated. The report was published prior to the availability of chromosome banding techniques, so the nature of the elongated X chromosome cannot be directly determined. However, such a rearrangement could involve a deletion of a portion of the short arm of the involved X. The elongated X chromosome was consistently the late replicating chromosome, with reported homogeneous labeling along its entire length. The patient was mentally retarded, a feature not ordinarily associated with the Turner syndrome. However, a review of 343 cytogenetically proven cases of the Turner syndrome [54–57] reveals that 18% of affected patients may be mentally retarded. When one notes the extent of the abnormal phenotypes of the other 13 patients with unbalanced X-autosome translocations (to be discussed later), it is interesting that 3 should have the phenotype of the Turner syndrome while the other 13 are much more severely affected with mental retardation-multiple anomaly syndromes. We suggest that the 3 patients with the Turner syndrome *do not* have X-autosome translocations at all, but instead have X-duplications or X-X translocations, therefore intrinsic structural alterations of the X chromosome. In all 3, the entire abnormal X chromosome was the late replicator, and in 2 the X-chromatin mass was reported as bipartite [43, 44]. Both of these factors favor an intrinsic structural X-alteration. Known X-duplications, which would contain 2 inactivation centers according to the hypothesis that a single inactivation center exists in proximal Xq, have been reported to produce bipartite X-chromatin masses [17, 58, 59]. Also, in an intrinsic structural alteration of the X chromosome, the structurally altered X *should* be the late replicator. In addition, the presence of mosaicism with a 45,X cell line probably favors an intrinsic structural X alteration.

The remaining 13 patients with unbalanced X-autosome translocations (Table 5) present varying mental retardation-multiple congenital anomaly syndromes and, for the most part, the phenotype of each patient is predictable on the basis of the specific translocation and its X-inactivation pattern [18–22, 25, 26, 41, 46–51].

1) In the KOP translocation reported by Allderdice and associates [20] and Grzeschik and co-workers [21], the Klinefelter syndrome results from the presence of an extra Xq translocated to 14q. The effect of 14q trisomy is mitigated by its inactivation which extends from the inactive Xq. It is also interesting that this case suggests that the Klinefelter phenotype is the result of excessive Xq, in contrast to the Xq– determination of the Turner phenotype. This

TABLE 4. Alleged Unbalanced X-Autosome Translocations With Features of the Turner Syndrome

Source	Karyotype	X-Breakpoint	Late Replicating X	Phenotype
Gouw [43]	46,XXp+ *may involve del (Xp)* *not banded*	Distal Xp	Entire Xt *Bipartite X-chromatin*	Turner syndrome, MR
Hugh-Jones [44]	45,X/46,XXq+ *mosaic – not banded*	Distal Xq	Entire Xt *Bipartite X-chromatin*	Turner syndrome
Thorburn [45]	45,X/46,XXq+ *mosaic – not banded*	Distal Xq	Entire Xt	Mild Turner syndrome

TABLE 5. Unbalanced X-Autosome Translocations With MR-MCA Phenotypes

Source	Karyotype	X-Breakpoint	Late Replicating X	Phenotype
Allderdice [20]	47,Xq−Y,der(14),t(X;14) (q13;q32), +der(14),(X;14) (q13;q32) *banded*	Xq13	One entire der(14)	Klinefelter syndrome and mild MR
Mukerjee [46]	46,X,t(Xp+;3q−),r(3) *not banded* *difficult to evaluate*	Distal Xp	Entire Xt	MR and MCA
Neuhäuser [47]	45,XXp+,−C *not banded* *difficult to evaluate*	Distal Xp	X-portion of Xt	MR, MCA
German [48]	46,XXq+ *not banded*	Distal Xq	X-portion of Xt	MR, FTT
Cohen [18]	46,X,der(X),t(X;8) (q28?;q11?) *banded*	Xq28	X-portion of Xt	MR, MCA *trisomy 8?*
Engel [49]	45,XXq+,−15 *not banded*	Distal Xq	Entire Xt 75% N1. X 11% Both 10%	MCA
Buckton [22]	45,X/46,X,der(X),t(Xp−;14q+) *mosaic – not banded*	Xp11	Entire Xp−	Short but otherwise normal
Summitt [19]	46,XX,der(21),rcp(X;21) (q11;p11)mat	Xq11	N1. X 100% N1. X and entire Xt 79% N1. X and X-portion of Xt 15%	21p− syndrome
Tipton [25]	46,X,der(X), t(X;8) (q26;q12)mat *banded*	Xq26	Entire Xt	MR and MCA, *some features of trisomy-8 syndrome*
Jenkins [41]	46,XY,−22,+t(X;22) (q28;q11) *banded*	Xq28	X-portion of Xt 77% Entire Xt 23%	MR and MCA, *? features of 22q− syndrome*
Leisti [26]	46,X,−X,+der(9),rcp(X;9) (q11;q32)mat *banded*	Xq11	Entire der(9) in most cases	Hypogonadism, primary amenorrhea, *some features of 9p− syndrome and Xp−*
Mikkelsen [50]	46,X,der(X),t(X;8) (p22;q21) *banded*	Xp22	Normal X	MR and MCA, *? 8-trisomy syndrome*
Crandall [51]	46,X,der(X),t(X;13) (q27;q12) *banded*	Xq27	X-portion of Xt, ? 13-portion	Some features of 13-trisomy syndrome

case also provides evidence that the X-inactivation center is located in the proximal long arm of X, distal to band q13.

2) The karyotype of the case of Mukerjee [46] which includes an X-3 translocation, and also includes a ring-3 chromosome, is complex. The lack of chromosome banding makes the karyotype even more difficult to interpret. At any rate, the entire Xt, including the 3-derived segment, replicates late.

3) The case reported by Neuhauser [47] involves an unidentified partial autosomal monosomy. Only the monosomy for the actually deleted segment would predictably be expressed since the autosome-derived portion of the Xt does not replicate late. The X-breakpoint is in the distal portion of the short arm of X.

4) The case described by German [48] involves an unidentified autosomal duplication, and that duplication would predictably be fully expressed since the autosome-derived portion of the Xt does not replicate late. In this case, the X-breakpoint is in distal Xq.

5) Case 2 of Cohen and associates [18] involves X and 8, the karyotype including the der(X) and resulting in partial duplication-8 since the 8-derived portion of the Xt is apparently not inactivated. Some features of the trisomy-8 syndrome are apparent.

6) The case of Engel [49] is an enigma. It involves an X-15 translocation, and autoradiography revealed late replication of the entire Xt in 75% of cells, of the normal X in 11% of cells, and of both in 10% of cells. To some extent inactivation thus seems random. This apparent random nature of X-inactivation might be explained, like that in Thelen's case [40], by the lack of negative selection against cells in which the normal X is inactivated if the segment of distal Xq which is deleted is minimal. The inexplicable feature is the survival of 10% of cells in which the normal X *and* the Xt are inactivated, producing effective nullisomy-X.

7) In the report of Buckton and associates [22] the patient has a mosaic karyotype, one cell line 45,X and the other, which includes the der(X), essentially 46,X,del(Xp) as a result of malsegregation of the parental X-14 translocation. Thus, she predictably is short. As a matter of fact, why she does not have the fullblown phenotype of the Turner syndrome is difficult to understand.

8) The patient reported by Summitt and associates [19] predictably exhibits the 21q– phenotype (Fig. 6). She is one of like-sex twins, and her co-twin is normal. Her karyotype (Fig. 7) includes 2 normal X chromosomes plus the Xt, the result of malsegregation of the balanced X-21 translocation in her mother (Fig. 1). Some cells of her buccal smear contain 2 X-chromatin masses. Autoradiography (Fig. 8) revealed late replication of the normal X in all cells, the inactivation also of the entire Xt in 79% of cells, and of the X-derived portion of the Xt in 15% of cells. Thus, the 21q– phenotype is the result of extension of inactivation to the 21-derived segment of Xt in the majority of cells. This

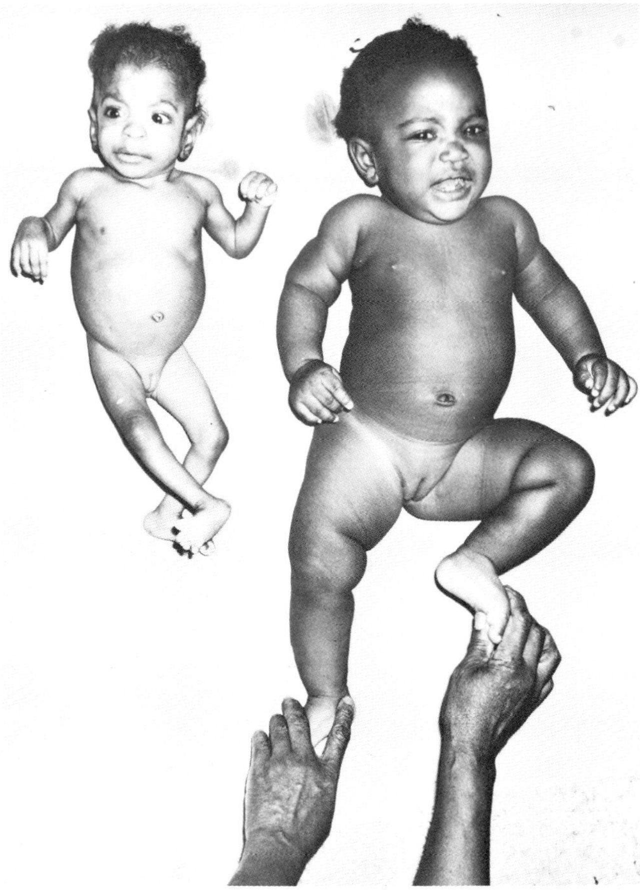

Fig. 6. Patient with karyotype 46,XX,der(21),rcp(X; 21) (q11; p11)mat, demonstrating 21 deletion phenotype, on left. At right is normal co-twin.

case also offers support for the location of the X-inactivation center in the long arm of X, distal to band q11.

9) The patient described by Tipton [25] predictably exhibits some features of the trisomy 8 syndrome (Fig. 9). The karyotype (Fig. 10) includes the der(X) of an X-8 translocation, the result of segregation of the balanced X-8 translocation in the mother (Fig. 3). Autoradiography reveals inactivation of the entire Xt in all cells (Fig. 11). The effect of the additional segment of chromosome 8 is thus apparently mitigated by its inactivation.

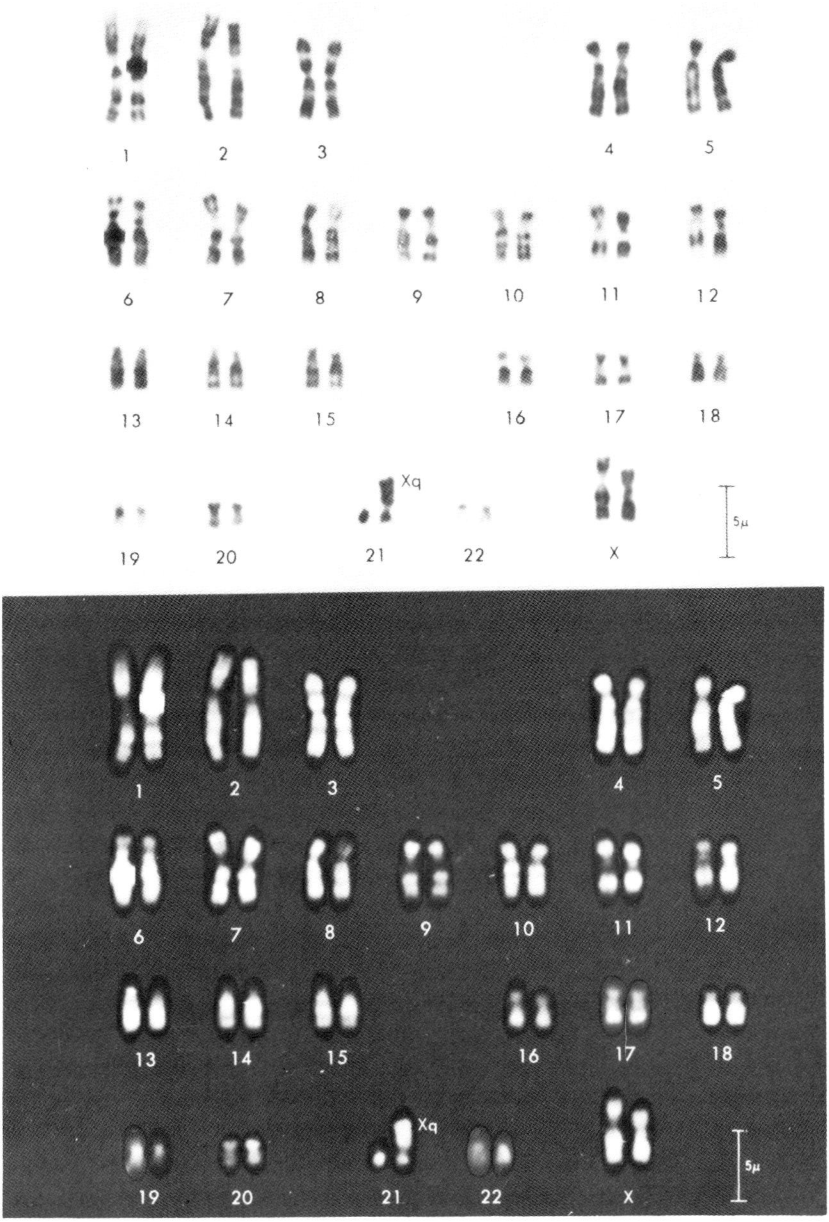

Fig. 7. Karyotype 46,XX,der(21),rcp(X;21) (q11;p11)mat, of patient in Fig. 6.

Fig. 8. Autoradiography of cell from patient in Fig. 6, showing late replication of 1 normal X and entire Xt.

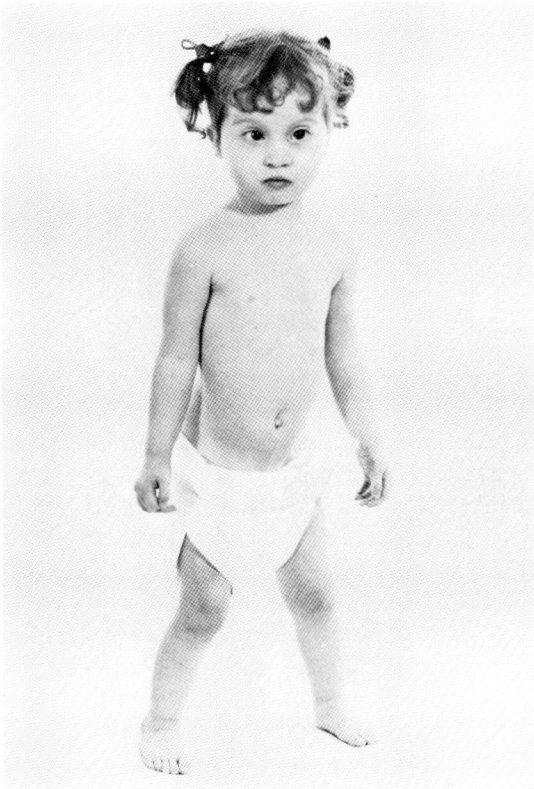

Fig. 9. Patient with karyotype 46,X,der (X),rcp (X;8) (q26; q12)mat, demonstrating features of trisomy 8 syndrome.

10) The patients of Jenkins [41] predictably exhibit a few features of the 22q− syndrome, attributable to inactivation of the 22-derived segment of the Xt in a proportion of cells. One would further predict that these patients have the Klinefelter syndrome since the karyotype includes the major portion of an X chromosome in the Xt, in addition to a normal X. In fact, the older of the 2 males has small testicles.

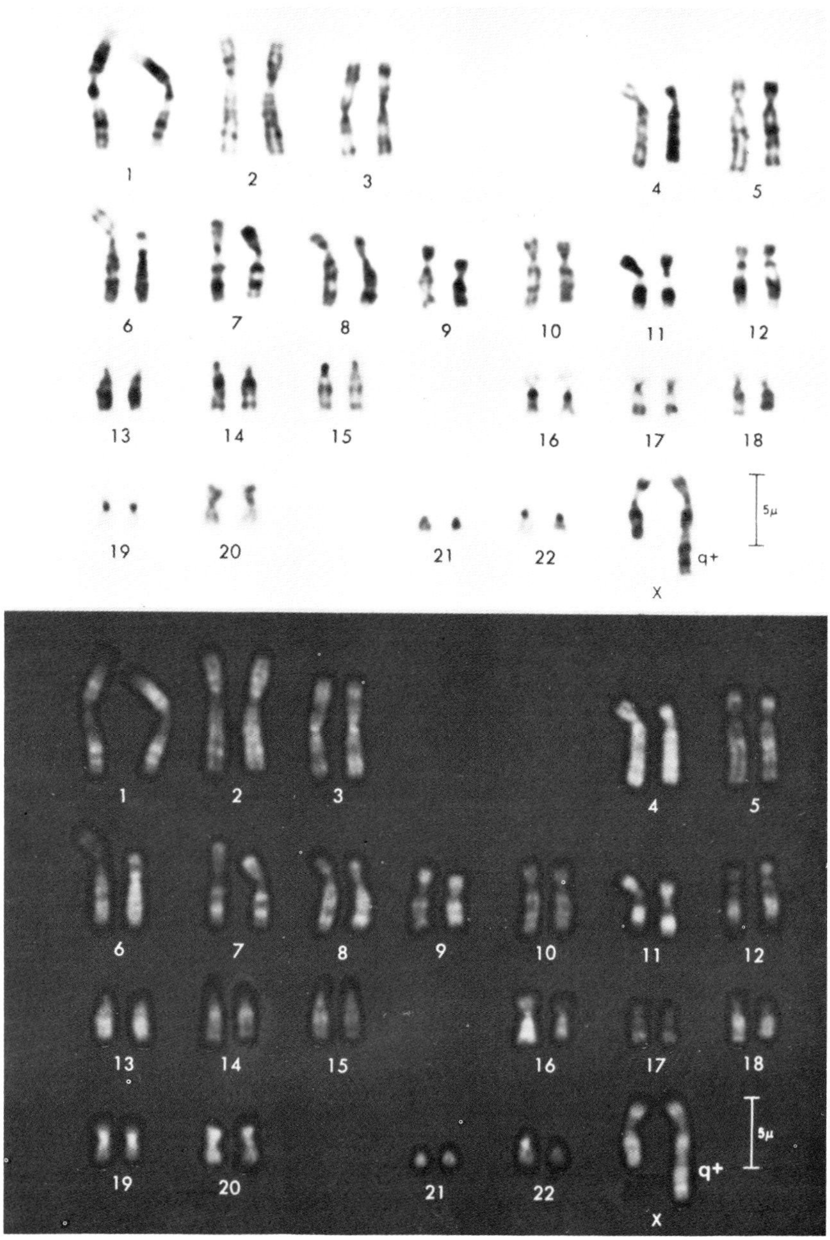

Fig. 10. Karyotype 46,X,der(X),rcp(X;8)(q26;q12)mat, of patient in Fig. 9.

11) The patient of Leisti [26] has an X-9 translocation, and her karyotype includes one normal X and two normal chromosomes 9 plus the derivative made up of virtually all of chromosome 9 and all of the long arm of X distal to band q11. She thus has monosomy for the short arm of X. The entire der(9) is inactivated. Thus, while the patient is trisomic for most of chromosome 9, she is monosomic for Xp. Furthermore, the patient predictably has gonadal dysfunction and features of the Turner syndrome because of the del(Xp). The pattern of X-inactivation, like that in the case of Summitt and associates [19], supports the location of the X-inactivation center in the long arm of the X chromosome, distal to band q11.

12) The case of Mikkelsen [50] is reported as a translocation between X and 8, with the X-breakpoint at p22 and the 8-break at q21. The der(X) includes all of Xq, and Xp to band 22, and that portion of 8q from band 21 to the terminus. The X-inactivation pattern here again is an enigma. It would appear that the single-active-X rule is broken, and the explanation is not apparent. The author claims that the Xp+ is due to translocation of a segment of 8q. While this may be open to challenge, it still does not explain the most unusual X-inactivation pattern. Even if the Xp segment deleted in the translocation is minimal, one would at least expect random inactivation, not *preferential* inactivation of the normal X. Why should there be selection against cells with an inactive Xt?

13) The patient reported by Crandall and associates [51] has an X-13 translocation. The karyotype includes a translocation chromosome made up of the major portion of an X chromosome from the p terminus to band q27, and the major portion of 13q. The patient predictably shows some features of the 13-trisomy syndrome.

Of interest is the fact that even though apparently inactivated as indicated by late replication, the autosome-derived segment of an X-autosome translocation chromosome, when it produces a duplication (or partial trisomy), is at least to some extent expressed. Could this be related in any way to the question as to why X-monosomy produces such severe effects whereas one X is inactivated in all somatic cells of the normal female? Is there an active segment in the inactive X [60] and this is the late replicating autosomal segment of an X-autosome translocation chromosome not totally inactive? Other remaining enigmas presented by X-autosome translocations are: 1) Apparent effective X-nullisomy in cells in which both the normal X and all or a portion of the Xt are inactivated. How is this possible? 2) Inactivation of the normal X in an unbalanced X-autosome translocation.

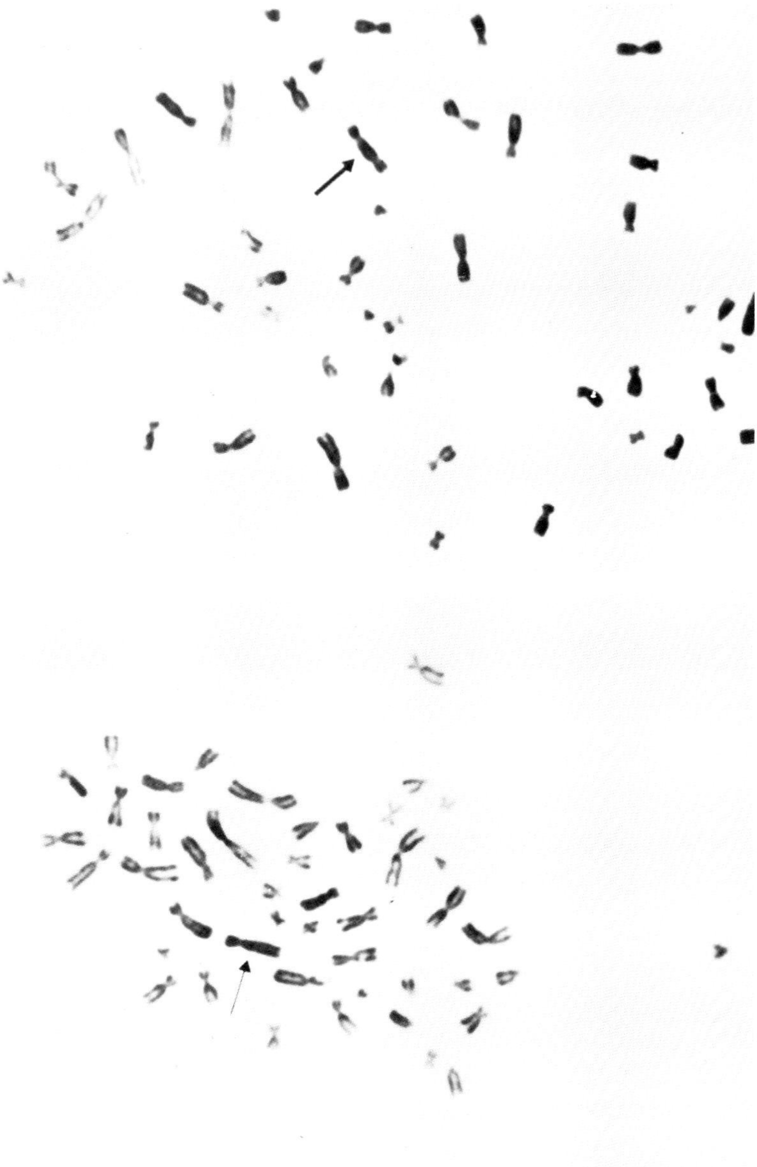

Fig. 11. Autoradiography of cell from patient in Fig. 9, showing late replication of entire Xt.

11

CONCLUSIONS

Conclusions possible on the basis of this review of X-autosome translocations are:

1) That in general the X replication/inactivation pattern IS that which preserves a single active X.
2) X-inactivation is governed and initiated in an "inactivation center" located in the proximal region of the long arm of X, probably the proximal 1/3, but apparently in or distal to band q13. Inactivation extends from this center to the remainder of the X chromosome. No convincing evidence exists for the presence of a "secondary inactivation center."
3) Extension of X-inactivation to the autosome-derived segment of an Xt varies between cases and to some extent even between cells of the same patient, with no readily apparent explanation. Whether or not the autosomal segment is inactivated is apparently *not* a function of its distance from the inactivation center in proximal Xq, or whether the X-breakpoint is in Xp or Xq. Inactivation *may* involve an autosomal segment as extensive as an *entire chromosome 9*. The availability of testable autosomal markers will considerably improve our ability to investigate directly the question of inactivation of autosomal loci. Its study through autoradiography or BRDU fluorescence is admittedly indirect and may be misleading.
4) In cases in which X-inactivation appears random, the X-imbalance in the Xt appears minimal and may be so minimal as not to exercise a negative selection force.
5) A critical region exists in the long arm of the X chromosome within which if a break occurs and is followed by translocation involving an autosome, gonadal dysgenesis results. This is best explained by some sort of position effect, not by effective hemizygosity for a mutant recessive gene or by loss of a small chromosome segment.
6) Reported unbalanced X-autosome translocations producing the Turner syndrome without other anomalies are NOT X-autosome translocations, but are, instead, X-duplications or X-X translocations.
7) The phenotypes of patients with UNBALANCED X-autosome translocations are in general attributable to effective autosomal imbalance – either monosomy which results from the extension of inactivation from the X-derived portion of the Xt to the autosome-derived portion, or to duplication of or partial trisomy for the autosome translocated to the Xt.

REFERENCES

1. Lyon MF: Gene action in the X-chromosome of the mouse (Mus musculus L). Nature 190:372–373, 1961.
2. Lyon MF: Sex chromatin and gene action in the mammalian X-chromosome. Am J Hum Genet 14:135–148, 1962.
3. Lyon MF: X-chromosome inactivation in mammals. In Woollam DHM (ed): "Advances in Teratology." New York: Academic Press, 1966, vol 1, pp 25–54.
4. Russell LB: Another look at the single-active-X hypothesis. Trans NY Acad Sci 26: 726, 1964.
5. Brown SW, Chandra HS: Inactivation system of the mammalian X chromosome. Proc Natl Acad Sci USA 70:195–199, 1973.
6. Hamerton JL: "Human Cytogenetics." New York: Academic Press, 1971, vol 1, pp 167–191.
7. Lindsten J, Tillinger KG: Self-perpetuating ring chromosome in a patient with gonadal dysgenesis. Lancet 1:593–594, 1962.
8. Rowley J, Muldal S, Lindsten J, Gilbert CW: H^3-Thymidine uptake by a ring chromosome in a human female. Proc Natl Acad Sci USA 51:779–786, 1964.
9. Atkins L, Santesson B, Voss H: Partial deletion of an X chromosome. Ann Hum Genet 29:89–95, 1965.
10. Hsu L, Hirschhorn K: Genetic and clinical considerations of long arm deletion of the X chromosome. Pediatrics 45:656–664, 1970.
11. Muldal S, Gilbert CW, Lajtha LG, Lindsten J, Rowley J, Fraccaro M: Tritiated thymidine incorporation in an isochromosome for the long arm of the X chromosome in man. Lancet 1:861–863, 1963.
12. Taft PD, Brooks S: Late labeling of an iso-X chromosome. Lancet 2:1069, 1963.
13. Miller OJ, Mukherjee BB, Bader S, Christakos AC: Autoradiographic studies of X-chromosome duplication in an XO/X-isochromosome X mosaic human female. Nature 200:918–919, 1963.
14. Gartler SM, Sparkes RS: The Lyon-Beutler hypothesis and isochromosome X patients with the Turner syndrome. Lancet 2:411, 1963.
15. Therman E, Patau K: Abnormal X chromosomes in man: Origin, behavior and effects. Humangenetik 25:1–16, 1974.
16. Lyon MF: Mechanisms and evolutionary origins of variable X-chromosome activity in mammals. Proc R Soc Lond [Biol] 187:243–268, 1974.
17. Therman E, Sarto GE, Patau K: Center for Barr body condensation on the proximal part of the human Xq: A hypothesis. Chromosoma 44:361–366, 1974.
18. Cohen MM, Lin C-C, Sybert V, Orecchio EJ: Two human X-autosome translocations identified by autoradiography and fluorescence. Am J Hum Genet 24:583–597, 1972.
19. Summitt RL, Martens PR, Wilroy RS Jr: X-autosome translocation in normal mother and effectively 21-monosomic daughter. J Pediatr 84:539–546, 1974.
20. Allderdice PW, Miller OJ, Klinger HP, Pallister PD, Opitz JM: Demonstration of a spreading effect in an X-autosome translocation by combined autoradiographic and quinacrine-fluorescence studies. Excerpta Med Int Cong Ser 233:14, 1971.

21. Grzeschik KH, Allderdice PW, Grzeschik A, Opitz JM, Miller OJ, Siniscalco M: Cytological mapping of human X-linked genes by use of somatic cell hybrids involving an X-autosome translocation. Proc Natl Acad Sci USA 69:69–73, 1972.
22. Buckton KE, Jacobs PA, Rae LA, Newton MS, Sanger R: An inherited X-autosome translocation in man. Ann Hum Genet 35:171–178, 1971.
23. de la Chapelle A, Schröder J: Apparently non-reciprocal balanced human (3q–;Xq+) translocation: Late replication of structurally normal X. In Wahrman J, Lewis KR (eds): "Chromosomes Today." New York: John Wiley & Sons, 1973, vol 4, pp 261–266.
24. Davis JR, Heine MW, Lightner ES, Giles HR, Graap RF: X-short arm deletion gonadal dysgenesis in 2 siblings due to unique translocation (Xp–;16q+). Clin Genet 10:202–207, 1976.
25. Tipton RE: "Cytogenetic Studies in Subjects With Mental Retardation and Congenital Anomalies Using Banding Techniques." Master of Science Thesis, University of Tennessee, 1974.
26. Leisti JT, Kaback MM, Rimoin DL: Human X-autosome translocations: Differential inactivation of the X chromosome in a kindred with an X-9 translocation. Am J Hum Genet 27:441–453, 1975.
27. Thorburn MJ, Martin PA, Pathak UN: Case report: Possible X/autosomal translocation in a girl with gonadal dysgenesis. J Med Genet 7:402–406, 1970.
28. Sarto GE, Therman E, Patau K: X inactivation in man: A woman with t(Xq–;12p+). Am J Hum Genet 25:262–270, 1973.
29. Mann JD, Valdmanis A, Capps SC, Puite RH: A case of primary amenorrhea with a translocation involving chromosomes of groups B and C. Am J Hum Genet 17:377–383, 1965.
30. Mann JD, Higgins J: A case of primary amenorrhea associated with X-autosomal translocation [46,X,t(Xq–;5q+)]. Am J Hum Genet 26:416, 1974.
31. Kallio H: Cytogenetic and clinical study of 100 cases of primary amenorrhea. Acta Obstet Gynecol Scand (Suppl) 24:7, 1973.
32. Lucas M, Smithies A: Banding patterns and autoradiographic studies of cells with an X-autosome translocation. Ann Hum Genet 37:9–12, 1973.
33. Gerald PS, Burns G, Monedjikova V: Localization of genes on the X chromosome by somatic cell hybridization. Pediatr Res 7:344, 1973.
34. Gilgenkrantz S, Mauuary G, Dutrillaux B, Masocco G: Translocation X sur autosome et replication tardive. Humangenetik 26:25–34, 1975.
35. Phelan JP, Upton RT, Summitt RL: Balanced reciprocal X-4 translocation in a female with early secondary amenorrhea. Am J Obstet Gynecol 129:607–613, 1977.
36. Sauer F, Greenstein RM, Reardon P, Riddick DH: Secondary amenorrhea associated with balanced X-autosome translocation. Obstet Gynecol 49:101–104, 1977.
37. Sujansky E, Hsu LY, Lucas M, Hirschhorn K: Nature of X-autosome translocation and choice of X-inactivation. Pediatr Res 7:343, 1973.
38. Dutrillaux B, Laurent C, Gilgenkrantz S, Frederic J, Carpentier S, Couturier J, Lejeune J: Les translocations du chromosome X. Etude apres traitement par le BUDR et coloration par lacridine orange. Helv Paediatr Acta (Suppl) 34:19–31, 1974.
39. Palmer C, Nance W, Cleary R, Dexter R: Structural abnormalities of the X chromosome in pure gonadal dysgenesis. Am J Hum Genet 25:57A, 1973.
40. Thelen TH, Abrams DJ, Fisch RO: Multiple abnormalities due to possible genetic inactivation in an X/autosome translocation. Am J Hum Genet 23:410–418, 1971.
41. Jenkins MB, Davis E, Thelen TH, Boyd L: A familial X-22 translocation with an extra X chromosome. Am J Hum Genet 26:736–745, 1974.

42. Punnett HH, Kistenmacher ML, Greene AL, Coriell LL: An (X;1) translocation, balanced, 46 chromosomes. Cytogenet Cell Genet 13:406–407, 1974.
43. Gouw WL, Coenegracht JM, Stalder G: A very large metacentric chromosome in a woman with symptoms of Turner's syndrome. Cytogenetics 3:427–440, 1964.
44. Hugh-Jones K, Wallace SJ, Thornber JM, Atkin NB: Gonadal dysgenesis with unusual abnormalities. Arch Dis Child 40:274–279, 1965.
45. Thorburn MJ, Miller CG, Dovey P: Anomalies of development in a girl with unusual sex chromosomal mosaicism. J Med Genet 4:283–287, 1967.
46. Mukerjee D, Burdette WJ: Multiple congenital anomalies associated with a ring 3 chromosome and translocated 3/X chromosome. Nature 212:153–155, 1966.
47. Neuhäuser G, Back F: X-autosom-translokation bei einem Kind mit multiplen Missbildungen. Humangenetik 3:300-311, 1967.
48. German J: Autoradiographic studies on human chromosomes. I. A review. In Crow JF, Neel JV (eds): "Proceedings 3rd International Congress of Human Genetics." Baltimore: The Johns Hopkins Press, 1967, pp 123–136.
49. Engel W, Vogel W, Reinwein H: Autoradiographische Untersuchungen an einer X-Autosomentranslokation beim Menschen: 45,X,15–,tan(15qXq+)+. Cytogenetics 10: 87–98, 1971.
50. Mikkelsen M, Dahl G: Unbalanced X/autosomal translocation with inactivation of the normal X chromosome. Cytogenet Cell Genet 12:357–366, 1973.
51. Crandall BF, Carrell RE, Howard J, Schroeder W, Muller H: Trisomy 13 with a 13-X translocation. Am J Hum Genet 26:385–392, 1974.
52. Funderburk SJ, Spence MA, Sparkes RE: Mental retardation associated with "balanced" chromosome rearrangements. Am J Hum Genet 29:136–141, 1977.
53. Tharapel AT, Summitt RL, Wilroy RS, Martens PR: Apparently balanced de novo translocations in patients with abnormal phenotypes: Report of 6 cases. Clin Genet 11: 255–269, 1977.
54. Simpson JL: "Abnormalities of Sexual Differentiation." New York: Academic Press, 1977.
55. Palmer CG, Reichmann A: Chromosomal and clinical findings in 110 females with Turner syndrome. Hum Genet 35:35–49, 1976.
56. Wilroy RS Jr, Summitt RL, Martens PR, Tharapel AT, Tipton RE, Givens JR: Phenotype-karyotype correlations in 81 patients with the Turner syndrome. Clin Res 25: 74A, 1977.
57. Summitt RL: Chromosomal abnormalities. In Jackson LG, Schimke RN (eds): "The Biology of Inherited Disease." New York: John Wiley & Sons (In press).
58. Sillesen I, Rasmussen K, Østerballe O, Nielsen J: Center for Barr body condensation. A case of Turner's syndrome with 45,X/46,X,dic(X) (Xqter→p22:p22→qter). Hum Genet 33:337–340, 1976.
59. Daly RF, Patau K, Therman E, Sarto GE: Structure and Barr body formation of an Xp+ chromosome with 2 inactivation centers. Am J Hum Genet 29:83–93, 1977.
60. Therman E, Sarto GE, Disteche C, Denniston C: A possible active segment on the inactive human X chromosome. Chromosoma 59:137–145, 1976.

Differences Between Rates of Trisomy 21 (Down Syndrome) and Other Chromosomal Abnormalities Diagnosed in Livebirths and in Cells Cultured After Second-Trimester Amniocentesis—Suggested Explanations and Implications for Genetic Counseling and Program Planning

Ernest B. Hook, MD

An apparent discordance has been noted between the rate of chromosome abnormalities diagnosed in 2nd-trimester fetuses after amniocentesis and the rate diagnosed in livebirths.

The pooled rate of Down syndrome in livebirths born to women aged 35–39 is about 3–4 per 1,000 in most large-scale studies to date [1–4], rising for women aged 40–44 to about 10–15 per 1,000 (Table 1). In contrast, the rates of Down syndrome found after intrauterine diagnosis of women aged 35–39 (who have been studied because of elevated age) have been reported by Ferguson-Smith to be as high as 8 per 1,000; in those 40 years and over it as high as 49 per 1,000 [5]. Thus, the rate diagnosed at the 2nd trimester of gestation after amniocentesis is apparently 2-fold greater than in livebirths, judging by his report. Others have also observed high rates for women undergoing amniocentesis because of older age. For example, Milunsky and Atkins report for the 35–39-year interval 2 instances of trisomy 21 in 318 studied (6 per 1,000) in their own experience, and 10 in 457 (22 per 1,000) in a literature review [6].

Such observations have stimulated a good deal of speculation as to possible explanations for the difference and have raised questions as to which figures are "valid" or "appropriate" for genetic counseling and program planning.

An immediate question is whether the difference between the prevalence in livebirths and that in amniotic fluid specimens in any quinquennium is simply due to sampling fluctuation, ie occurs by chance alone.

Birth Defects: Original Article Series, Volume XIV, Number 6C, pages 249–267

TABLE 1. Comparison of Rates per 1,000 Births of Down Syndrome in Livebirths by Maternal Age Quinquennia

Maternal age (years)	Upstate N.Y. (from regression analysis) [1][a,b]	Upstate N.Y. (direct estimate) [1][a,c]	Australia [2]	Sweden [3]	Massachusetts [4][a]
< 20	–	0.63	$\left[\frac{15}{35,555}\right] = 0.42$	$\left[\frac{18}{30,325}\right] = 0.59$	$\left[\frac{46}{67,490}\right] = 0.68$
20–24	0.66	0.65	$\left[\frac{128}{207,931}\right] = 0.62$	$\left[\frac{87}{117,593}\right] = 0.74$	$\left[\frac{173}{261,091}\right] = 0.66$
25–29	0.94	0.96	$\left[\frac{208}{253,450}\right] = 0.82$	$\left[\frac{96}{108,746}\right] = 0.88$	$\left[\frac{187}{237,281}\right] = 0.79$
30–34	1.44	1.38	$\left[\frac{194}{170,970}\right] = 1.13$	$\left[\frac{72}{49,487}\right] = 1.45$	$\left[\frac{201}{159,264}\right] = 1.26$
35–39	4.33	4.37	$\left[\frac{297}{86,046}\right] = 3.45$	$\left[\frac{73}{19,522}\right] = 3.74$	$\left[\frac{314}{84,521}\right] = 3.72$
40–44	12.96	13.45	$\left[\frac{240}{24,498}\right] = 9.80$	$\left[\frac{73}{4,880}\right] = 14.96$	$\left[\frac{288}{21,881}\right] = 13.16$
⩾ 45	36.39[d]	43.26[d]	$\left[\frac{37}{1,707}\right] = 21.68$	$\left[\frac{19}{306}\right] = 62.09$	$\left[\frac{38}{1,078}\right] = 35.25$
N.S.	–	$\frac{0}{476}$	$\left[\frac{0}{11}\right]$	–	$\left[\frac{3}{195}\right] = 15.38$
TOTAL	–	1.36	$\left[\frac{1,119}{780,168}\right] = 1.45$	$\left[\frac{438}{330,859}\right] = 1.32$	$\left[\frac{1,250}{832,531}\right] = 1.50$

[a] White livebirths only. See Table 3 for discussion of possible racial variation in frequency.

[b] Estimated rates by single-year interval were obtained from a regression analysis, then divided by 0.375 to correct for under reporting. The rates in each quinquennium were derived by weighing the number of mothers in each single-year interval within a quinquennium.

[c] Observed numbers of cases in each quinquennium divided by 0.375 to correct for estimated under reporting.

[d] 45–49 only

ANALYSIS AND INTERPRETATION

In the 40–44-year interval enough data have been gathered to allow crude comparisons, which suggest an apparent statistically significant difference between the rates diagnosed at time of amniocentesis and in livebirths. In the pooled results of 4 studies which have presented data by single-year maternal age intervals ([7–13]) Table 2) there were 28/959 cases observed (29.2 per 1,000). The lower 95% confidence limit of this proportion is 19.5 per 1,000, which is higher than any of the rates for this interval in Table 1. (Excluding the results of Ferguson-Smith, the rate was 21/816 = 25.7 per 1,000, with a lower 95% confidence limit of 16.0 per 1,000, again higher than any rates for this interval in Table 1.) For the 35–39- and 45–49-year intervals there are not yet sufficient data to document a difference, but it appears likely that as data accumulate, the difference will become significant. The latter inference is based, not only on the comparison in the 40–44-year interval, but also on the data of Milunsky and Atkins [6] for the 35–39-year interval and on the findings reported by others [14]. These latter reports, unfortunately, do not provide sufficient detail to allow breakdown by single-year interval in all quinquennia for those studied because of elevated maternal age.

Some explanation other than chance alone must be sought for the reported differences between the rates of Down syndrome in livebirths and in cells cultured from amniotic fluids obtained in the 2nd trimester. There are at least 6 possible explanations. These are listed briefly here and discussed in detail below.

1) The reported rates in livebirths in Table 1 are falsely low.

2) The reported rates in amniotic-fluid specimens in some studies in Table 2 are falsely high.

3) Women undergoing amniocentesis because of advanced maternal age have a higher risk of Down syndrome birth than pregnant women of the same age who do not choose this procedure.

4) Crude comparisons by 5-year maternal age intervals between the 2 types of studies will produce artifactual differences, unless appropriate age adjustments are done (see below).

5) Even comparisons by single-year intervals will result in artifactual differences, albeit small ones, unless differences in gestational age are also adjusted.

6) There is a high (spontaneous) death rate of fetuses with Down syndrome and other chromosomal abnormalities between the usual time of amniocentesis (16th–18th week of gestation) and livebirth (40th week).

I emphasize that these considerations are not mutally exclusive. The first three are possible contributing factors. The last three are definite factors, although only two of these appear likely to make major contributions to the difference.

Table 2. Reported Rates of Karyotypes Associated With Down Syndrome Diagnosed in Amniotic Fluid Cells Obtained at 2nd Trimester Amniocentesis, in Published Studies Where the Indication for Procedure Was Advanced Maternal Age Only and Where Data by Single-Year Maternal Age Interval Are Available

Maternal age	Ferguson-Smith [5]	Polani et al [7]	USA collaborative [8–10]	Canadian collaborative [11-13][a]	Combined USA and Canadian collaborative
35	0/17	0/23	0/23	0/38	0/61
36	0/16	0/27	0/35	0/30	0/65
37	1/26	0/33	0/50	1/54	1/104
38	0/35	1/78	0/60	0/46	0/106
39	0/36	0/79	0/71	0/80	0/151
(35 – 39)	(1/130) = 7.7[d]	(1/240) = 4.2[d]	(0/239) = 0[d]	(1/248) = 4.0[d]	(1/487) = 2.1[d]
40	0/42	2/134	1/80	1/75	2/155
41	2/37	2/101	1/60[b]	3/48	4/108
42	2/34	2/81	1/44[b]	1/40	2/84
43	1/21	2/48	1/21	1/24	2/45
44	2/9	1/24	1/28	1/8	2/36
(40 – 44)	(7/143) = 49.0[d]	(9/388) = 23.2[d]	(5/233) = 21.5[d]	(7/195) = 35.9[d]	(12/428) = 28.0[d]
45	0/6	1/14	0/6	0/11	0/17
46	0/3	0/0	0/7	0/3	0/10
47	0/0	0/2	0/2	1/2	1/4
48	0/1	0/0	0/1	0/0	0/1
49	0/0	06	0/0	0/0	0/0
(45 – 49)	(0/10) = 0[d]	(1/14) = 71.4[d]	(0/16)[c] = 0[d]	(1/16) = 62.5[d]	(1/32) = 31.3[d]
⩾35	(8/283) = 28.3[d]	(11/642) = 15.6[d]	(5/488) = 10.2[d]	(8/459) = 19.6[d]	(14/947) = 14.8[d]
⩾40	(7/153) = 45.8[d]	(10/402) = 24.9[d]	(5/249) = 20.1[d]	(8/211) = 37.9[d]	(13/460) = 28.3[d]
⩾45	(0/10) = 0[d]	(1/14) = 71.4[d]	(0/16) = 0[d]	(1/17) = 58.8[d]	(1/33) = 30.3[d]
⩾50	–	–	–	(0/1) = 0[d]	(0/1) = 0[d]

[a]Data abstracted from appendix of Canadian collaborative report [13] pp. 109–118. Instances which a culture failed and no follow-up data on fetus are available are exluded here from the totals given in report by Simpson et al.

[b]Missed by amniocentesis, diagnosed at birth.

[c]Assuming no mothers 50 or over.

[d]Rate per 1,000.

Are Rates in Livebirths Falsely Low?

There is some variation between the studies of livebirths presented in Table 1, and it is possible that the reported rates in any study may be lower than the "true" quinquennial rates in the population, reflecting such factors as sampling fluctuations and temporal differences. It is of interest, however, that the Australian study of Collman and Stoller [2], which Christianson [15] has characterized as "generally accepted as providing the most reliable rates [in livebirths] available," reported quinquennial rates lower than most of the others.* Never-

*As far as community studies are concerned, that of Collman and Stoller [2] has been the most frequently cited regarding rates for genetic counseling, although the more recent studies of Fabia cited in reference 4, and Lindsjo [3], which used the same methodology, may be more authoritative.

theless, Ferguson-Smith declared that such studies of livebirths, the results of which are generally used for genetic counseling, have resulted in "gross" underestimates [5]. He attributes this to the fact that they are retrospective studies, ie have involved ascertainment some time after birth, and therefore may have missed many affected individuals. Studies such as those of Collman and Stoller [2], Lindsjo [3], and Fabia [4] are subject to these possible criticisms.

The New York State study [1], the results of which are also presented in Table 1, was based on notifications of affected individuals upon birth certificates and differs from the other livebirth studies cited with regard to time of ascertainment.† The results of this study are not at great variance with those from the others listed. However, birth-certificate reporting of defects is incomplete, and while this was corrected for – using estimates of completeness derived from a separate investigation of the data source – the resulting adjustment introduced a measure of uncertainty. Thus, the estimates do not provide strong evidence concerning the completeness of the other studies.

Very recently data have become available from 4 studies which involved thorough cytogenetic or phenotypic evaluation at birth of all individuals in the liveborn populations investigated and appear likely to have achieved almost complete ascertainment of all Down syndrome instances. Summary results are presented in Table 3 [15, 17–20]. (An extensive analysis by single-year maternal age interval and by study will be presented elsewhere.) Many fewer data are available, and thus the results are subject to a larger sampling variation than the data for livebirths. Nevertheless, comparison of the data in Tables 1 and 3 indicates no reason to believe that the studies summarized in Table 1 have resulted in gross underestimates of the true quinquennial rates in older mothers. In fact, for the 35–39-year interval the observed rates in all studies in Table 1 are higher than those in Table 3. This may of course be due to sampling fluctuation.

While it is still possible that some of the community studies (in particular that of Collman and Stoller [2]) may have slightly underestimated the true rates of Down syndrome in the populations studied, the suggestion that their rates are too low by, say, 2-fold appears highly implausible.

Are Falsely High Rates Obtained in Studies at Amniocentesis?

In evaluating amniocentesis studies at least 2 factors must be considered. First of all, the chromosomes 21 and 22 are indistinguishable using conventional stains. Unless banding studies are done, 47, trisomy 22 may be misdiagnosed as 47, trisomy 21 in cytogenetic analysis of cultured amniotic fluid cells. Trisomy

†If one defines "retrospective" as referring to the perspective of the investigator, the New York study is retrospective. If one defines this from the perspective of the age of ascertainment of those in the data source and the nature of the investigation, it fits better in a "cross-sectional" or perhaps a "cohort" classification. The issue is primarily a semantic one, and there may not be universal agreement as to the appropriate descriptive term [16].

Table 3. Summary of Prevalence Rates per 1,000 Livebirths of Down Syndrome in Four Systematic Studies of Liveborn Infants*

Maternal age		Down syndrome		Livebirths		Rate per 1,000	
< 16	< 20	1	18	1,700	27,476	0.59	0.66
16–19		17		25,776		0.66	
20–24		43		63,461		0.68	
25–29		44		54,130		0.81	
30–34		40		22,909		1,74	
35–39		28		8,280		3.38	
40–44		31		2,041		15.19	
45–49		6		114		52.63	
⩾50		0		2		0	
N.S.		0		1 97		0	
TOTAL		210		178,570		1.18	

*These data are pooled from observations in Atlanta (131,900 white livebirths) [17], Oakland (12,424 white livebirths) [18], the US collaborative study (available data on 22,484 white singleton livebirths) [19], and Edinburgh, Scotland (11,702 livebirths) [20]. The Edinburgh study involved cytogenetic analysis, the others involved phenotypic evaluation. The methods of these studies are discussed elsewhere [15, 21–26]. An analysis by single-year maternal age interval is in progress. Available data on black livebirths in the same US studies indicate no significant difference from the rates presented here and will be presented in detail elsewhere. They are not included here because of a suggestion in the literature of differences in maternal age specific rates of Down syndrome between the races [27]. Exclusion of data on black livebirths will also enable comparison with the data in Tables 1 and 2. The livebirth data in Table 1 are derived from studies among whites, and the individuals participating in the prenatal studies summarized in Table 2 were predominantly whites in the US, Canada, and Europe. For the same reason data from a systematic study of newborns in Jerusalem which found higher rates [28, 29] have not been included because a very large proportion of the group were born in North Africa and the Orient and may have a higher risk of Down syndrome than persons of European origin. It should be noted that despite the report of Stark and White [27] a recent extensive review [30] of the literature found no definite evidence for consistent differences in maternal age-specific differences between races. Even the most suggestive observations, such as reports of very low rates in Maoris [31] and of high rates in Israelis of North African and Oriental origin [28, 29], appear to warrant further confirmation.

N.S. = Not stated.

22 is relatively frequent among fetuses studied in the 1st trimester but extremely rare in livebirths [32]. Trisomy 22 has also been observed among later fetal deaths and affected fetuses may well survive to the time of amniocentesis or later. For example, among fetuses spontaneously lost before the 27th week, Creasy and Crolla found 9 with 47, trisomy 22, of whom 1 had survived to the 16th week [33]. Kuleshov et al observed one instance of 47, trisomy 22 among "antenatally dead" infants (presumably macerated stillbirths) [34].

Many reports of prenatal diagnosis do not indicate whether banding has been done upon all cases observed with 47 chromosomes and an extra group G chromosome. Unless this is clearly stated, it is possible that at least some fetuses diagnosed as having Down syndrome (47, trisomy 21) after amniocentesis actually had 47, trisomy 22. (It is also possible but less likely, that even 47,XYY fetuses may have been diagnosed as 47,XY,+21 Down syndrome in the absence of banding studies, although the Y chromosome should be morphologically distinguishable.) In the Canadian study listed in Table 2 [11–13], for example, instances diagnosed prenatally as trisomy-21 Down syndrome (and other cytogenetic abnormalities) were not always confirmed by banding, particularly in the 1st half of the study [12].

Second, the reporting in the literature may be selective. Relatively small numbers of individuals have undergone prenatal diagnosis at any one center, and it is not unexpected that some laboratories have observed high rates purely because of sampling fluctuation. The studies of Ferguson-Smith [5], Polani et al [7], and others [14] have been published *because* they observed apparently high rates in their prenatal studies compared to those reported in liveborns. It is possible that there are laboratories which have observed lower rates but have not reported them.

Data pertinent to this question are provided by the 2 studies which give rates by single-year maternal age interval and whose results would have been published regardless of what the findings were. These are the USA and Canadian collaborative studies. The data (Table 2) indicate that at least for ages 40–44 there is a difference between the livebirth and amniocentesis studies, although the rates in the collaborative studies are lower than those in the studies of Ferguson-Smith [5] and Polani et al [7]. Thus, while selective reporting contributes to some of the apparent difference, it cannot explain all of it.

Are There Higher Maternal Age-Specific Risks of Down Syndrome Birth to Women Undergoing Amniocentesis?

Women who choose prenatal diagnosis are a small sample of the population of pregnant women at high risk (because of advanced age) and may be unrepresentative even with regard to the age-specific risk.

Polani et al [7] have suggested that women undergoing prenatal diagnosis because of advanced maternal age have a higher risk of Down syndrome than other women of the same age, ie that there are factors associated both with a) choice of the diagnostic procedure and b) a higher age-specific rate of Down syndrome livebirth. They state that in women encountered in their practice who have chosen prenatal diagnosis, there is a "clear suggestion" of histories of "increased" x-ray exposure, "subfertility," and spontaneous miscarriage. No data on matched controls or other comparison groups are presented. It appears plausible that, as they suggest, women with such histories might be more likely to consult with obstetricians at major referral centers prior to pregnancy and thus be more likely to be aware of and to elect amniocentesis. But the relationships of such histories to a markedly increased risk of a Down syndrome conception, although suggestive [35–37], is not definitive.

There is still disagreement as to the role of irradiation as a possible risk factor for Down syndrome, for example. (See [38 and 39] for constrasting evidence.) In the Canadian study an attempt was made to separate out those with histories of multiple miscarriage or known extensive x-ray exposure from the remainder of the older women, particularly where such a history appeared to be related to choice of amniocentesis [12]. Such individuals are excluded from the summary of the Canadian data presented in Table 2. There were 17 women aged 35 or over in whom these additional factors were known to be present, and no known cytogenetic abnormality was detected in the 16 women in whom cytogenetic analysis was successful [11, 12].

If the suggestion of Polani et al [7] concerning the significance of previous x-ray exposure, "subfertility," and fetal loss is correct, rates derived from studies of a selected group of the women undergoing amniocentesis could not be extrapolated to the remainder of the population who do not share the additional putative risk factors. These additional putative factors would, of course, be important for accurate genetic counseling in individual instances and for identifying subgroups who are at a risk over and above that estimated from maternal age alone. These factors are not pertinent, however, to the issue of purported underestimates of the "true" rates in livebirth studies.

Comparison by 5-Year Intervals Will Produce Artifactual Differences in Rates

After age 33 the birth rate drops very rapidly, while the rate of Down syndrome livebirths increased markedly (Table 4). These 2 factors may result in artifacts if rates by 5-year maternal age intervals are compared without appropriate age adjustment [43]. For example, in the 35–39-year interval the largest group having infants is age 35, the second largest is 36, etc. Within this interval the younger women have the lowest rate of Down syndrome occurrences but contri-

Table 4. Comparison of Total Numbers of Livebirths and Estimated Rates of Down Syndrome in Livebirths by Single Year Interval*

Maternal age	USA livebirths, 1975 [41]	Estimates of rates of Down syndrome per 1,000 livebirths: New York State study [1]	Swedish study [42]
< 16	46,092	--	--
16	74,138	--	--
17	119,682	--	--
18	162,281	--	--
19	192,687	--	--
20	207,891	0.52	0.64
21	216,818	0.59	0.67
22	221,662	0.65	0.71
23	225,125	0.71	0.75
24	222,180	0.77	0.79
25	217,680	0.83	0.83
26	208,515	0.89	0.87
27	197,780	0.95	0.92
28	184,371	1.01	0.97
29	128,440	1.07	1.02
30	106,576	1.13	1.08
31	88,782	1.21	1.14
32	77,649	1.38	1.25
33	57,225	1.69	1.47
34	45,268	2.15	1.92
35	35,724	2.74	2.51
36	26,932	3.49	3.28
37	21,986	4.45	4.28
38	17,520	5.67	5.60
39	13,247	7.21	7.32
40	10,150	9.19	9.57
41	6,615	11.71	12.51
42	4,665	14.91	16.36
43	3,041	19.00	21.39
44	1,841	24.20	27.96
45	922	30.84	36.55
46	411	39.28	47.79
47	190	50.04	62.47
48	78	63.75	81.67
49	27	81.21	106.76
Total	3,144,198		

*Excluding persons aged ⩾ 50 or for whom the age was not stated.

bute the bulk of livebirths. Among those undergoing amniocentesis because of advanced maternal age, however, there are proportionally fewer aged 35 and 36 but more aged 37, 38, and 39, in whom the risk of Down syndrome birth is greater. (See Table 5 for data illustrating the difference in distribution within the 35–39 and 40–44 quinquennia.) For this reason, comparisons by 5-year intervals between rates in livebirths and in amniotic fluid cells obtained at the 2nd trimester would result in higher rates in the latter, even if there were identical rates by single-year maternal age interval. This factor may explain a significant portion of the discrepancy in quinquennial rates between the 2 types of studies.

Comparison by Single-Year Intervals Without Adjustment for Gestational Age Will Also Result in an Artifactual Difference

Almost all analyses of maternal age are age-truncated, that is, the ages are those of each woman involved at her last birthday at the time of observation. Thus the average age of all women reported as having livebirths at "age 35," for example, is about 35.4. (It is not 35.5 because of the rapid fall in birth rate between ages 35 and 36.) These women have an average age at conception of about 34.7 years, assuming an average gestation of 40 weeks (0.7 years). Those undergoing prenatal diagnosis at "age 35" are also, on the average, 35.4 at the time of observation. But amniocentesis usually occurs at 16–18 weeks of gestation (0.3 years from conception). The average age at conception is therefore 35.1, and the average age at time of expected childbirth is 35.8. In other words, women whose (truncated) age is 35 at amniocentesis are on the average 0.4 years *older* at the time of conception than those having a delivery at this age.

Unless, prior to all comparisons, all ages are corrected for this difference, the rates observed in studies of amniocentesis by single-year maternal age interval should be *higher* than those observed in livebirths. This would be the difference associated with an average 0.4-year greater age. Based on regression equations estimating the rate of Down syndrome as a function of maternal age [1, 42], 0.4 year is associated with a 10% difference in rate. This artifact is present irrespective of any of the other 5 factors discussed and can be eliminated by adding an average of 0.4 year to the ages of those undergoing amniocentesis before undertaking a comparison with rates in livebirths.*

*Since maternal ages are usually reported to at most a single year, the approach would have to involve plotting (or regression analysis) of rates for those undergoing amniocentesis as if they occurred 0.4 year later than reported, ie at 35.4, 36.4, 37.4, etc. Rates at 35.0, 36.0, 37.0, etc would be derived from this analysis by interpolation (or from the regression equation) and compared with the rates in mothers reported as having livebirths at the latter ages.

TABLE 5. Comparison of Proportion of Births by Single-Year Maternal Age Interval and Those Undergoing Amniocentesis in 35–39 and 40–44 Quinquennia

	Livebirths			Amniocentesis	
Maternal age	USA 1975 [41]	New York State 1975 [44]	US collaborative study [8]	All four studies in Table 2	Ratios for USA observations[a]
35	31.0%	30.4%	9.6%	11.8%	0.31
36	23.3%	23.5%	14.6%	12.3%	0.63
37	19.1%	19.1%	20.9%	19.1%	1.09
38	15.2%	15.6%	25.1%	25.6%	1.65
39	11.5%	11.4%	29.7%	31.6%	2.56
Total 35–39	100.1% (115,409)	100.0% (11,694)	99.9% (239)	100.0% (853)	1.00
40	38.6%	40.0%	34.3%	34.3%	0.89
41	25.1%	25.7%	25.8%	25.8%	1.03
42	17.7%	18.5%	18.9%	20.8%	1.07
43	11.6%	10.8%	9.0%	11.9%	0.78 } 1.25
44	7.0%	5.0%	12.0%	7.2%	1.71 }
Total 40–44	100.0% (26,319)	100.0% (2,497)	100.0% (233)	100.0% (959)	1.00

[a]Ratios of percents within maternal-age quinquennium of women having amniocentesis (US collaborative study data) to percents of women having livebirths in the US in 1975.

An Increased Fetal Death Rate Is Associated With Chromosome Abnormalities

While most fetal losses occur in the 1st trimester, a certain proportion of fetuses undergo spontaneous fetal death after the 16th week of gestation [45]. There are no complete cytogenetic studies of fetuses that have been lost after this point, but 2 small studies provide data on stillbirths, which are usually defined as spontaneous fetal deaths occurring after the 28th week of gestation [46]. Machin and Crolla found 1 trisomy 21 case in 156 stillbirths, or 0.6% [47], and in an apparently similar study, Kuleshov et al, found 1 case in 83 stillborn infants or 1.2% [34]. The age composition of the women in these studies was not presented, so comparisons by quinquennium with rates for livebirths (Tables 1 and 3) are not possible. Nevertheless, it is noteworthy that the observed rate of trisomy 21 among the entire group in these 2 small studies of stillbirths is 5–10 times the rates of 1.2 to 1.4 per 1,000 observed among total livebirths in the studies in Tables 1 and 3.

These data are consistent with a high spontaneous fetal death rate from Down syndrome in the latter part of gestation. This would obviously explain some of the differences between the rates observed in livebirths and those diagnosed in amniotic fluid cells obtained at the 16th–18th week of fetal life.

Regarding the period from the 16th to the 27th week of gestation, no direct data are available. Creasy and Crolla, however, reported that of 8 trisomy 21 and 7 trisomy G (ie trisomy 21 or 22) fetuses they studied who had been lost between the 10th and 26th week of gestation, 3 had been lost after the 16th week (at the 19th, 20th and 22nd week) [33].

The exact proportion of Down syndrome zygotes that results in fetal death is unknown, but appears to be at least 65% [33, 48]. As the studies cited above indicate, a significant fraction of this loss apparently occurs in the 2nd trimester or later. On this interpretation, many instances of 47, trisomy 21 diagnosed at amniocentesis, perhaps as many as half, will undergo spontaneous fetal death if there is no prior medical termination.

The discussion here has been limited to Down syndrome because only for this abnormality are stable rates on livebirths available by single-year maternal age interval or even by 5-year interval. Nevertheless, for other cytogenetic abnormalities – in particular for 47, trisomy 18 and 47, trisomy 13 – there also appears to be a higher rate diagnosed after amniocentesis than in livebirths. The extent of this difference can only be estimated using indirect methods (see Appendix). The possible explanations for this difference are of course similar to those considered for Down syndrome. In particular, a high fetal death rate from these conditions after the 18th or 20th week of gestation appears likely to make a strong contribution.

CONCLUSION

The factors discussed to this point indicate that a difference may be expected between rates of Down syndrome (and other chromosomal disorders) in livebirths and in cultured amniotic fluid cells obtained at 2nd trimester amniocentesis. Even when appropriate age corrections are performed, differences are likely to remain in view of the apparently high spontaneous fetal death rate from Down syndrome after the 16th week of gestation. If the factors mentioned by Polani et al [7] contribute significantly to the empiric risk of a Down syndrome birth, these must also be adjusted for, if possible, in making comparisons between women having livebirths and those undergoing amniocentesis.

Assuming that accurate age-adjusted rates are available, which data should be used for genetic counseling and for program planning – those derived from studies of livebirths or those from studies of amniotic fluid cells?

I can see no alternative to using data derived from studies of livebirths.* It is the occurrence of a disorder in a liveborn infant that results in prolonged human toll and long-term social effects. Occurrences of a spontaneous fetal death or stillbirth to a woman seeking a livebirth also involve a personal loss, but the presence of a chromosome abnormality in the fetus would not be expected to increase this sense of loss for the parent. Indeed, if anything, it may make the event easier to accept.

The opposite view has been voiced by a group who believe a 47, trisomy 21 stillbirth or early fetal loss should be accorded the same weight in estimating projected social toll and in counseling as an affected liveborn infant [49]. It may not be coincidental that this view originates from a pathology unit, whose primary clinical experience is apparently derived at the autopsy table. The distinction between the human toll engendered by a spontaneous stillbirth and that by a liveborn individual who expires and eventually comes to necropsy may not be as readily apparent to pathologists as to the parents themselves and the physicians and clinical geneticists who counsel them.

---

*Of course in a small minority of instances there is a much higher empiric risk of an affected child than is predicted solely on the basis of parental age, eg if there is a 21-translocation, a mosaic trisomy 21 line, or a history of a previous Down syndrome birth to one parent. This discussion applies to the vast bulk of individuals in whom no additional known risk factors are present. If, in the future, the nature and magnitude of putative additional risk factors, such as those discussed by Polani et al [7] are better characterized, these also must be considered in genetic counseling and program planning concerning risks of affected livebirths, where possible.

APPENDIX

Other Cytogenetic Abnormalities

For abnormalities other than Down syndrome it is possible to do only indirect comparisons of rates in livebirths and in those diagnosed prenatally, unlike the direct analysis carried out for Down syndrome. This approach is required because exact data on the association of maternal age with the risk of conditions other than Down syndrome in livebirths are unavailable.

The following discussion is limited to trisomy 13, trisomy 18, 47,XXY, and 47,XXX genotypes. Data on cases diagnosed in four prenatal studies are presented on Table 6. In a total of 1,859 women studied there were 5 instances of trisomy 18, 2 of trisomy 13, and 10 with extra X chromosomes, resulting in rates of 2.7, 1.1, and 5.4 per 1,000, respectively. Stable estimates of livebirth rates of all these conditions are available, but not, unfortunately, rates by single-year interval or by quinquennium as for Down syndrome.

These other conditions are known, however, to be associated with advanced maternal age, and it appears that this association is not stronger for any of them

TABLE 6. Chromosome Abnormalities Other Than Those Associated with Down Syndrome Diagnosed Prenatally in Four Studies by Single-Year Maternal Age Interval

Maternal age	Ferguson-Smith (5)	Polani et al [7]	US collaborative [8–10][a]	Canadian collaborative [11–13]
35	–	–	–	–
36	–	–	–	47,+18;47,XY,7q+
37	–	–	–	–
38	–	XXY	XX/XO; 47, +18; 47, XXY	46, XX, t(5, 16)
39	XXX	–	47, XXX	–
40	–	47,XY, +5		47, +13
41	XXY	46,XX/47,XX, +9	47, +13/46	47,XXY/46,XY
42	–	XXY		47, +18
43	–	–	47,+18	47, +18
44	–	–	47, XXX	–
45	–	XXY	–	–
46	–	XXX	–	–
47	–	–	–	–
48	–	–	–	–
49	–	–	–	–
⩽50	–	–	–	–
NS	–	–	–	–
Total studied[b]	283	642	488	459

[a]Also balanced translocation carriers on whom data are not now available.
[b]From Table 2.

than for Down syndrome. This allows some inferences to be drawn by indirect analysis.

One starts with the crude rates (ie rates unadjusted for maternal age) of various cytogenetic conditions in the total populations of livebirths studied [50]. One then compares these rates with the rates in women diagnosed prenatally because of advancing maternal age. As Table 7 shows, the ratios of the latter to the former rates are much greater than 1.0. There are two reasons for this. The first is that populations of older women studied prenatally have high risks, whereas the livebirths used for comparison were born to women of all fertile ages. The second reason relates to some of the factors discussed in the text, particularly the observations regarding fetal deaths after the 18th week, that explain differences between maternal age-specific rates in livebirths and in those diagnosed prenatally.

The ratio of the crude rates diagnosed prenatally in older mothers to the crude liveborn rates for Down syndrome is 13.7 (Table 7), but the ratios for trisomy 13 and trisomy 18 are 22.0 and 20.4. Because the latter conditions do not appear to have stronger association with maternal age than Down syndrome, it appears that, for these two conditions, there must be a difference between maternal age-specific rates in livebirths and in those diagnosed prenatally, as for Down syndrome. Moreover, since the ratios are greater than for Down syn-

TABLE 7. Comparisons of Crude Rates of Chromosome Abnormalities Detected in Livebirths and in Prenatal Studies of Women in Whom the Indication for Study was Advanced Maternal Age

Statistical parameter	Down syndrome	47,+18	47,+13	XXY and XXX
Number of affected livebirths observed in six cytogenetic studies [50]	71 (+1)[a]	7	3	70
Rate per 1,000 livebirths[b]	1.264	0.123	0.053	1.229
Rate per 1,000 in women studied prenatally because of advanced maternal age (Table 6)	17.304	2.704	1.082	5.408
Ratio of rates	13.7	22.0	20.4	4.4
Relative ratio normalized to a value of 1.0 for Down syndrome	1.0	1.6	1.5	0.3

[a]71 47, trisomy 21 instances plus one interchange trisomy.
[b]56,952 total livebirths [50].

drome, this difference should be greater. For genotypes with an extra X chromosome, however, the ratio is 4.4, lower than for Down syndrome. This could mean that there is less discrepancy than for Down syndrome (or none at all) between maternal age-specific rates of these conditions in livebirths and in those diagnosed prenatally. It could also be solely due to the fact that the association of XXX and XXY genotypes with advancing maternal age is not as strong as for Down syndrome.

Thus, for trisomy 18 and trisomy 13, this analysis supports the existence of an effect similar to that discussed in the body of the text concerning Down syndrome, but of even stronger magnitude. For XXX or XXY genotypes no decision is possible one way or the other.

I believe the most plausible explanation for the observations concerning trisomy 13 and trisomy 18 is that affected fetuses have a higher associated relative risk of fetal death after the 18th week than do those with trisomy 21. There is some direct evidence for this suggestion, although it is based on sparse data. In the cytogenetic studies of stillbirths already cited, Machin and Crolla [47] reported 3 trisomy 18 instances, one trisomy 13 instance, and one 47,XY/47,XXY in 156 stillbirths, compared to one case of 47, trisomy 21. Kuleshov et al [34] found one trisomy 18 instance in 83 stillbirths, and one trisomy 21. Comparisons of pooled data from these two studies with rates in livebirths appear in Table 8. For trisomy 13 and trisomy 18 the apparent relative likelihood of a stillbirth compared with a livebirth is much higher than for trisomy 21, which in turn is higher than for XXX or XXY.

TABLE 8. Comparison of Crude Rates of Chromosome Abnormalities in Studies of Stillbirths and Livebirths

Statistical parameter	Down syndrome	47,+18	47, +13	XXX or XXY
Rate per 1,000 in two stillbirth studies (pooled) [34, 47]	(2/239) = 8.37	(4/239) = 16.74	(1/239) = 4.184	(1/239) = 4.184
Rate per 1,000 in six cytogenetic studies of livebirths (pooled) [50]	1.264	0.123	0.053	1.229
Ratio of rates	6.62	136.10	78.94	3.40
Ratio normalized to a value of 1.0 for Down syndrome	1.0	20.5	11.9	0.5

This is not unexpected, since trisomy 13 and trisomy 18 are much more severe conditions than Down syndrome, while the sex chromosome disorders are milder. Nevertheless, the apparent relative excess for trisomy 18, a 20-fold greater rate, is somewhat surprising. (Trisomy 18 occurs only one-tenth as frequently as Down syndrome in livebirths [50] but twice as frequently in stillbirths [34, 47] . The size of this excess may be due to sampling fluctuation, however. Only sparse numbers have been studied, and the magnitude of the difference may turn out not to be as great as it appears from this comparison.

REFERENCES

1. Hook EB, Chambers GM: Estimated rates of Down syndrome in livebirths by one year maternal age intervals for mothers aged 29–49 in New York State Study – Implications of the risk figures for genetic counseling and cost benefit analysis of prenatal diagnosis programs. In Bergsma D, Lowry RB (eds): "Numerical Taxonomy of Birth Defects and Polygenic Disorders." New York: Alan R Liss for The National Foundation–March of Dimes, BD:OAS: XIII(3A):123–141, 1977.
2. Collman RD, Stoller A: A survey of mongoloid births in Victoria, Australia, 1942–1957. Am J Public Health 52:813–829, 1962.
3. Lindsjo A: Down's syndrome in Sweden. Acta Paediatr Scand 63:571–576, 1974.
4. Hook EB, Fabia, J: Frequency of Down syndrome in livebirths by single year maternal age interval: Results of a Massachusetts study. Teratology. (In press)
5. Ferguson-Smith MA: Prospective data on risk of Down's syndrome in relation to maternal age. Lancet 2:252, 1976.
6. Milunsky A, Atkins L: The frequency of chromosomal abnormalities diagnosed prenatally. In Hook EB, Porter IH (eds): "Population Cytogenetics – Studies in Humans." New York: Academic Press, 1977, pp 11–26.
7. Polani PE, Alberman E, Berry AC, Blunt S, Singer JD: Chromosome abnormalities and maternal age. Lancet 2:516–517, 1976.
8. The NICHD National Registry for Amniocentesis Study Group: Midtrimester amniocentesis for prenatal diagnosis – Safety and accuracy. JAMA 236:1471–1476, 1976.
9. Lubs HA: Incidence of abnormal findings from amniocentesis. Report presented at Symposium on the National Registry for Amniocentesis: Results and Implications of the study. October 20, 1975, Washington, DC.
10. Bryla D: Personal communication.
11. Simpson NG, Dallaire L, Miller JR, Siminovich L, Hamerton JL, Miller J, McKeen C: Prenatal diagnosis of genetic disease in Canada: Report of a collaborative study. Can Med Assoc J 115:739–746, 1976.
12. Simpson NG: Personal communication.
13. Anonymous: Diagnosis of Genetic Disease by Amniocentesis – A Canadian Study. Report No. 5, Ottawa: Medical Research Council, 1977, pp 109–118.
14. Sachs ES, Jahoda MGJ, Niermeijer AF, Galjaard H: An unexpected high frequency of trisomic fetuses in 229 pregnancies monitored for advanced maternal age. Hum Genet 36:43–46, 1977.
15. Christianson RE: Down syndrome and maternal age. Lancet 2:1198, 1976.
16. White C, Bailor JC III: Retrospective and prospective methods of studying association in medicine. Am J Public Health 46:35–44, 1956, cited by MacMahon B, Pugh TF, "Epidemiology: Principles and Methods." Boston: Little Brown, 1970, p 44.

17. Erickson JD: Personal communication.
18. Christianson RE: Personal communication.
19. Rosenberg L: Personal communication.
20. Ratcliffe S: Personal communication.
21. Erickson JD: Racial differences in the incidence of congenital malformations. Ann Hum Genet 39:315–320, 1976.
22. Center for Disease Control: Metropolitan Atlanta Congenital Defects Program: Report No. 1, April 1, 1969.
23. Flynt JW, Ebbin AJ, Oakley GP, Falek A, Heath CW: Metropolitan Atlanta Congenital Defects Program. In Hook EB, et al (eds): "Monitoring, Birth Defects and Environment – The Problem of Surveillance." New York: Academic Press, 1971, pp 155–158.
24. Slone D, Shapiro S, Heinonen OP, Monson RR, Hartz SC, Siskind V, Rosenberg L, Mitchell AA: Maternal drug exposure and birth defects. In Kelly S, Hook EB, Janerich D, Porter IH (eds): "Birth Defects: Risks and Consequences." New York: Academic Press, 1976, pp 265–274.
25. Heinonen O, Slone D, Shapiro S: "Birth Defects and Drugs in Pregnancy." Littleton: Publishing Sciences Group, 1977, pp 8–10.
26. Jacobs PA, Melville A, Ratcliffe S, Keay AJ, Syme J: A cytogenetic study of 11,680 newborn infants. Ann Hum Genet 37:359–376, 1974.
27. Stark CR, White ND: Cluster analysis and racial differences in risk of Down's syndrome. In Hook EB, Porter IH (eds): "Population Cytogenetics – Studies in Humans." New York: Academic Press, 1977, pp 275–284.
28. Wahrman I, Fried K: The Jerusalem prospective newborn survey of mongolism. Ann NY Acad Sci 171:341–360, 1970.
29. Harlap S: Down's syndrome in West Jerusalem. Am J Epidemiol 97:225–232, 1973.
30. Hook EB, Porter IH: Human population cytogenetics: Comments on racial differences in frequency of chromosome abnormalities, putative clustering of Down's syndrome, and radiation studies. In Hook EB, Porter IH (eds): "Population Cytogenetics – Studies in Humans." New York: Academic Press, 1977, pp 353–365.
31. Morris R; Down's syndrome in New Zealand. NZ Med J 73:195–198, 1971.
32. Hamerton JH, Canning N, Ray M, Smith S: A cytogenetic survey of 14,069 infants. Clin Genet 8:223–245, 1975.
33. Creasy MR, Crolla JA: Prenatal mortality of trisomy 21 (Down's syndrome). Lancet 1: 473, 1973.
34. Kuleshov NP, Alekhin NI, Egolina NA, Karetnikova NA: Frequency of chromosome abnormalities among infants dead (sic) during the perinatal period. Genetica 11:(No. 11) 107–113, 1975.
35. Alberman E, Elliot M, Creasy M, Dhadial R: Previous reproductive history in mothers presenting with spontaneous abortions. Br J Obstet Gynaecol 82:366–373, 1975.
36. Alberman E, Polani PE, Fraser-Roberts JA, Spicer CC, Elliot M, Armstrong E: Parental exposure to X-irradiation and Down's syndrome. Ann Hum Genet 36:195–208, 1972.
37. Alberman E, Polani PE, Fraser-Roberts JA, Spicer CC, Elliot M, Armstrong E, Ohadal RK: Parental X-irradiation and chromosome constitution in their spontaneously aborted fetuses. Ann Hum Genet 36:185–194, 1972.
38. Uchida IA: Maternal radiation and trisomy 21. In Hook EB, Porter IH (eds): "Population Cytogenetics – Studies in Humans." New York: Academic Press, 1977, pp 285–299.
39. Cohen BH, Lilienfeld AM, Kramer S, Hyman LC: Parental factors in Down's syndrome – Results of the second Baltimore case-control study. In Hook EB, Porter IH (eds):

"Population Cytogenetics – Studies in Humans." New York: Academic Press, 1977, pp 301–352.
40. Anonymous: Diagnosis of Genetic Disease by Amniocentesis – A Canadian Study. Report No. 5, Ottawa: Medical Research Council, 1977, p 120.
41. Provisional data from National Center for Health Statistics. "US Vital Statistics, 1975." Washington, DC.
42. Hook EB, Lindsjo A: Down syndrome in livebirths by single year maternal age interval in a Swedish study: Comparison with results from a New York state study. Am J Hum Genet 30:19–27, 1978.
43. Hook EB: Risk of Down's syndrome in relation to maternal age. Lancet 2:465, 1976.
44. Office of Biostatistics, New York State Department of Health, Special tabulation, 1977.
45. Abramson FD: Spontaneous fetal death in man: A methodological and analytical evaluation. Doctoral dissertation, University of Michigan, 1971, University Microfiles, p 86.
46. MacNalty AS (ed): "British Medical Dictionary." London: Caxton, 1963, p 1354.
47. Machin GA, Crolla JA: Chromosome constitution of 500 infants dying during the perinatal period. Humangenetik 23:183–198, 1974.
48. Kajii T, Ohama K, Niikawa N, Ferrier A, Avirachan S: Banding analysis of abnormal karyotypes in spontaneous abortion. Am J Hum Genet 24:539–547, 1973.
49. Bain AD, Smith II, Bauld R, Riley SB, Watson J: Prenatal diagnosis, stillbirths, and the macerated fetus. Lancet 2:375, 1976.
50. Hook EB, Hamerton JH: The frequency of chromosome abnormalities detected in consecutive newborn studies – Differences between studies – Results by sex and severity of involvement. In Hook EB, Porter IH (eds): "Population Cytogenetics – Studies in Humans." New York: Academic Press, 1977, pp 63–80.

The Interregional Cytogenetic Register System (ICRS)*

Gerald H. Prescott, DMD, Marian L. Rivas, PhD, Linda Shanbeck, MS, David W. Macfarlane, MD, Herman E. Wyandt, PhD, W. Roy Breg, MD, Herbert A. Lubs, MD, R. Ellen Magenis, MD, Robert L. Summitt, MD, Catherine G. Palmer, PhD, Frederick Hecht, MD, William Kimberling, PhD, and Dorothy Clow

The Interregional Cytogenetic Register System (ICRS) was initiated in 1973 and had been in the process of development until 1976 when the Register became operational [1]. The goal was to design and maintain a registry of clinical and cytogenetic information which would be useful in providing better health care and delivery in the area of cytogenetics directly through the implementation of genetic services (diagnosis, counseling, and follow-up) and indirectly through genetic research.

When fully operational, the Registry could be instrumental as a resource in the following areas of research and better health care services (the research approach to each general area will require detailed protocols that are not discussed in this report):

A. GENETICS OF CHROMOSOMAL ABERRATIONS
1. Recurrence risk estimates
2. Mutation rate estimates
3. Natural selection
    a. Prenatal: fertility, segregation distortion, etc
    b. Postnatal: morbidity/mortality
4. Frequency and location of breakpoints

*This project is funded by Contract NOI-HD-32713 awarded by the National Institute of Child Health and Human Development.

Birth Defects: Original Article Series, Volume XIV, Number 6C, pages 269–279

5. Gene mapping (family studies/deletion-duplication/somatic cell hybrid mapping)
6. Population dynamics

B. CLINICAL
1. Phenotype/karyotype correlations
2. Natural history of clinical cytogenetic disorders

C. ETIOLOGIC-DEMOGRAPHIC
1. Temporal and geographic clustering
2. Environmental factors
3. Parental age effects

D. HEALTH CARE NEEDS/SERVICES
1. Geographic distribution of referrals
2. Sources of referrals
3. Natural history of cytogenetic disorders
4. Follow-up studies
5. Financial/social burdens of cytogenetic diseases

E. GENETIC COUNSELING
1. Identification of individuals at risk
2. Improved accuracy of information used in genetic counseling (all of the above, A–D, contribute to this end)
3. Permit follow-up counseling as new information becomes available

F. EDUCATION OF HEALTH PERSONNEL

G. INTERNAL HOUSEKEEPING AND RECORD KEEPING

To achieve these objectives the system gathers, on three separate forms designed for a computer data base, information on each patient who has had a cytogenetic study done.

A Procedure Manual was developed to instruct on how items on all three forms are to be filled in. The manual and forms have been copyrighted and are available from the University of Oregon Health Sciences Center at cost. The computer program is also available.

The core of the Registry is the information derived from the Cytogenetics Laboratory Form (Fig. 1), which is divided into three sections. Section I deals with "Sex Chromatin Studies," which include X-chromatin, Y-chromatin, and Y-antigen results, respectively. Section II, entitled "Chromosome Studies," deals with the laboratory procedure and the type of tissue used to obtain each karyotype. We feel the matrix under part D of this section is particularly useful, as it allows one to record different results on a patient 1) from the same tissue, 2) from tissues taken at different times, and 3) from different procedures used on the same cells.

Section III, entitled "Chromosome Results," is broken down into six parts: 1) general karyotype results, 2) sex chromosome complement (including mosaicism), 3) autosomal mosaicism, 4) breakage and somatic rearrangements, 5) the Paris chromosome nomenclature, and 6) a matrix to delineate the specific chromosome results. If the origin of the chromosomal abnormality is known from family or meiotic studies, this information is also recorded. The matrix is flexible and allows essentially any known aberration to be recorded on computer. The breakpoints in a particular band in a chromosome can be specified.

A second form is the Master Information Form (Fig. 2) and is for demographic data. The form includes information about the proband identified in the study as well as the parents.

A third form used for each patient is the Clinical Information Form (Fig. 3). It allows one to list the phenotypic manifestations and/or diagnosis. The form uses a modified ICDA (International Classification of Diseases, Adapted) system, which we have developed. This modified ICDA system is a six-digit code rather than a four-digit code as found in volume 2 of the ICDA Manual [2]. Fifteen phenotypic descriptions can be recorded on this page as well as up to three single gene diagnoses using the code numbers in McKusick's *Mendelian Inheritance in Man* [3]. There are also lines to record source of referral, ascertainment, and the sources of medical information.

The ICRS data base currently is processed on a 128K Digital Equipment Corporation PDP-10 computer. The computer is located at the University of Oregon (Eugene campus) and is accessed via remote terminal from the Genetics Division of the University of Oregon Health Sciences Center in Portland, 100 miles away. The system organization consists of three primary indexed-sequential files corresponding to the three entry forms. Each individual has one demographic record of 304 characters (Master Information Form). No restriction is placed on the number of 365 characters on the second form (Cytogenetic Laboratory Form) for each individual, which are linked by pointers in the file to facilitate retrieval of complete information on each patient. A third record (Clinical Information Form) of 142 characters of phenotypic data is similarly stored on each patient.

This information is readily available through interactive COBOL [4] programs (COBOL conforms to American National Standards Institute specifications as described in the document USA Standard COBOL, X3.23-1968); this allows access to individual patient records for verification, data modification, and complete batch processing for statistical analysis. The files are thoroughly cross-referenced for ease in retrieval of requests for research information.

A special aspect of the Register is the ability to enter and retrieve data on families as well as individuals, since the system identifies all individuals with a unique family member.

ICRS
LABORATORY CYTOGENETICS FORM

*1. REGION: ___ *CENTER: ___

*2. FAMILY NO.: ___ *INDIV. NO. ___

*3. LAB NO.: ___

4. HOSP. NO.: ___ UNIT NO.: ___

*5. TYPE OF FORM: [1] New [2] Add [3] Change [4] Delete (Item): ___

*6. DATE SAMPLE RECEIVED: __/__/__ (Mo. Day Year) 7. DATE OF REPORT: __/__/__ (Mo. Day Year)

I. SEX CHROMATIN STUDIES

A. X–CHROMATIN

[1] Not Tested
[2] Positive
[3] Negative
[4] Unsatisfactory

___ No. Cells Studied
___ % Cells 1 Body
___ % Cells > 1 Body

B. Y–CHROMATIN

[1] Not Tested
[2] Positive
[3] Negative
[4] Unsatisfactory

___ No. Cells Studied
___ % Cells 1 Body
___ % Cells > 1 Body

C. Y–ANTIGEN

[1] Not Tested
[2] Tested
 [1] Positive
 [2] Negative
 [3] Inconclusive

II. CHROMOSOME STUDIES

A. TYPE OF SPECIMEN:

[1] Blood
[2] Bone Marrow
[3] Amniotic Fluid
[4] Skin
[5] Other ___ (Specify)

B. CULTURE FAILURE?

[1] No [2] Yes

C. CHROMOSOME COUNTS:

Number of Cells Observed

≤44	45	46	47	≥48	>2n	Total

D. PROCEDURES USED:

Procedure	Culture Duration	Hour / Day / Week	Mitogen	Mitotic Inhibitor	Stain 1	Sequential Stain 2	Stain 3	No. Cells Karyotyped	Best Quality
1.		H D W							
2.		H D W							
3.		H D W							
4.		H D W							
5.		H D W							
CODES			00-No Mitogen 01-PHA 02-Con A 03-PPD 04-LPS 05-Pokeweed 99-Other	00-Not used 01-Used	00-Unbanded 01-Q 02-G	03-R 04-C 05-G-11	06-Ac. Orange 07-T 08-Ag stain 99-Other		1-Excellent 2-Average 3-Poor

A-268-Rev. 1–10/76

A Fig 1. ICRS laboratory cytogenetics form

III. CHROMOSOMAL RESULTS

A. GENERAL KARYOTYPE RESULTS:
- 1 Normal Karyotype
- 2 Abnormal Karyotype
  - 1 Unbalanced
  - 2 Balanced
  - 3 Not Certain

B. **SEX CHROMOSOME COMPLEMENT:**

1 XX	1 XY
2 XO	2 XXY
3 XXX	3 XYY
9 Other: Specify in matrix	4 XXXY
	9 Other: Specify in matrix

C. AUTOSOMAL MOSAICISM?
- 1 Detected (If so, Number of Cell Lines: ____)
- 2 Not Detected
- 3 Suspected

D. BREAKAGE / SOMATIC REARRANGEMENTS?
- 1 Data Not Available
- 2 Data Available
  - 1 Increased Breakage Observed
  - 2 Somatic Rearrangements Observed
  - 3 Neither Observed

E. CURRENT INTERNATIONAL NOMENCLATURE:

____

____

____

F. SPECIFIC CHROMOSOME RESULTS:

	Aberration Code	Chrom.	Arm	Region	Band	Chrom.	Arm	Region	Band	Origin of Chromosome Abnormality
1.										
2.										
3.										
4.										
5.										
6.										
7.										

N = Not Tested
M = Maternal
P = Paternal
D = DeNovo
U = Unknown
A = Acquired
W = Maternal first
X = Maternal second
Y = Paternal first
Z = Paternal second

ABERRATION CODES

01 Polymorphism

Aneuploidy
- 10 Monosomy, entire chrom.
- 11 Trisomy, entire chrom.
- 12 Tetrasomy, entire chrom.
- 13 Polysomy, > 4
- 14 Double aneuploidy
- 15 Partial trisomy
- 16 Partial monosomy

Polyploidy
- 23 Triploidy
- 24 Tetraploidy
- 25 Polyploidy (> 4n)

Duplications
- 30 Duplication, unspec.
- 31 Direct duplication
- 32 Inverted duplication

Deletions
- 40 Deletion, unspec.
- 41 Terminal
- 42 Interstitial
- 43 Ring chromosome

Isochromosomes
- 50 Isochromosome

Translocations
- 60 Translocation, unspec.
- 61 Centric fusion, Robertson.
- 62 Reciprocal translocation
- 63 Complex translocation
- 64 Philadelphia chromosome

Inversions
- 70 Inversion, unspec.
- 71 Paracentric inversion
- 72 Pericentric inversion

Insertions
- 80 Insertion, within a chrom., unspec.
- 81 Insertion between chromosomes, unspec.
- 82 Direct insertion, within a chrom.
- 83 Direct insertion, between two chroms.
- 84 Inverted insertion, within a chrom.
- 85 Inverted insertion, between two chroms.

Other
- 90 Polycentric (> 2) chromosome
- 91 Fragments
- 92 Polyradials
- 93 Gaps and /or breaks
- 94 Marker chromosome

G. CELLS STORED?
- 1 No
- 2 Yes
  - 1 Local
  - 2 Camden
  - 3 Other ____________ Specify

Trouble Coding? 2 No 1 Yes

____ Recorder

__ / __ / __ Mo. Day Year Entered

B

ICRS
MASTER INFORMATION FORM

*1. REGION: ____ *CENTER: ____

*2. FAMILY NO.: ____ *INDIV. NO. ____

*3. LAB NO.: ____

4. HOSP. NO.: ____ UNIT NO.: ____

*5. TYPE OF FORM: [1] New [2] Add [3] Change [4] Delete (Item): ____

PATIENT

6. NAME: ____ (Last / First / Middle)

6.A. MAIDEN SURNAME ____

7. SEX: [1] Male [2] Female [3] Ambiguous [4] Not Recorded

8. BIRTHDATE: __/__/__ (Mo. Day Year) Place of Birth: ____ (City / State / Country)

9. **MULTIPLE BIRTH?** [1] **No** [2] **Twin-Monozygous** [3] **Twin-Dizygous** [4] **Twin-Zygosity Unknown** [5] **Other** ____

10. RACE: [1] Not Recorded [2] Caucasian [3] Black [4] Oriental [5] Am. Indian [6] Other ____

11. **RELIGION:** [1] **Not Recorded** [2] **Protestant** [3] **Catholic** [4] **Jewish** [5] **Mormon** [6] **Other** ____

12. DECEASED? [1] No [2] Yes [3] Not Known DATE OF DEATH: __/__/__ (Mo. Day Year)

13. MARITAL STATUS: [1] Not Recorded [2] Single [3] Married [4] Divorced [5] Other: ____

14. PEDIGREE AVAILABLE? [1] Yes [2] No

15. ADOPTED? [1] No [2] Yes [3] No Information Available

16. LEGAL GUARDIAN OR RESPONSIBLE AGENT: ____

FATHER

17. NAME: ____ (Last / First / Middle) *INDIV. NO. ____

18. BIRTHDATE: __/__/__ (Mo. Day Year) PLACE OF BIRTH: ____ (City / State / Country)

19. RACE: [1] Not Recorded [2] Caucasian [3] Black [4] Oriental [5] Am. Indian [6] Other ____

20. **RELIGION:** [1] **Not Recorded** [2] **Protestant** [3] **Catholic** [4] **Jewish** [5] **Mormon** [6] **Other** ____

21. DECEASED? [1] No [2] Yes [3] Not Known DATE OF DEATH: __/__/__ (Mo. Day Year)

MOTHER

22. NAME: ____ (Last / First / Middle) *INDIV. NO. ____

23. MAIDEN NAME: ____ (Last)

24. BIRTHDATE: __/__/__ (Mo. Day Year) PLACE OF BIRTH: ____ (City / State / Country)

25. RACE: [1] Not Recorded [2] Caucasian [3] Black [4] Oriental [5] Am. Indian [6] Other: ____

26. RELIGION: [1] Not Recorded [2] Protestant [3] Catholic [3] Jewish [5] Mormon [6] Other: ____

27. DECEASED? [1] No [2] Yes [3] Not Known DATE OF DEATH: __/__/__ (Mo. Day Year)

3/9/76

____ Recorder

__/__/__ (Mo. Day Year) Entered

Fig. 2. ICRS master information form

ICRS
CLINICAL INFORMATION FORM

*1. REGION: ___ CENTER: ___

*2. FAMILY NO.: ___ INDIV. NO. ___

*3. LAB NO.: ___

4. HOSP. NO.: ___ UNIT NO.: ___

* 5. TYPE OF FORM: [1] New [2] Add [3] Change [4] Delete (Item): ___

6. Ascertainment: [1] Proband [2] Not Proband

6.A. Referred by ___ Code ___

___ Local Code ___

7. Reason Referral ___

___

___ Code ___

___ Local Code ___

8. Relationship ___ 9. Clin. Phenotype: [1] Clinically Normal [2] Affected [3] Unknown

10. Clinical Manifestations / Diagnosis Code: ICDA

11. ___
12. ___
13. ___
14. ___
15. ___
16. ___
17. ___
18. ___
19. ___
20. ___
21. ___
22. ___
23. ___
24. ___
25. ___

McKusick Code

26. ___ McK ___
27. ___ McK ___
28. ___ McK ___

29. Source Medical Information: [1] Hospital Records [2] Genetic Clinic Records
[3] Private Physician [4] Specialist's Report [5] Autopsy Data [6] Family "Hearsay"
[7] Church Records [8] State Vital Statistics [9] Other / Unknown ___
Specify

30. Comments/ References ___

___

31. Trouble Flag [1]

A-267-Rev. 1–10/76

Recorder ___ Entered ___

Fig. 3. ICRS clinical information form

TABLE 1. Chromosomal Aberrations of Heteromorphisms on File (November 1, 1977)

Aberrations	Center 1	Center 2	Center 3	Center 4	Center 5	Center 6	Total
Monosomy, entire chromosome	7	2	0	2	14	1	26
Trisomy, entire chromosome	382	60	24	263	579	58	1,366
Tetrasomy, entire chromosome	2	0	0	0	5	1	8
Polysomy (>4) (extra chromosomes)	1	0	0	0	0	0	1
Double aneuploidy	5	0	0	3	2	0	10
Partial trisomy	33	7	0	29	32	6	107
Partial monosomy	3	5	0	22	12	0	42
Triploidy	1	0	1	2	1	0	5
Tetraploidy	2	0	0	0	0	0	2
Polyploidy (>4N)[a]	0	0	0	0	0	0	0
Duplication: unspecified	1	1	0	0	0	0	2
Direct duplication	0	1	0	0	0	0	1
Inverted duplication[a]	0	0	0	0	0	0	0
Deletion: unspecified	21	1	0	1	1	0	24
Deletion: terminal	13	6	2	18	23	3	65
Deletion: interstitial	2	0	0	2	0	1	5
Ring chromosome	4	1	1	8	14	1	29
Isochromosome	11	2	0	10	20	3	46
Translocation: unspecified	37	3	1	7	7	1	56
Translocation: centric fusion	29	5	3	27	37	53	154
Translocation: reciprocal	22	6	1	72	29	10	140
Translocation: complex	1	0	0	0	0	1	2

Philadelphia chromosome	37	1	0	0	39	9	86
Inversion: unspecified	1	0	0	0	1	0	2
Inversion: paracentric	0	1	0	0	0	1	2
Inversion: pericentric	17	16	0	43	27	1	104
Inversion: unspecified	0	0	0	0	0	1	1
Insertion between chromosomes	0	0	0	0	1	0	1
Direct insertion within a chromosome	0	0	0	0	0	1	1
Direct insertion between two chromosomes	0	0	0	0	1	0	1
Inverted insertion within a chromosome[a]	0	0	0	0	0	0	0
Inverted insertion between two chromosomes[a]	0	0	0	0	0	0	0
Polycentric (>2) chromosome	0	0	0	0	0	0	0
Fragments	0	1	0	1	2	0	4
Polyradials	0	0	0	0	0	0	0
Gaps and/or breaks	0	0	0	0	2	3	5
Marker chromosome	23	1	0	4	14	0	42
Abnormal sex chromosome	106	24	13	64	91	18	316
Sex chromosome mosaic	35	4	2	38	49	4	132
Autosomal mosaic	54	5	2	22	58	3	144
Increased breakage	29	7	0	9	19	9	73
Somatic rearrangements	58	14	32	0	50	11	165
Heteromorphism	363	302	0	237	89	4	995
TOTALS:	1,300	476	82	884	1,219	204	4,165

[a]When detected, the ICRS will have the capacity to list these types of aberrations.

Total number of studies = 10,349.

Some patients may have more than one aberration or heteromorphism.

Confidentiality measures have been instituted at several levels. Patients' names are recorded only one time, and are mailed under separate cover from cytogenetic and phenotypic data. Similarly, the computer files preserve this separation. Entry to the data base is limited to the primary individuals involved in data processing and is further insured by a system of passwords. Data are checked for accuracy during data entry for identifying information and consistency for each individual. Files are stored on magnetic tape with two back-up copies in case of malfunction, and are restored to the disk only when necessary for processing.

At present the participating centers include Yale University School of Medicine (Dr. W. Roy Breg), University of Colorado School of Medicine (Dr. Herbert A. Lubs), University of Oregon Health Sciences Center (Dr. R. Ellen Magenis), University of Tennessee Center for the Health Sciences (Dr. Robert L. Summitt), Indiana University School of Medicine (Dr. Catherine G. Palmer), New York State Birth Defects Institute Laboratory (Drs. Ian H. Porter and Ernest B. Hook). The director of the Register is Dr. Gerald H. Prescott of the University of Oregon Health Sciences Center.

Each of these centers records patient information on the three forms, which are then sent to Oregon for processing and computerization. Normal and abnormal studies, as well as culture failures, are recorded. Each center receives an updated monthly print-out of its own cases in the following order: 1) laboratory number, 2) alphabetical lists of patients, 3) by chromosomal aberration. It is anticipated that in the near future enough data will be accumulated to permit testing of some of the Register goals regarding patient care and research.

The Register conforms to the rules laid down by each participating institution as well as the federal government in regard to maintaining patient confidentiality [5]. Each patient is referred to by an individual number and, when transporting material, names are not included with the diagnosis. Due to local regulations, some centers do not send patient names at all, but only individual numbers. Much effort has been expended to conform to the Privacy Act which was enacted in 1974 [6].

Another area of concern is quality control of the participating laboratories. Each karyotype is judged excellent, average, or poor according to standardized karyotypes agreed upon by the laboratory directors for each staining technique used. Approximately every 12 weeks a random sample of 10 karyotypes is requested by the Oregon center from each participating laboratory. These samples are checked for uniformity in quality and errors in coding the forms. It has been reassuring and surprising to some that within a few months' experience, uniformity of quality control between laboratories was achieved. Also, the human error involved in handling thousands of cases is less than that predicted.

Given the broad range of problems in research and health care provision, the Register will not have the time or funds to seek solutions to the more complex problems listed at the beginning of this article. Investigators seeking access to data compiled by the Register must submit a request in writing to the director. This request will be considered by the directors of the participating laboratories for permission to use their material. The request should contain the research hypothesis, cost estimates for computer expenses, funding sources, and curriculum vitae and bibliography of the investigators.

Investigators seeking information about individual cases or selected subsets of cases will be referred to the individual directors who performed the studies. We feel this will insure maximum use of the material and allow each center to control with whom and to what extent they collaborate on individual cases. As of December 1, 1977, there were 10,349 cases in the Registry, including 6,184 normal cases and 4,165 cases with a chromosome abnormality or heteromorphism. A summary of the data is given in Table 1.

Cytogenetic laboratories interested in contributing to the Register should contact the project director for the criteria and requisites for membership.

ACKNOWLEDGMENTS

We acknowledge the exceptional efforts of the following people: Susan Bennett, David Dwyer, Joanne Gruen, Helen Howells, Carolyn Fox, and Robin Schwartz.

REFERENCES

1. Emery AEH, Miller JR (eds): "Registers for the Detection and Prevention of Genetic Disease." New York: Stratton Intercontinental Medical Book Corporation, 1976.
2. US Department of Health, Education and Welfare: "International Classification of Diseases (Adapted for Use in the United States)." Public Health Service Publication No. 1693. Washington, DC: US Government Printing Office, 1968, vol 2 (8th rev).
3. McKusick VA: "Mendelian Inheritance in Man," 4th Ed. Baltimore: John Hopkins University Press, 1975.
4. Digital Equipment Corporation: "Cobol Language Handbook," 2nd Ed. Maynard, Massachusetts, 1973.
5. Westin AF: "Computers, Health Records, and Citizen Rights." US Department of Commerce, National Bureau of Standards Monograph 157. Washington, DC: US Government Printing Office, 1976.
6. Bushkin AA, Schaen SI: "The Privacy Act of 1974: A Reference Manual for Compliance." McLean, Virginia: System Development Corporation, 1976.

Are Balanced Translocations Really Balanced? Preliminary Cytogenetic Evidence for Position Effect in Man*

Frederick Hecht, MD, Barbara Kaiser-McCaw, PhD, Shivanand Patil, PhD, and Herman E. Wyandt, PhD

Certain types of chromosome changes are known to be deleterious to Man. Among "bad" chromosome changes are those in which there is extra or missing material. Total or partial duplication or deletion of a chromosome is microscopically detectable. (It is quite possible that some duplications or deletions are still below the level of the microscope, even with fine banding of well-extended chromosomes in prophase or early metaphase.)

Paradoxically, certain submicroscopic chromosome rearrangements resulting in duplication or deletion of genetic material have been found. These involve the Lepore class of abnormal hemoglobins, which are fusion products due to nonhomologous pairing and unequal crossover between globin genes.

Point mutations like the one that results in hemoglobin S are now well known, even though no human eye has seen the change in the DNA of the chromosome.

Certain breaks in chromosomes may also interfere with transcription or translation, but this is as yet unproved.

Various mechanisms thought to affect the human phenotype are summarized in Table 1. The subject of this paper is position effect, the "bottom line" of Table 1.

*Supported by grant CA16747 from the National Cancer Institute for work on the "Cytogenetics of Clonal Neoplasias" and by grant HD08236 from the National Institute of Child Health and Human Development for work on "Chromosome Structure, Function and Behavior." Research was initiated while the authors were in Oregon and continues in Arizona with collaboration from Drs. H. S. Kaplan and A. Epstein, Stanford University.

Birth Defects: Original Article Series, Volume XIV, Number 6C, pages 281–286

TABLE 1. Chromosomal Mechanisms Affecting the Human Phenotype

Mechanism	Microscopically visible	Status
Extra material		
Duplication, total (trisomy)	Yes	Proved
Duplication, partial	Yes	Proved
Duplication, partial	No	Proved
Missing material		
Deletion, total (monosomy)	Yes	Proved
Deletion, partial	Yes	Proved
Deletion, partial	No	Proved
Mutation		
Point mutation	No	Proved
Break site mutation	Yes	Unproved
Position effect	?	Unproved

DEFINITION OF POSITION EFFECT

The term "position effect" is used here to designate a change in the phenotype brought about by a change in the position of genetic material. The phenotype may be affected on a clinical level, resulting in malformations or retardation, or on a cellular or biochemical level. The effect on the phenotype is not due to any net gain or loss of genetic material. *The effect on the phenotype is due solely to repositioning of genetic material.*

Repositioning of genetic material is presumed to affect gene activity; ie genes that are active are inactivated or genes which are inactive are activated, when chromosome material is relocated.

POSITION EFFECT IN OTHER ORGANISMS

The rules regarding position effect come from studies of two genetically favorite animals: the fruit fly (Drosophila) and the mouse. "Position effect" is the name accorded to the suppression of gene activity by a chromosome rearrangement. The term "position effect variegation" is also used.

The rules for position effect are as follows:

1) One break must occur in a heterochromatic region – ie in a region along the chromosome that stains differently and that, presumably, contains repetitive sequences of DNA.

2) A second break must occur in euchromatin. This break must be near a euchromatic gene and result in a rearrangement that brings it into proximity to the broken heterochromatin.

3) The two breaks must result in a rearrangement in which the gene is moved near broken heterochromatin in the cis position on the same chromosome.

This implies that heterochromatin may normally play a role in gene regulation.

4) When reversion of the chromosome rearrangement occurs, returning the chromosome situation to normal, there is reversion of the phenotype from variegated to normal, wild type.

5) Virtually all genes are subject to suppression by position effect from heterochromatin.

This implies that with position effect, gene suppression occurs at the level of transcription or translation, not at the protein level.

6) The intercession of undisturbed (unbroken) heterochromatin can protect a gene from position effect.

This implies that heterochromatin normally derepresses gene activity by working linearly along the chromosome for only a short distance. Derepression and repression of gene activity presumably do not involve a long-range diffusible effect, but rather a short-range intrachromosomal effect.

7) The euchromatic regions of chromosomes tend to lose their capacity to stain as sharply banded regions when they are brought into close proximity to heterochromatin. This cytologic change has also been observed by measuring the relative amounts of ultraviolet-absorbing substances in specific euchromatin bands brought next to heterochromatin. Analogous changes in the uptake of Feulgen stain have also been noted in interband regions involved in the position effect phenomenon.

These and other key aspects of position effect are reviewed lucidly in detail, with references by Lewis [1] and Baker [2].

CIRCUMSTANTIAL EVIDENCE FOR POSITION EFFECT AS A CAUSE OF BIRTH DEFECTS

Before 1974 case reports appeared that describe abnormal children with "balanced" chromosome rearrangements. It was assumed that these cases reflected chance, that is, simply biased ascertainment. How could a chromosome rearrangement in which no material had been lost or gained cause mental retardation or malformations?

In 1974 Jacobs [3] noted that in large population chromosome studies there is an excess of phenotypically subnormal individuals among people with euploid (balanced) structural chromosome rearrangements.

The current evidence indicating that some balanced chromosome rearrangements may result in abnormal phenotypes has been reviewed by Tharapel et al [4] and by Summitt (in this volume).

TRANSLOCATIONS IN CANCER CELLS: CYTOLOGIC EVIDENCE FOR POSITION EFFECT?

A number of human neoplasms are marked by a relatively consistent cytogenetic abnormality. The best known example is the Philadelphia chromosome, which is present in 85–90 percent of typical cases of chronic myelocytic leukemia (CML). The Philadelphia chromosome is usually the result of a translocation from the long arm (q) of a chromosome 22 to a 9q. The CML cells presumably contain the normal amount of genetic information. The chromosome change is not quantitative, but merely morphologic.

This knowlege has led us to suggest that chromosome rearrangements in tumor cells may correlate with changes in genetic expression due to position effect [5]. Genetic information that regulates cell proliferation may be, we feel, situated on the rearranged chromosomes. This information may be lost by chromosome deletion, or its function may be altered by juxtaposition to heterochromatin. This could occur by translocating the genetic information to a heterochromatic region or to a position near heterochromatin that has been broken.

Another equally consistent chromosome rearrangement is the 14q+ chromosome in both African and North American Burkitt lymphoma. The extra material on 14q in Burkitt lymphoma is the result of a balanced translocation from 8q. This represents another case in which a population of neoplastic cells characterized by a nonrandom, apparently balanced translocation appears to have a proliferative advantage. We investigated this translocation to determine whether it was balanced and whether any cytologic or histochemical changes could be detected.

Materials and Methods

Cell lines. The characteristics of two newly established North American Burkitt lymphoma cell lines have been described [6]. These lines have been designated SU-AmB-1 and SU-AmB-2. (SU-AmB = Stanford University–American Burkitt).

African Burkitt lymphoma cell lines were also studied karyotypically; these lines have been described elsewhere [5, 6].

Peripheral blood from two clinically normal individuals was used also to establish short-term (3-day) lymphocyte cultures.

Chromosome preparations. Cytogenetic analysis was done on six cultures: two North American Burkitt lymphomas, two African Burkitt lymphomas, and two normal controls. Metaphase cells were accumulated by adding colchicine (0.01 μg/ml medium) to cultures 1.5 hours before harvest. The cells were then treated with hypotonic KCl (0.075 M), fixed in methyl alcohol:acetic acid (3:1) solution, dropped onto clean glass slides, and air dried. The entire harvest procedure was completed at 20°C. Slides were stained by quinacrine mustard for Q-banding by the acetic-saline Giemsa method for G-banding, and by acridine orange for R-banding [5, 7].

Chromosome analysis. At least ten cells from each line were photographed under oil immersion with a Zeiss photomicroscope equipped for light and fluorescent microscopy with an HBO-200 mercury light source. Photographic enlargements (X 2,400) were scored for chromosome number and morphology. Ten cells were selected on the basis of quality of metaphase and location of the chromosomes. Ten cells from each individual were analyzed for a total of 60 cells.

Semiautomated microphotodensitometry was done using a Zeiss photomicroscope II, equipped with a photodensitometer and scanning stage interfaced to a Tektronix 31 programable calculator. Programs were devised for obtaining standardized densitometric profiles from banded chromosomes [7].

RESULTS AND DISCUSSION

The chromosomes 14 from the two North American and two African Burkitt lymphomas were compared visually. In all four cases the extra chromosomal material on 14q appeared similar in amount and had identical staining characteristics. Using the acridine orange R-banding methods [7], this rearrangement was found to be a translocation from 8q to 14q [5]. This observation was confirmed by semiautomated photodensitometry, which showed that the addition of material from 8q in every case (both African and North American Burkitt lymphomas) was equal to approximately 20% of the total length of a normal 14q.

With acridine orange we have found a color shift in the 8q−:14q+ translocation in all four cases of Burkitt lymphoma examined. We have confirmed this color shift by densitometry and have also noted a change in staining intensity by Q-banding (unpublished results).

Green on acridine orange means native, double-stranded DNA; red means denatured, single-stranded DNA. In classic cytogenetic terms, our findings mean that the heterochromatic material on 8q became euchromatic when it was translocated to 14q.

This does not constitute evidence on a gene level for position effect, since we did not examine gene products. This is merely preliminary cytologic evidence.

It is of interest that chromosome rearrangements leading to position effect in Drosophila are usually accompanied by cytologic changes – namely, by a change in chromaticity.

Our findings are consistent with the concept that one consequence of the 18q−:14q+ translocation in Burkitt lymphoma is position effect. We have also found that the rearrangement of chromosome 14 in ataxia-telangiectasia lymphocyte clones results in similar changes by Q-banding. The details of these findings will be reported elsewhere.

Other chromosome rearrangements currently under study in this regard include the Philadelphia translocation in CML and X-autosome translocations. These observations on chromosome 14 appear to parallel those on the Philadelphia (Ph_1)

chromosome. The Ph_1 chromosome in CML has been shown to be a balanced translocation between chromosome 22 and the long arm of chromosome 9 or, less frequently, between 22 and one of several other chromosomes. The 14q+ translocation may thus represent a comparable tissue-specific, nonrandom chromosomal abnormality, which is characteristic of lymphocytic neoplasia.

How do balanced translocations affect normal cell growth and differentiation? The answer will shed important light on cell growth in cancer. The answer may also, we believe, bear on one cause of mental retardation and physical malformations, namely position effect.

REFERENCES

1. Lewis EB: The phenomenon of position effect. Adv Genet 3:73–115, 1950.
2. Baker WK: Position-effect variegation. Adv Genet 14:133–169, 1968.
3. Jacobs PA: Correlation between euploid structural chromosome rearrangements and mental subnormality in humans. Nature 249:164–165, 1974.
4. Tharapel AT, Summitt RL, Wilroy RS Jr, Martens P: Apparently balanced de novo translocations in patients with abnormal phenotypes. Clin Genet 11:255–269, 1977.
5. Kaiser-McCaw B, Epstein AL, Kaplan HS, Hecht F: Chromosome 14 translocation in African and North American Burkitt's lymphoma. Int J Cancer 19:482–486, 1977.
6. Epstein AL, Henle W, Henle G, Heweston J, Kaplan HS: Surface marker characteristics and Epstein-Barr virus studies of two established North American Burkitt's lymphoma cell lines. Proc Natl Acad Sci USA 73:228–232, 1976.
7. Wyandt HE, Vlietinck RF, Magenis RE, Hecht F: Colored reverse-banding of human chromosomes with acridine orange following alkaline/formalin treatment: Densitometric validation and applications. Humangenetik 23:119–130, 1974.

Duplication (Partial Trisomy) of the Distal Long Arm of Chromosome 17: A New Clinically Recognizable Chromosome Disorder*

Martha S. Berberich, MD, John C. Carey, MD, Helen J. Lawce, BS, and Bryan D. Hall, MD

INTRODUCTION

We have evaluated a family (Fig. 1) in which three cousins had a distinct malformation syndrome. The proposita had a duplication (partial trisomy) of the distal long arm of chromosome 17. The son of a paternal aunt and the daughter of a paternal uncle had clinical features identical to those of the proposita. The aunt, uncle, and father of the proposita had karyotypes that included a balanced reciprocal translocation between the distal long arm of chromosome 17 and the distal long arm of chromosome 21. To our knowledge no previous reports have identified individuals with only trisomy or duplication of the distal long arm of chromosome 17.

CASE REPORTS

Case 1

The proposita (*III-13*), is a 17-year-old white female (Fig. 2). Her nonconsanguineous parents were 24 years old when she was born. The mother had had a previous pregnancy that had resulted in a spontaneous abortion at 3 months. She had a normal daughter by a previous marriage. The proposita was born after

*This paper was supported in part by NIH Center grant GM-19527, MCH Project No. 445, and the Birth Defects Center, National Foundation–March of Dimes grant CE-47.

Birth Defects: Original Article Series, Volume XIV, Number 6C, pages 287–295

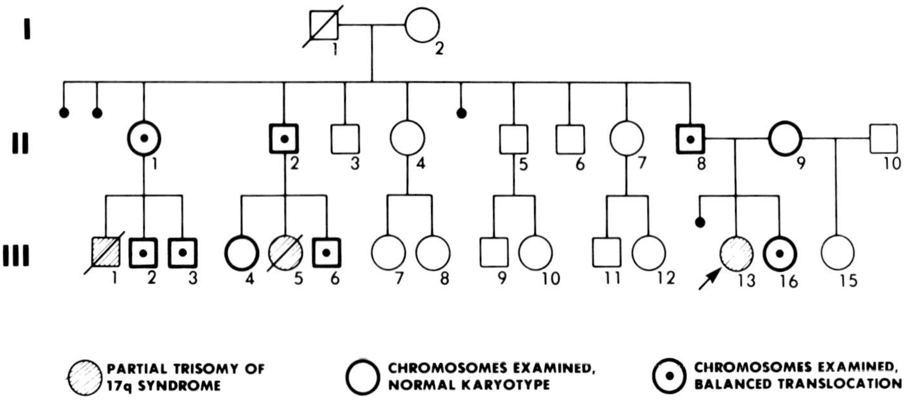

Fig. 1. Family pedigree of the 17q trisomy family. All family members not examined live in the Pacific Northwest and will be studied in the future.

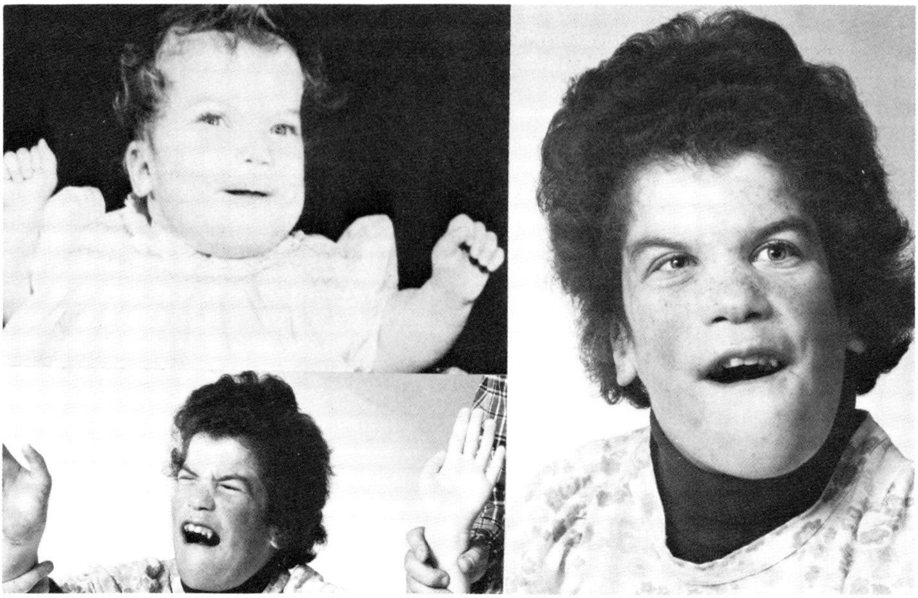

Fig. 2. Top left: Proposita as an infant. Bottom left: Proposita at 17 years of age. Note small ulnar deviated thumb. Right: Note temporal hair, widow's peak, narrow bifrontal diameter, thin upper lip.

an uncomplicated 40-week pregnancy with a birthweight of 3,090 gm (25th percentile) and birth length of 45.7 cm (3rd percentile). No neonatal problems occurred. Her subsequent development was delayed; she held up her head at 1 year, sat alone at 2½ years and walked at 6 years. Her developmental age was less than 1 year when she was 6 years of age. Her height was significantly below the 3rd percentile by 2 years. She developed secondary sexual characteristics and menstruated irregularly beginning at age 13.

Physical examination at age 17 years showed her height (130 cm), weight (36.5 kg), and head circumference (50.75 cm) to be less than the 3rd percentile. Asymmetric microcephaly was present. The forehead was short and had a narrow bifrontal diameter. Scalp hair showed a widow's peak and excessive temporal hair. Her low-set ears were posteriorly rotated with a poorly formed antitragus. The normally spaced eyes gave a "squinty" appearance due to decreased distance between the eyelids and persistent epicanthal folds. Her fundi could not be visualized owing to high myopia. The nasal bridge was depressed, wide, and built-up along its lateral borders. There was malar hypoplasia, broad mouth with thin upper lip and downturned corners, cleft palate, and cleft uvula. The neck was webbed and had a low posterior hairline. She had severe kyphoscoliosis and a resultant right thoracic curve. A soft systolic heart murmur was heard.

Multiple abnormalities of the limbs were present with limited (160°) elbow extension, while her other joints were hyperextensible. The palms and fingers were short, the 5th metacarpals being relatively shorter. There was a mild 2-3-4 cutaneous syndactyly of the right hand and mildly hypoplastic thumbs, which were maintained in ulnar deviation. The fingernails were vertically short and the finger creases poorly formed. Muscle mass was decreased in all limbs. Nonpitting edema was present on the dorsa of the feet.

Results of laboratory studies done prior to the above evaluation included normal amino acid chromatography, bridged sella turcica on skull films, ventricular dilatation on pneumoencephalogram, and a normal intravenous pyelogram. Nonbanded chromosomes had been previously interpreted as normal; however, trypsin-pretreated, Giemsa-banded chromosomes (Fig. 3) done at this evaluation showed 46,XX,der(21),rcp(17;21)(q23;q22)pat, which represented a partial trisomy, or more properly a duplication of the distal portion of the long arm of chromosome 17. This was verified when the karyotype of the proposita's father (Fig. 4) revealed 46,XY,rcp(1721)(q23:q22). The paternal aunt *(II-1)* and the paternal uncle (*II-2*), who were balanced carriers (Fig. 1), had similar karyotypes. Four other paternal relations *(III-2, III-3, III-6, III-14)* were identified as carrying the same balanced translocation. Neither of the other two abnormal

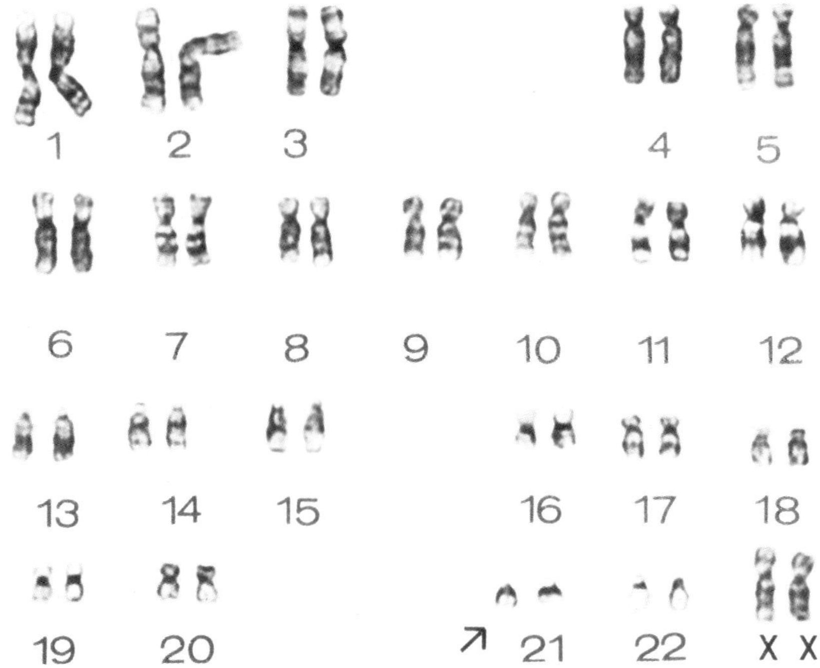

Fig. 3. Giemsa-banded (trypsin-pretreated) karyotype on proposita *(III-13)*, showing extra material on the long arm of chromosome 21. This represents a duplication of the distal long arm of chromosome 17.

paternal relatives *(III-1, III-5)* was alive to karyotype; however, it is assumed they had the same karyotype as the proposita, since they had identical clinical features and because a parent of each *(II-1, II-2)* was a balanced carrier of the same rearrangement.

Case 2 (III-1)

This subject, who was a first cousin of the proposita and the eldest child in his family, weighed 3,005 gm at birth (50th percentile) following a 36-week pregnancy complicated by persistent vaginal bleeding. He was flaccid and unresponsive at birth. His development was extremely limited and he was never able to sit. He died at 10 years of age of subacute bacterial endocarditis. By history, his height, weight, and head circumference fell markedly below the 3rd percentile during the first year of life. His clinical features, derived from photographs (Fig. 5) and history, are strikingly similar to the proposita's (Table 1). He had a cardiac murmur that was proven at autopsy to represent an atrial septal

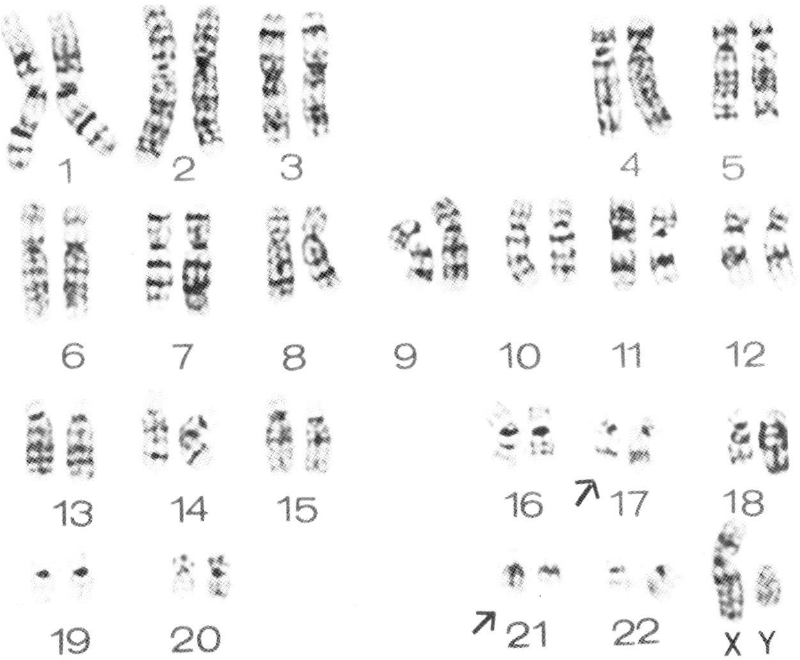

Fig. 4. Giemsa-banded (trypsin-pretreated) karyotype on the father *(II-8)* of the proposita, showing a balanced reciprocal translocation between the long arm of chromosome 17 and the long arm of chromosome 21.

defect. Other autopsy findings included left hydronephrosis, cavernous hemangioma of the left kidney, dilated ureters, bilateral inguinal hernias, intra-abdominal testes, holoprosencephaly, micropolygyria, and brain heterotopias.

Case 3 (III-5)

This subject was the other similarly affected first cousin of the proposita. Her mother had preeclampsia during the third trimester of a 36-week pregnancy. The infant weighed 2,800 gm (50th percentile) at birth. She died at 3 months of age from pulmonary edema secondary to congenital heart defects, which at autopsy were found to be an atrial septal defect and subaortic stenosis. Hospital records and photographs (Fig. 6) show her clinical features (Table 1) to be similar to those of the proposita. Additional autopsy findings included hiatus hernia, hypoplastic and cystic ovaries, cystic kidneys, right hydronephrosis, right tortuous ureter, small bladder diverticulum, and dilatation of the lateral ventricles of the brain.

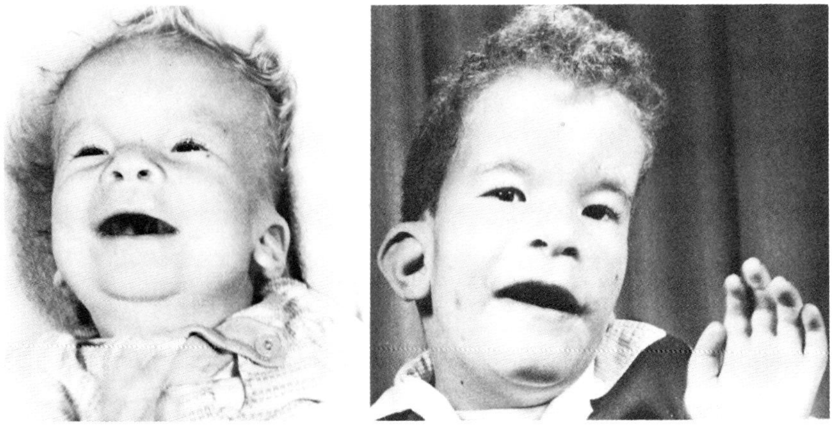

Fig. 5. Left: *Case 2 (III-1)* as an infant. Note cleft palate, thin upper lip. Right: Subject at 8 years. "Squinty" eyes, narrow bifrontal diameter, abnormal ears, and small ulnar deviated thumbs are evident.

DISCUSSION

Although the amount of genetic material present in this family's duplication of the distal long arm of chromosome 17 is relatively small, the clinical features are marked, unusual, and distinctive. For this reason duplication of the distal long arm of chromosome 17 can be classified as a clinically recognizable syndrome. Table 1 reviews the major clinical features found in the proposita (*Case 1,* III-13) and her two similarly affected first cousins (*Case 2, III-1* and *Case 3, III-5*). Additional features found in at least one of three affected individuals but not listed in the table include excessive temporal hair, high myopia, cleft uvula, hiatus hernia, cystic ovaries, bladder diverticulum, intraabdominal testes, dislocated hip, decreased muscle mass, mild unilateral 2-3-4 cutaneous finger syndactyly, short and tapering fingers, short palms, and vertically short fingernails.

Abnormalities of chromosome 17 appear to be relatively rare [1]. This is the first report of duplication (trisomy) of only the distal long arm of chromosome 17 and the fourth to describe a break in the long arm of chromosome 17. One previous report noted possible monosomy for the distal long arm of chromosome 17 in a spontaneously aborted fetus [1]. Another report mentioned an apparently normal newborn with a balanced translocation similar to that in the parents of the patients reported here [2]. In neither of these cases was chromosome banding utilized. A third article presented a patient with an isochromosome 17 who had some features that were similar to our patient's; however, that patient [3] was probably monosomic for the short arm of chromosome 17 as well as trisomic for some portion of the long arm of chromosome 17.

It should be noted that the family presented herein is also at risk of having offspring with deletion of the distal portion of the long arm of chromosome 17. Why this event has not occurred in this family may be a matter of chance, selection, or nonviability of a 17q– fetus. Additionally, the proposita may lack a

TABLE 1. Major Clinical Features of Partial Trisomy 17q

Features	Case 1	Case 2	Case 3	No.	%
Growth deficiency (postnatal)	+	+	+	3/3	100
Mental retardation (severe)	+	+	+	3/3	100
Microcephaly (asymmetric)[a]	+	+	+	3/3	100
Ventricular abnormalities (brain)	0	+	+	2/3	66
Facies					
Narrow bifrontal diameter	+	+	+	3/3	100
Squinty eyes[b]	+	+	+	3/3	100
Epicanthal folds (persistent)	+	+	+	3/3	100
Malar hypoplasia	+	+	+	3/3	100
Flat, wide nasal bridge	+	+	+	3/3	100
Large mouth (down-turned corners)	+	+	+	3/3	100
Thin upper lip	+	+	+	3/3	100
Cleft palate	+	+	+	3/3	100
Ear abnormalities[c]	+	+	+	3/3	100
Neck webbed	+	+	0	2/3	66
Kyphoscoliosis	+	+	0	2/3	66
Congenital heart defects[c]	0?	+	+	2/3	66
Kidney anomalies[c]	0	+	+	2/3	66
Limbs					
Hypoplastic thumbs (mild)	+	+	0	2/3	66
Ulnar deviated thumbs	+	+	+	3/3	100
Pedal edema (feet)	+	+	0	2/3	66
Hyper- and hypoextensible joints	+	+	+	3/3	100

[a]None of the three cases had head circumference measurements as newborns.
[b]Result of persistent epicanthal folds, decreased distance between upper and lower eyelids in presence of normal palpebral fissure length.
[c]See text for description and listing of specific defects.

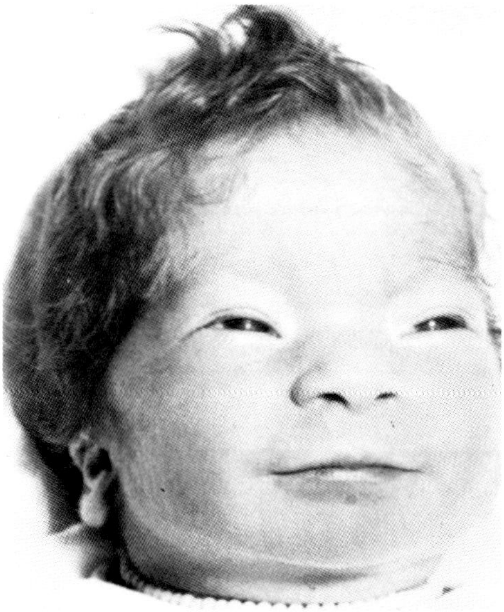

Fig. 6. *Case 3* (III-5) as a newborn. Note "squinty" eyes, temporal hair, and flat wide nasal bridge with built-up lateral borders.

distal segment of the long arm of chromosome 21. In view of the lack of clinical features in the proposita of 21q– we feel that the small amount of missing material in the long arm of 21 may not be genetically active (or is minimally active) and does not contribute to the phenotypic abnormalities.

We hope the clinical delineation of duplication (partial trisomy) of the distal long arm of chromosome 17 will lead to a more rapid identification of other affected persons, so that involved families can receive accurate genetic counseling. This will also allow detection of any balanced carriers before they embark upon having children without knowing the potential and detectable problems that may be present in the future.

ACKNOWLEDGMENTS

We would like to thank Mary Dessau for typing this manuscript, and the family reported herein for their constant support and interest while we were attempting to elucidate the nature and genetics of their problems.

REFERENCES

1. Ikeuchi T, Fujimoto S: A familial transmission of heterozygous no. 17 chromosomes in relation to spontaneous abortion. Jpn J Hum Genet 43:383, 1968.
2. Bochkov NP, Kuleshov NP, Chebotarev AN, Alekhin VI, Midian SA: Population cytogenetic investigation of newborns in Moscow. Humangenetik 22:139, 1974.
3. Salamanca-Gómez F, Armendares S: Identification of isochromosome 17 in a girl with mental retardation and congenital malformations. Ann Génét (Paris) 18:235, 1975.

Trisomy 18 in Sibs and Maternal Chromosome 9 Variant*

Richard M. Pauli, MD, PhD, Roberta A. Pagon, MD, and Judith G. Hall, MD

We recently observed the first instance, to our knowledge, of classic trisomy 18 in two live-born sibs. The mother has a variant of chromosome 9. We raise the question of whether the mother's seemingly unrelated chromosome variant predisposed to nondisjunction.

Family H (Fig. 1) was first seen after the birth of their third child in 1968, who had typical stigmata of trisomy 18 and a 47,XX,+18 karyotype, without evidence of translocation or mosaicism in peripheral blood. The father was 25 and the mother 26 years old at the time of the proband's birth. A subsequent pregnancy resulted in a spontaneous abortion and was followed by an ectopic pregnancy (with no examination of tissue in either case and no examination of ovarian tissue in the latter). Prior to the birth of the next child the family declined the option of amniocentesis for religious and moral reasons. At birth, this child also demonstrated physical findings characteristic of trisomy 18 (Fig. 2) and was found to have a 47,XX,+18 karyotype in leukocyte culture. An older boy and an older girl are normal. The father had an unremarkable karyotype, but the mother was found by C-banding to be heterozygous for a chromosome 9 variant resembling a pericentric inversion of the entire heterochromatin region inv(9qh) (Fig. 3). Chromosome counts of more than 100 cells from each parent revealed no evidence of trisomy 18 mosaicism in either lymphocyte or skin fibroblast cultures.

Recurrence of both identical and dissimilar chromosomal aneuploidies within sibships has been repeatedly reported [1]; these have included the association of trisomy 18 and trisomy 21 [1–5] and the presence of mosaic trisomy 18 [6]

*This study was supported by Gene Center grant GM 15253 from the USPHS and Fellowship Training grant CA-90, The National Foundation–March of Dimes 6-44.

Birth Defects: Original Article Series, Volume XIV, Number 6C, pages 297–301

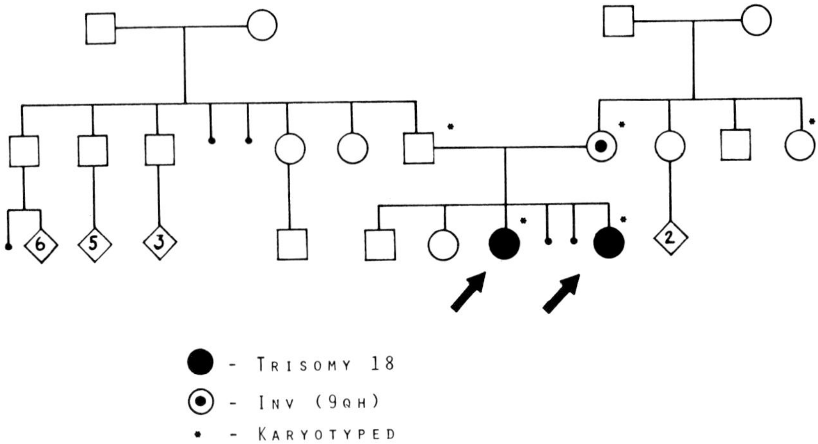

Fig 1. Pedigree of *Family H*

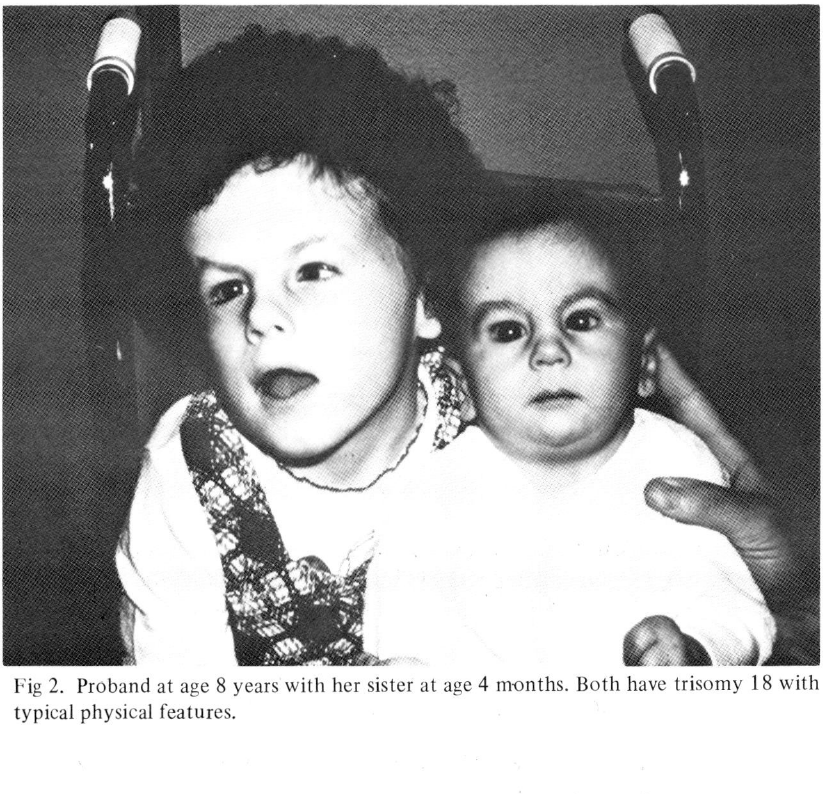

Fig 2. Proband at age 8 years with her sister at age 4 months. Both have trisomy 18 with typical physical features.

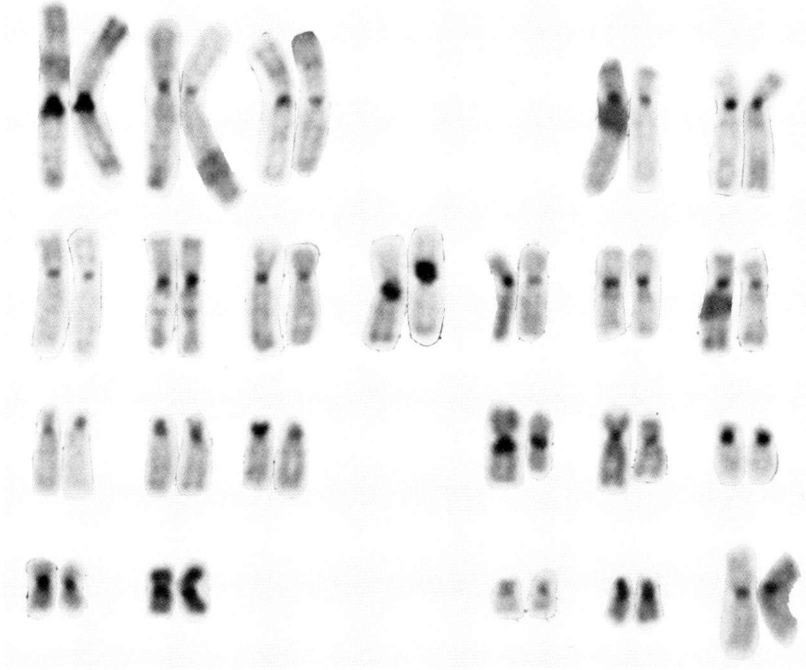

Fig 3. C-banded karyotype of the mother, showing the apparent pericentric inversion of chromosome 9.

in sibs. In one instance, amniocentesis detected a fetus with trisomy 18 in a woman with a previous trisomy 18 child [7]. In that case, attempts were made to detect mosaicism in the mother, but the parental karyotypes were not adequately documented.

Postulated mechanisms for familial recurrence of nondisjunction include parental germinal mosaicism [8, 9], the existence of a recessive gene predisposing to nondisjunction [10], genetic predisposition to autoimmunity resulting in aneuploidy [11], and structural chromosome abnormality of one parent [12]. None has been proved in man.

The apparent pericentric inversion inv (9qh) is a common heritable polymorphism, occurring in 1–2% of the population [13–16]. Individuals heterozygous [13, 17] or homozygous [17, 18] for inv (9qh) do not have abnormal phenotypes, but some investigators have associated inv(9qh) with nondisjunction of chromosomes 8 [19], 9 [20–22], 13 [23], and 21 [19]. What role, if any, the presence of this pericentric inversion in the mother played in the recurrence of trisomy 18 in these sibs is speculative. Possibly this inversion favors nonhomologous pairing from a distributive pool of nonexchange chromosomes in meiosis (as hypothesized but not proved in Drosophilia) [12, 24]. Alternative explanations include a position effect due to the inversion of heterochromatin influencing a gene important in disjunction, or some sort of association of chromosome 18 and the inverted region during meiosis when a unique configuration of the chromosome 9 heterochromatin is observed [25].

Regardless of the mechanism, the recurrence of regular trisomy 18 in two offspring of young parents has several implications. Not only should prenatal diagnosis be offered during subsequent pregnancies to any couple with one trisomic child, but until a prospective study defines the clinical significance of inv(9qh), consideration should be given to the possibility that individuals with this polymorphism may be at increased risk for nondisjunctional events. Banded karyotypes of a series of parents of more than one trisomic child may give insight into structural chromosomal aberrations which may predispose to nondisjunction.

ACKNOWLEDGMENTS

The authors wish to acknowledge Drs David W. Smith, James W. Hanson, Neil Duncanson, and E. Franklin Stone for their contribution of clinical information; Dr Holger Hoehn and Mr Bradley Gong for the thorough cytogenetic investigation; and Mrs Bonnie McLoughlin for editorial and clerical assistance.

REFERENCES

1. Hecht F, Bryant JS, Gruber D, Townes PL: The nonrandomness of chromosomal abnormalities: Association of trisomy 18 and Down's syndrome. N Engl J Med 271:1081, 1964.
2. Turner B, denDulk GM, Watkins G: The 17-18 trisomy and 21 trisomy syndromes in siblings. J Pediatr 64:601, 1964.
3. Holmgren G, Ånséhn S: The trisomy 21 and the trisomy 17-18 syndromes in siblings. Hum Hered 21:577, 1971.
4. Crandall BF, Ebbin AJ: Trisomy 18 and 21 in two siblings. Clin Genet 4:517, 1973.
5. Soudek D, McCreary B, Laraya P: Trisomies 21 and 18 in two sibs; another case. J Pediatr 87:326, 1975.
6. Shih L-Y, Hsu LYF, Sujansky E, Kushnick T: Trisomy 18 mosaicism in two siblings. Clin Genet 5:420, 1974.
7. Atkins L, Milunsky A, Shahood JM: Prenatal diagnosis: Detailed chromosomal analysis in 500 cases. Clin Genet 6:317, 1974.

8. Beratis NG, Kardon NB, Hsu LY, Grossman D, Hirschhorn K: Parental mosaicism in trisomy 18. Pediatrics 50:908, 1972.
9. Hartl DL: Recurrence risks for germinal mosaics. Am J Hum Genet 23:124, 1971.
10. Kwiterovich PO, Cross HE, McKusick VA: Mongolism in an inbred population. Bull Johns Hopkins Hosp 119:268, 1966.
11. Fialkow PJ: Autoimmunity: A predisposing factor to chromosomal aberrations? Lancet 1:474, 1964.
12. Grell RF: Distributive pairing in man? Ann Genet 14:165, 1971.
13. De La Chapelle A, Schröder J, Stenstrand K, Fellman J, Herva R, Saarni M, Anttolainen I, Tallila I, Tervilä L, Husa L, Tallqvist G, Robson EB, Cook PJL, Sanger R: Pericentric inversions of human chromosomes 9 and 10. Am J Hum Genet 26:746, 1974.
14. Madan K, Bobrow M: Structural variation in chromosome No. 9. Ann Genet 17:81, 1974.
15. McKenzie WH, Lubs HA: Human Q and C chromosomal variations: Distribution and incidence. Cytogenet Cell Genet 14:97, 1975.
16. Müller HJ, Klinger HP, Glasser M: Chromosome polymorphism in a human newborn population. II. Potentials of polymorphic chromosome variants for characterizing the ideogram of an individual. Cytogenet Cell Genet 15:239, 1975.
17. Wahrman J, Atida J, Goitein R, Cohen T: Pericentric inversions of chromosome 9 in two families. Cytogenetics 11:132, 1972.
18. Vine DT, Yarkoni S, Cohen MM: Inversion homozygosity of chromosome no. 9 in a highly inbred kindred. Am J Hum Genet 28:203, 1976.
19. Boué J, Taillemite JL, Hazael-Massieux P, Leonard C, Boué A: Association of pericentric inversion of chromosome 9 and reproductive failure in ten unrelated families. Humangenetik 30:217, 1975.
20. Bowen P, Ying KL, Chung GSH: Trisomy 9 mosaicism in a newborn infant with multiple malformations. J Pediatr 85:95, 1974.
21. Schinzel A, Hayashi K, Schmid W: Mosaic-trisomy and pericentric inversion of chromosome 9 in a malformed boy. Humangenetik 25:171, 1974.
22. Martin AO, Turk KB, Macintyre MN: An analysis of chromosome 9 inversions in four families. Am J Hum Genet 26:58A, 1974.
23. Dill FJ, Poland B, MacLeod P: Trisomy 13 in an abortus from a woman heterozygous for an inversion in chromosome 9. Am J Hum Genet 26:26A, 1974.
24. Grell RF, Valencia JI: Distributive pairing and aneuploidy in man. Science 145:66, 1964.
25. Stahl A, Luciani J, Devitor M, Capodano A, Gagne R: Constitutive heterochromatin and micronuclei in the human oocyte at the diplotene stage. Humangenetik 26:315, 1975.

Rob(14q15q) Translocation in Several Members of a Family

Patricia L. Monteleone, MD, Sister Leo Rita Volk, MS, Gurbax S. Sekhon, MD, Joan Grzegocki, BS, Michael Tietjens, BS, James A. Monteleone, MD, and Hari Sekhon, BS

Robertsonian translocations involving two D-group chromosomes are among the commonest structural chromosome rearrangements in man, occurring with a frequency of 1 in 1,500 live births [1]. These translocations may be sporadic or inherited. Cohen [2] estimated that 77% of these D-D translocations involve chromosomes 13 and 14, 9% involve chromosomes 13 and 15, 5% involve chromosomes 14 and 15, 8% are 13-13 translocations, and 1% are 15-15 translocations. The combined mutation rate for all balanced Robertsonian translocations of the D-D variety is of the order of 2.5×10^{-5} per gamete per generation with the proportion of these mutations which occur between chromosomes 14 and 15 unlikely to be more than one-third of the total, and perhaps even less. Most individuals with any type of D-D translocation are phenotypically normal [3–5]; however, a few individuals with abnormal phenotypic features have been found with D-D translocation but with no consistent clinical pattern. Zeuthen [5] reported a 14-15 translocation in five members of a family. Except for reading difficulties in two of these individuals, the translocation carriers were all healthy. A healthy 26-year-old male with 14-15 translocation was reported by Jacobs [6]. He was diagnosed only because his parents had undergone chromosomal analysis as part of an evaluation in a subfertility clinic in which the mother was found to have a 14-15 translocation.

We report here a family of 7 members, 4 of whom have a 14-15 translocation. The proposita is a 6½-year-old female (Fig. 1) who was referred for chromosomal analysis because of multiple congenital anomalies and mental retardation. Family history revealed 4 healthy male sibs, no spontaneous abortions or stillbirths, and no parental consanguinity. Her parents were 41 and 46 years old at the time of her birth. There is a family history of 5 mentally retarded paternal nephews and

Birth Defects: Original Article Series, Volume XIV, Number 6C, pages 303–308

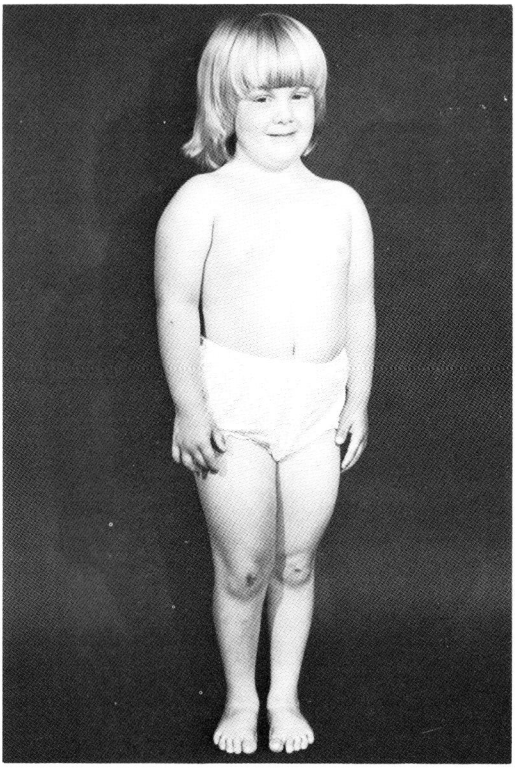

Fig. 1. Proposita.

nieces. Physical examination revealed an alert, well-nourished white female with slight epicanthal folds, hypertelorism, flat nasal bridge, a large head suggestive of arrested hydrocephalus, small rounded ears, and flat occipital area of the head. The child's chest was shield-like with widely spaced nipples. The hairline at the nape of the neck was quite low, and the 5th digits on the hands were very short. Her height has remained in the 3rd percentile, and she has a history of left hydronephrosis and congenital hip dysplasia with subluxation of the left hip. Speech and language skills tested when she was 5 years old are abnormal with functioning at the 3-year-old level. Overall mental age is in the educable range.

Chromosomal analysis with Giemsa banding on peripheral blood revealed a 45,XX,rob(14q15q) constitution appearing to be balanced (Fig. 2). There was no evidence of deletion or duplication in the karyotype. Chromosomal analysis was then done on the parents and sibs, and karyotypes of the father and 2 male sibs revealed a 45,XY,rob(14q15q) chromosomal constitution similar to the proposita. Karyotypes of the mother and 2 sibs were normal. The parents would not permit analysis of other members of the paternal side of the family.

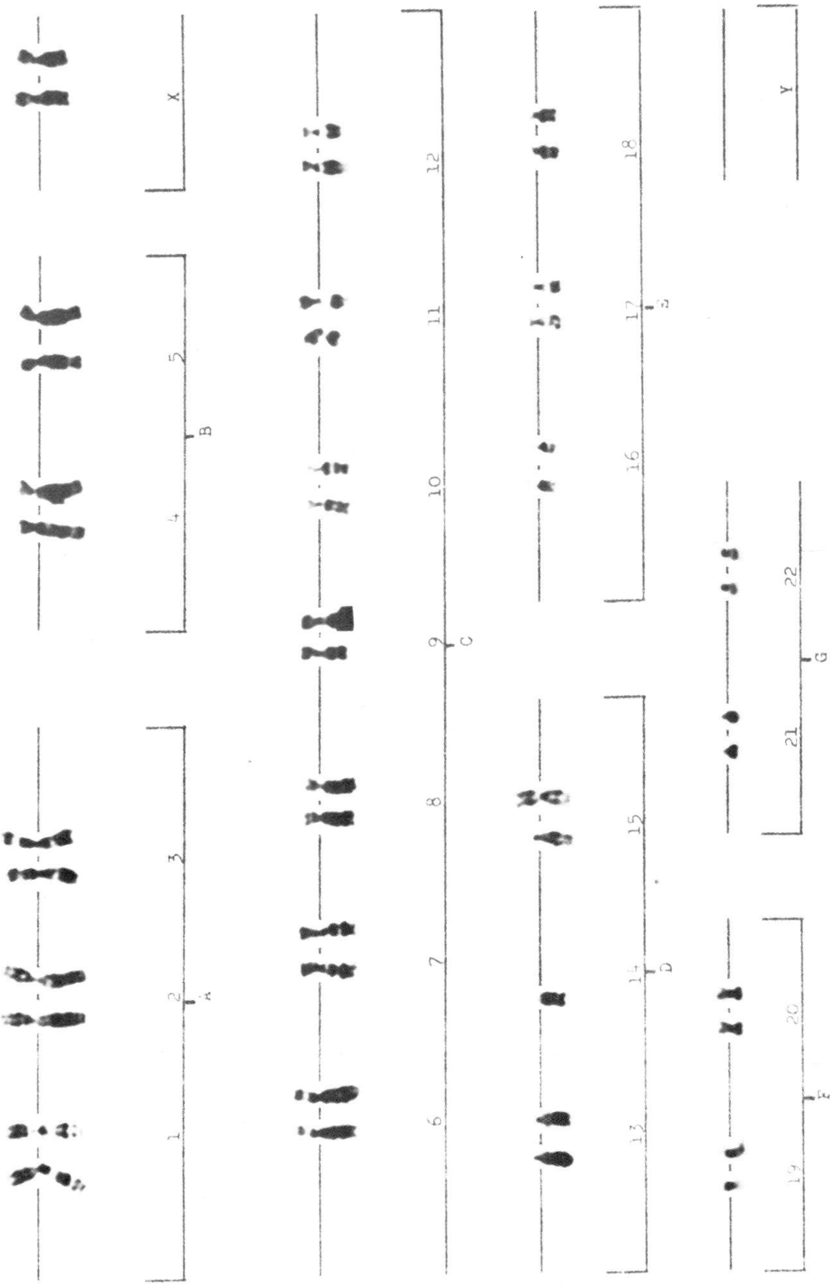

Fig. 2. Karyotype of proposita.

DISCUSSION

We describe here an apparently balanced translocation in an abnormal appearing child and the same translocation in 3 other family members who are clinically normal.

There are several possible explanations for the possible effect of the balanced 14-15 translocation in this child. It may be due to a small deletion or duplication at the site of the chromosome break which is undetectable using available analytic techniques. It is also possible that a gene mutation has occurred at the site of the break. One can also consider a possible detrimental effect from arrangement of gene position – the so-called position effect. It is possible that the child's clinical findings have no association with the apparently balanced 14-15 translocation. Interesting is the finding for 13-14 translocations that a considerably higher number of males with a balanced translocation inherit the same translocation compared to offspring of females with this translocation. This has not been documented for 14-15 translocations, but in our case the father was the individual with 14-15 translocation. Hamerton [7] postulated that sperm carrying the translocation chromosome might have some selective advantage over sperm with normal chromosomes.

Genetic counseling of this family is especially difficult because of several unanswered questions about 14-15 translocations. A problem which arises in counseling is the question of whether one structural abnormality of the chromosomes increases the risk of offspring with structural abnormalities in other chromosome pairs. Marsden et al [8] reported a child with G-G translocation Down syndrome whose mother had a D-D translocation, and there have been other structural abnormalities of non-D group chromosomes reported in offspring of D-D translocation carriers [9–17]. More data are required before this question can be answered. It is controversial, and analysis of randomly ascertained propositi is necessary in order to make reliable calculations concerning the risk of other translocations, Down syndrome, and nondisjunction in general, in the progeny of persons with D-D translocation. The studies of Fernhoff et al [18] suggested a greater risk of various types of aneuploidy among offspring of carriers of D-D translocations than among persons with normal chromosomes. However, these data are from 13-14 translocation carriers and there are no data specifically for 14-15 translocation carriers. There is known to be an increased incidence of de novo non-Robertsonian balanced translocations in surveys of mentally retarded persons, but no increase of Robertsonian translocations among this same group [14–21]. Part of the problem encountered in determining the genetic risks associated with any of the D-D translocations is that all types have been grouped together. Autoradiographic labeling studies have shown the nonrandomness of D-D translocations within the D group [2] and have shown that the 13-14 type is by far the most common. These may not even be a homogeneous group since the origin of the centromere may be from either chromosome [22]. Likewise, the location of break points may vary [23], and thus clinical manifestations might vary. Some involved chromosomes may have small fragments of the original short arm of the acrocentric chromosome present. The centromeres may be fully or partially fused. Heterogeneity would result from any of these mechanisms and contribute to the difficulty of counseling. Daniel and Lam-Po-Tang [24] have proposed that Robertsonian translocations be defined as stable dicentric whole arm translocations with both centromeres in a median position. The mechanism for the stability of the D-D translocations in such an event could be the proximity of their centromeres but suppression of one centromere by the other has also been proposed. Both Fernhoff et al [18] and Hamerton [7] showed no evidence of increased fetal wastage among 13-14 translocation carriers. A prospective study is needed to compare the intrafamily rate of spontaneous abortions among D-D translocation carriers with that among the members of kindreds who have normal chromosomes. Studies of spontaneous abortuses would be helpful and, likewise, meiotic studies may be helpful in determining if D trisomy (specifically trisomy 14 or 15) is a common pattern among aborted offspring of D-D translocation carriers. Theoretically, these have been thought to be nonviable.

If we are to improve genetic counseling and amniocentesis interpretation, we need to refine our techniques for diagnosis, delineate clinical manifestations of 14-15 translocations, detail the family pedigrees, and define the exact nature of each D-D translocation; ie whether it is 13-13, 13-14, 13-15, 14-14, 14-15, or 15-15.

ACKNOWLEDGMENT

The authors wish to express appreciation to Miss Doris Weiland for secretarial assistance.

REFERENCES

1. Jacobs PA, Frackiewicz A, Law P: Incidence and mutation rates of structural rearrangements of the autosomes in man. Ann Hum Genet 35:301–319, 1972.
2. Cohen MM: The chromosomal constitution of 165 human translocations involving D group chromosomes identified by autoradiography. Ann Genét 14:87–96, 1971.
3. Hamerton JL: "Human Cytogenetics." London and New York: Academic Press, 1971, vol 1.
4. Levine H: "Clinical Cytogenetics." Boston: Little, Brown, 1971.
5. Zeuthen E, Nielson J: D/D translocation in males examined for military service. J Med Genet 10:356–361, 1973.
6. Jacobs PA, Buckton KE, Christie S, Newton M, Matthew D: A family with two translocations and a polymorphism involving chromosome 14. J Med Genet 11:65–68, 1974.
7. Hamerton JL: Robertsonian translocations; evidence on segregation from family studies. In: "Human Population Cytogenetics." Edinburgh: University Press, 1970, pp 63–80.
8. Marsden HB, Mackay RI, Murray A, Ward HE: Down's syndrome with a familial D/D reciprocal translocation and a G/G chromosome. J Med Genet 3:56–58, 1966.

9. Chaptal J, Jean R, Emberger JM: Translocation entre deux chromosomes du group 13-15 chez la mere d'un mongolien par trisomie 21 réguliere. Arch Fr Pediatr 22:35, 1965.
10. Hamerton JL, Gianelli G, Carter CO: A family showing transmission of a D/D reciprocal translocation and a case of regular 21-trisomic Down's syndrome. Cytogenetics 2:194, 1963.
11. Hustinx TJ: Two cases of Down's syndrome with a translocation. Hum Chromosome Newsl 9:17, 1963.
12. Orye E, Delire C: Familial D/D and D/G_1 translocation. Helv Paediatr Acta 22:36, 1967.
13. Stahl A, Aubert L, Arroyo H, Luciani JM: Translocation familiale D/D. Association à un sydrome de Turner XO/XX. Ann Endocrinol (Paris) 27:847, 1966.
14. Tiepolo L, Fraccaro M, Hulten M, Lindsten J: Double aneuploidy (46,XXY,D-, t(DqDq)+). Ann Genet 10:114, 1967.
15. Yunis JJ, Alter M, Hook EB, Mayer M: Familial D/D translocation. Report of a pedigree and DNA replication analysis. N Engl J Med 271, 1133, 1964.
16. Zergollern J, Hoefnagel D, Benirschke K, Corcoran PA: A patient with trisomy 21 and a reciprocal translocation in the 13-15 group. Cytogenetics 3:148, 1964.
17. Palmer CG, Conneally PM, Christian JC: Translocations of D chromosomes in two families: t(13q14q) and t(13q14q)+(13p14p). J Med Genet 6:166, 1969.
18. Fernhoff PM, Singh DH, Hanson J, Trusler S, Dumont CR, Chen AT: Association of D/D translocations with fetal wastage and aneuploidy. J Med Genet 13:389–393, 1976.
19. Funderbunk SJ, Spence MA, Sparks RS: Mental retardation associated with balanced chromosome rearrangements. Am J Hum Genet 29:136–141, 1977.
20. Jacobs PA: Correlation between euploid structural chromosome rearrangements and mental subnormality in humans. Nature 249:164–165, 1974.
21. Breg WR, Miller DA, Allderdice PW, Miller OJ: Identification of translocation chromosomes by quinacrine fluorescence. Am J Dis Child 123:561–564, 1972.
22. Sparkes RS, De Chiere PR: Inherited 13/14 chromosome translocation as a cause of human fetal wastage. Obstet Gynecol 35:601–607, 1970.
23. Palmer CG, Morris JL, Thompson BH, Nance WE: Fertility and 13/14 translocation. Lancet 1:728, 1973.
24. Daniel A, Lam-Po-Tang PR: Structure and inheritance of some heterozygous Robertsonian translocations in man. J Med Genet 13:381–388, 1976.

Familial Translocation t(8;14) With a Case of Tertiary Trisomy, +14q–*

Dushan Soudek, MD, PhD, Patricia Hunter, BA, Susan O'Shaughnessy, RT, Nancy E. Simpson, PhD, FCCMG, and Vera Soudek, MD, DPH

INTRODUCTION

The number of defective offspring caused by the 3:1 meiotic disjunction is less common than that with chromosomal imbalance caused by 2:2 disjunction in parents carrying a balanced reciprocal translocation. The first type of disjunction results in a chromosome number of 45 or 47, and the second type in a count of 46 in the offspring. The most common representative of the first group is tertiary trisomy in which one of the two derivative chromosomes is added to a normal chromosome set. This chromosome is always the shorter of the two derivative chromosomes. Tertiary trisomy, in typical cases, consists of a partial trisomy of two different chromosomes. In most cases the origin of tertiary trisomy can be traced to a translocation in one of the parents; less frequent cases have been described with de novo tertiary trisomy [1]. Before banding techniques were introduced, many cases of tertiary trisomy were listed as cases of small accessory chromosomes of unknown origin. Even so, with contemporary banding methods the identification of the extra chromosome often requires a chromosome analysis of the parents' karyotypes.

The clinical features will be described for a child with tertiary trisomy 14q– and the following account will illustrate the use of several banding techniques (quinacrine fluorescence [QFQ], Giemsa trypsin and constitutive heterochromatin banding [GTC, CBG], parental karyotypes and the dosage effect of two enzymes

*This study was assisted by the Ontario Mental Health Foundation

Birth Defects: Original Article Series, Volume XIV, Number 6C, pages 309–315

(nucleoside phosphorylase [NP] and glutathione reductase [GR]) for indentification of a supernumerary chromosome. The locus for NP has been assigned to 14q11→14q21 [2] and GR to 8pter→8 cen [3]. Segregation ratios among the offspring of the sibs of the father for carriers of the balanced reciprocal translocation t(8;14) (q24;q21), normals, and spontaneous abortions will be given and discussed.

CASE REPORT

A female child was born in 1973 after an uneventful pregnancy and delivery. Her mother was 25 years old and her father was 27 years-old. Her birthweight was 2,720 gm. As a newborn she was an unusual looking infant, inactive, with a small head, flexion contractures of both 5th fingers, and an occasional dusky spell when sucking the bottle. Later she developed blue and choking spells and was admitted to an institution for the mentally retarded when 15 months old. Her mother could no longer cope with her extreme feeding difficulties and lack of responsiveness. At that time her developmental age was 6–7 months.

The patient is now 4 2/12 years old. Her height (90 cm) and weight (11 kg) are below the 3rd percentile and her head circumference is 46.5 cm (< 2%). According to the Denver Development screening test she functions at a 9–13-month level. She has an abnormal electroencephalogram (EEG) and is taking phenobarbital for minor motor seizures. Her eyes have a very slight upward slant; her mouth corners are turned down. She has hypotelorism and low-set posteriorly rotated ears (Fig. 1). Her occiput is prominent and thumbs set proximally. She rarely establishes eye contact; she neither talks nor walks yet. She crawls, feeds herself with a spoon, and can play with toys in a simple way.

The mother of the patient requested prenatal diagnosis for her next pregnancy because of the above patient. A report from another laboratory stated that the patient had a karyotype of 47,XX,+21 after receiving the cells by mail. This karyotype was done using G banding with very little clinical description and the parents' chromosomes were not analyzed.

Because the patient did not appear to have the Down syndrome clinically, her chromosomes were reanalyzed in our laboratory; those in the lymphocytes of both her parents as well as those from the cells of the fetus in the amniotic fluid.

The chromosomes of the mother were normal. The karyotype of the fetus was 46,XY and a normal male was delivered.

METHODS

QFQ, GTC, and CBG banding were done on lymphocyte cultures from the child and her father. About 50 of their cells were analyzed visually and 15 full and partial karyotypes were prepared. At least 5 karyotypes of the mother and of 19 relatives of the father were made using the GTC banding technique.

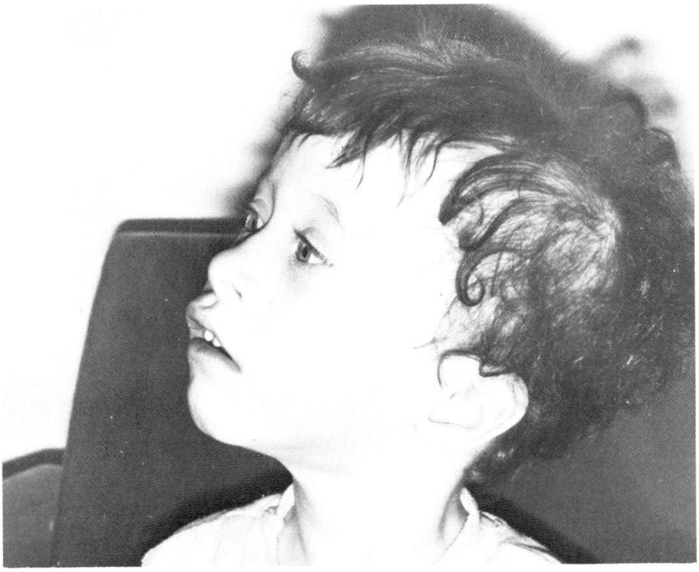

RESULTS

Cytogenetics

The chromosome analysis of the father showed that he was a carrier of a balanced reciprocal translocation t(8;14) (q24;q21). The detailed formula of the translocation is 46,XY,t(8;14) (8pter→8q24::14q21→14qter;14pter→14q21::8q24→8qter). In view of this father's karyotype and the reanalysis using the 3 banding techniques in our laboratory, the patient's karyotype was interpreted to be tertiary trisomy 47,XX,+der(14),t(8;14) (q24;q21)pat (Figs. 2 and 3).

On the other hand, from the banding, the translocation could have been t(8;14) (q21;q11) which, if true, would have implied that the patient's karyotype is 47,XX,+der(8),t(8;14) (q21;q11)pat; ie that more of chromosome 8 was trisomic than chromosome 14. The +der(8) interpretation was, in fact, made by another laboratory.

Dosage Effect of the Enzymes

The activity for nucleoside phosphorylase (NP) in the red cells of the patient was about 50% greater than that of normal and other trisomic controls, whereas for glutathione reductase (GR) it was normal [4]. Details of the enzyme studies will be published elsewhere. The enzyme studies, therefore, supported the +der(14) hypothesis.

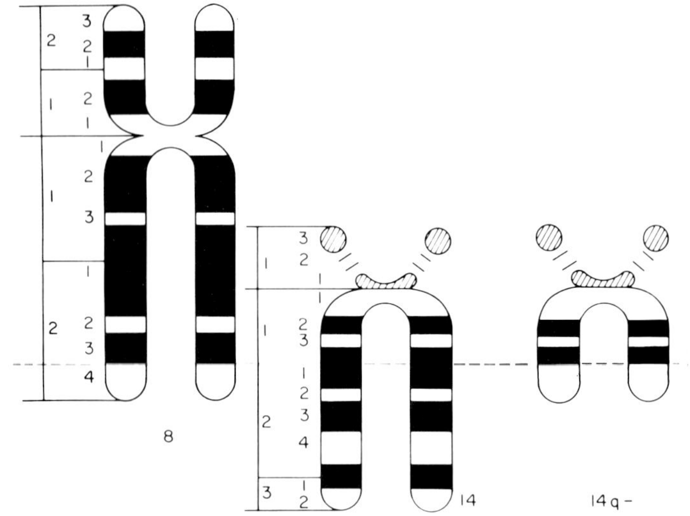

Fig. 2. A diagrammatic drawing of the reciprocal translocation and of the derivative chromosome.

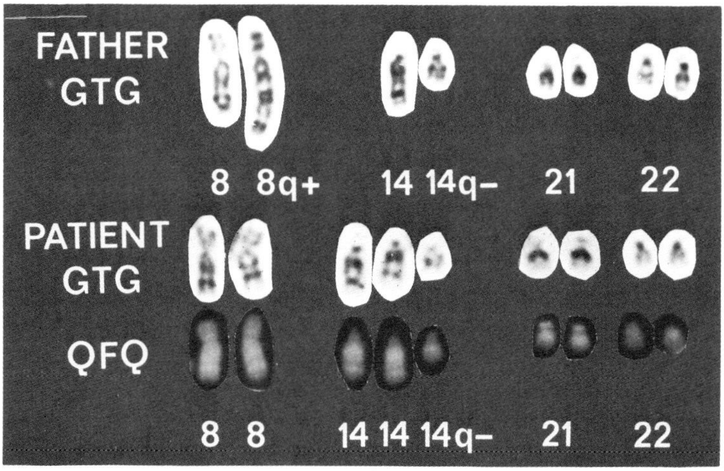

Fig. 3. Partial karyotypes of the balanced carrier (father) and of the patient.

Segregation Ratios

The pedigree in Figure 4 summarizes the cytogenetic analyses in the father's family. Three of his brothers were balanced carriers of t(8;14) and, although we could not study their father (II:1) because he was deceased, he likely carried the translocation, the mother (II:2) having normal chromosomes.

Assuming that II:1 had the translocation and excluding II:3 and the sibs of the patient, there were 8 normal offspring, 7 balanced carriers, and 3 spontaneous abortions among the offspring of balanced carriers.

The sex ratio of those with the translocation was compared with that of those with a normal karyotype in the offspring of balanced carriers, including the offspring of II:1. The sex ratios were 8 males: 1 female for the carriers and 6 males: 3 females for the normals.

DISCUSSION

Several authors [5–8] refer to the difficulties in distinguishing partial D chromosomes even when banding techniques are used. It is well known that such identification is facilitated when a parent is a balanced carrier, but even so there are sometimes difficulties. More than one banding technique is also helpful.

The use of known dosage effect of NP [2] and GR [9], whose loci were known to be in chromosomes 14 and 8, respectively, substantiated the identification of +der(14) in the patient. Neither of the enzymes had a dosage effect in other trisomies [Frecker and Simpson, in preparation]. The use of dosage effect of enzymes whose loci are mapped should prove useful in identifying small segments of chromosomes in trisomy or monosomy as more data on dosage and mapping become available.

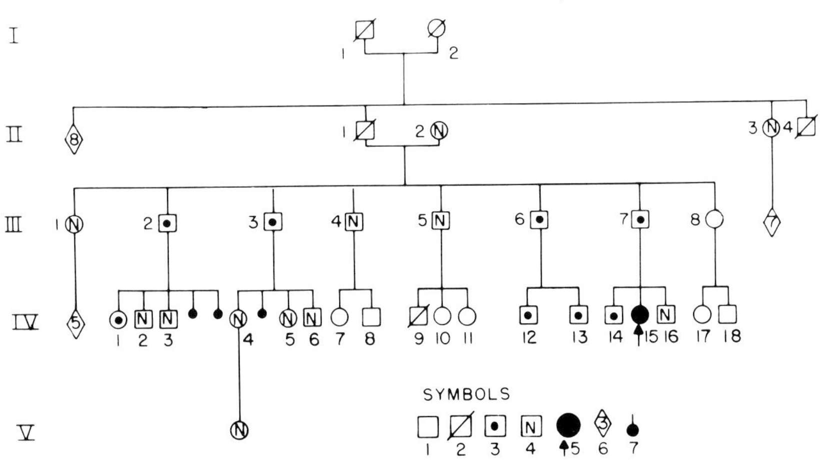

Fig. 4. The pedigree of the studied family. Symbols: 1, chromosomes not examined; 2, deceased; 3, carrier of balanced translocation; 4, normal chromosomes; 5, tertiary trisomy; 6, 3 persons of undisclosed sex; 7, spontaneous abortion.

Complete trisomy 14 is likely lethal and rare, but partial trisomy 14 is more viable; the severity of the defect decreases with the size of the triplicated 14q segment [6]. Therefore, relatively many cases of partial trisomy 14q− have survived and have been reported. Eight cases were discussed and tabulated by Raoul [10]. Pfeiffer's case [11] has a different chromosome structure, and cannot be compared with the other cases. Since Raoul's report, 9 other cases have been published [5–8, 12–16], one of them with uncertain chromosomal diagnosis [7]. Two of the reported cases (Laurent, quoted by Raoul [10], [14]) are combinations of partial trisomy 14q− and partial monosomy of another chromosome with a chromosome count of 46. Two of the cases are familial [12, 14]; 4 of them originated de novo (Muldal, quoted by Raoul [10], [7, 13, 15]). The differences in clinical characteristics are not surprising and are likely due to the different sizes of the 14q− triplicated segment and the differences in other chromosomes taking part in the chromosomal rearrangement either as partial trisomy or partial monosomy. In the reported cases chromosomes 1–3, 8–10, 12, 13, 16, 19–21 have been involved. On the other hand, common features of all cases are growth and psychomotor retardation, and a variable spectrum of nonspecific malformations frequently associated with chromosomal aberrations. Thus, partial trisomy 14q− syndrome can be defined only in very broad clinical terms.

The ratio of normal to balanced carrier karyotypes among the offspring of carriers was 8:7; the numbers are too small to be very meaningful. The 3 spontaneous abortions, among 14 conceptuses of balanced carriers of course, could have had the imbalanced karyotypes, but were no more frequent than the 2 abortions and 1 early death among the 13 conceptuses with normal parents (III:1 had 2 abortions).

The balanced translocation may be more viable in males (there were 8 males to 1 female) but more data would be needed to establish that the 8:1 ratio is different from the 6:3 ratio for those children with normal karyotypes born to the balanced carriers.

The lack of cases of an imbalanced karyotype other than for our patient may reflect a presence of factors limiting the viability of male germ cells carrying an unbalanced chromosomal set. Such differential viability is well known for nondisjunction resulting in full trisomies.

ACKNOWLEDGMENTS

The authors thank Dr. J. R. Miller (Department of Medical Genetics, University of British Columbia, Vancouver, British Columbia) and Dr. R. M. Greenstein (Division of Cytogenetics, University of Connecticut, Farmington), for supplying us with the cytogenetics data regarding some members of the studied family. We also thank the family members for their cooperation.

REFERENCES

1. Lindenbaum RH, Bobrow M: Reciprocal translocations in man. 3:1 Meiotic disjunction resulting in 47-or 45-chromosome offspring. J Med Genet 12:29–43, 1975.
2. George DL, Francke U: Gene dose effect: Regional mapping of human nucleoside phosphorylase on chromosome 14. Science 194:851–852, 1976.
3. McKusick VA, Ruddle FH: The status of the gene map of the human chromosomes. Science 196:390–405, 1977.
4. Simpson NE, Soudek D, Hunter P, Frecker M: Dosage effect of nucleoside phosphorylase (NP) in a patient with tertiary trisomy der(14)pat. Can J Genet Cytol 18:573, 1976.
5. Fawcett WA, McCord WK, Francke U: Trisomy 14q–. In Bergsma D (ed): "New Chromosomal and Malformation Syndromes." Miami: Symposia Specialists for The National Foundation–March of Dimes, BD:OAS Xi(5):223–228, 1975.
6. Young SR, Donovan DM, Greeg HA, Burch K, Potter DG: Tertiary trisomy, 47,XX, +14q–, resulting from maternal balanced translocation 46,XX,t(14;16) (q11;q24). Hum Genet 33:331–334, 1976.
7. Cohen MM, Rosenman A, Dagan J, Legum C: Partial trisomy D; a diagnostic and cytogenetic dilemma. J Med Genet 13:535–538, 1976.
8. Simpson J, Zellweger H: Partial trisomy 14q– and parental translocation of No. 14 chromosome. J Med Genet 14:124–127, 1977.
9. de la Chapelle A , Vuopio P, Icen A: Red cell glutathione reductase activity in patients with bone-marrow trisomy 8. In Bergsma D (ed): "Human Gene Mapping 3." Basel: S Karger for The National Foundation-March of Dimes, BD:OAS XII (7):277, 1976.
10. Raoul O, Rethore M-O, Dutrillaux B, Michon L, Lejeune J: Trisomie 14q partielle I. Trisomie 14q partielle par translocation maternelle t(10;14) (p15.2;q22). Ann Genet 18:35–39, 1975.
11. Pfeiffer RA, Buettinghaus K, Struck H: Partial trisomy 14 following a balanced reciprocal translocation t(14q–;21q+). Humangenetik 20:187–189, 1973.
12. Surana RB, Conen PE, Braudo M, Keith JD: Familial tertiary trisomy with t(14q–; 1+?). In Bergsma D (ed): "Clinical Cytogenetics and Genetics." Miami: Symposia Specialists for The National Foundation–March of Dimes, BD:OAS X(8):53–65, 1974.
13. LoCurto F, Maraschio P, Milanesi P, Severi F, Ugazio AC, Zuttardt O: The syndrome of partial trisomy 14q. Eur J Pediatr 123:237–241, 1976.
14. Pena SDJ, Ray M, McAlpine PJ, Ducasse C, Briggs J, Hamerton JL: Tertiary trisomy 14; Is there a syndrome? In Bergsma D, Schimke RN (eds): "Cytogenetics, Environment and Malformation Syndromes." New York: Alan R Liss for The National Foundation–March of Dimes, BD:OAS XII(5):113–118, 1976.
15. Yeatman GW, Riccardi VM: Partial trisomy of chromosome 14:(+14q–). In Bergsma D, Schimke RN (eds): "Cytogenetics, Environment and Malformation Syndromes." New York: Alan R Liss for The National Foundation– March of Dimes, BD:OAS XII(5): 119–124, 1976.
16. Coco R, Penchaszadeh VB: Partial trisomy 14q and familial translocation t(2;14) (q12;q13). Ann Genet 20:41–44, 1977.

A Complex Translocation Involving Chromosomes 3,11, and 14 With an Interstitial Deletion, del(14) (q13q22) in a Child With Congenital Glaucoma and Cleft Lip and Palate*

Philip D. Buchanan, PhD, Kathleen W. Rao, Carolyn L. Doerr, MA, and Arthur S. Aylsworth MD

Interstitial deletions and complex translocations are rare in man and may be very difficult to diagnose. With the application of trypsin and quinacrine banding, we have identified an interstitial deletion from the long arm of a chromsome 14 in a child with congenital glaucoma and cleft lip and palate. This deletion has been identified as resulting from a complex translocation involving chromosomes 3, 11, and 14.

CASE REPORT

The patient is a black male infant whose birth size was appropriate for his estimated gestational age of 38 weeks. He was initially referred for respiratory distress on the first day of life. He was noted to have a bilateral cleft lip, complete cleft palate, and cloudy corneas. Evaluation under anesthesia revealed congenital glaucoma. Trabeculotomies were then performed.

The pregnancy was uneventful, without exposure to drugs or cytotoxic agents, and the delivery was spontaneous. The mother was 29 years old and in good health. The father was 43 years old and has a history of hypertension. The patient has three older sisters and one older brother, all of whom are alive and well. There is no family history of abortions or stillbirths.

*This work was supported in part by grants from the North Carolina Department of Human Resources and The National Foundation–March of Dimes.

Birth Defects: Original Article Series, Volume XIV, Number 6C, pages 317–322

CYTOGENETIC STUDIES

Chromosome studies were performed on peripheral blood cultures from the patient and his parents. Chromosome analysis of 44 lymphocyte metaphases using both trypsin and quinacrine banding show the patient has a chromosome count of 46 (Figs. 1 and 2). A number 3, 11, and 14 chromosome are missing and a complex, unbalanced translocation involving these chromosomes is present in all of the cells studied. Part of the short arm of a chromosome 3 is translocated to the long arm of a chromosome 11 (Fig. 3). Part of the long arm of a chromsome 11 is translocated to the long arm of a chromosome 14. Two chromosome breaks in the long arm of chromosome 14 have resulted in an interstitial deletion of chromosome material between bands q13 and q22. In addition, part of the long arm of a chromosome 14 has been translocated to the short arm of a chromosome 3. Chromosome studies on both parents are normal.

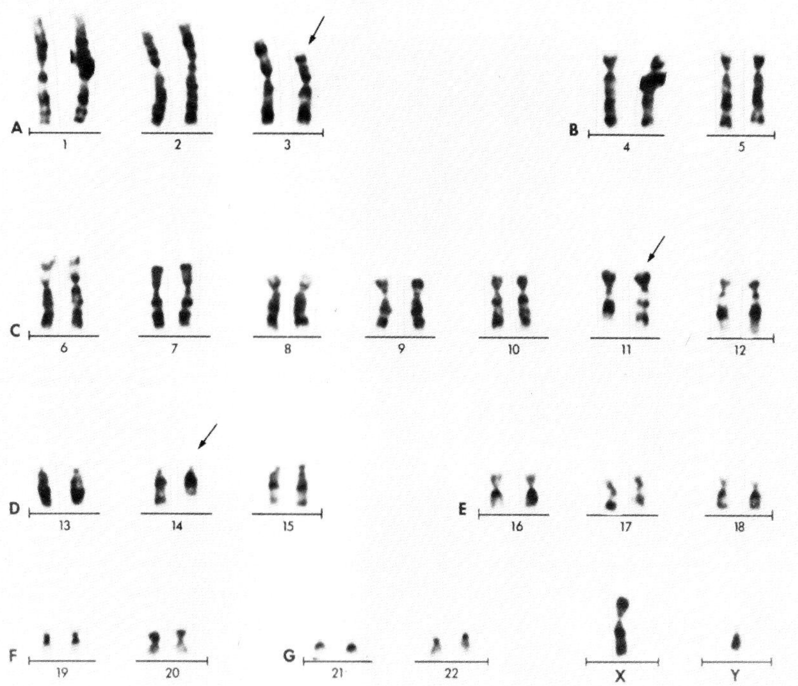

Fig. 1. Lymphocyte karyotype prepared with trypsin banding. Arrows point to the translocation chromosomes.

Fig. 2. (top) Lymphocyte karyotype prepared with quinacrine banding. Arrows point to the translocation chromosomes.

Fig. 3. (bottom) Diagrammatic presentation of the complex rearrangement and interstitital deletion. The idiogram is based on the Paris Conference nomenclature [19].

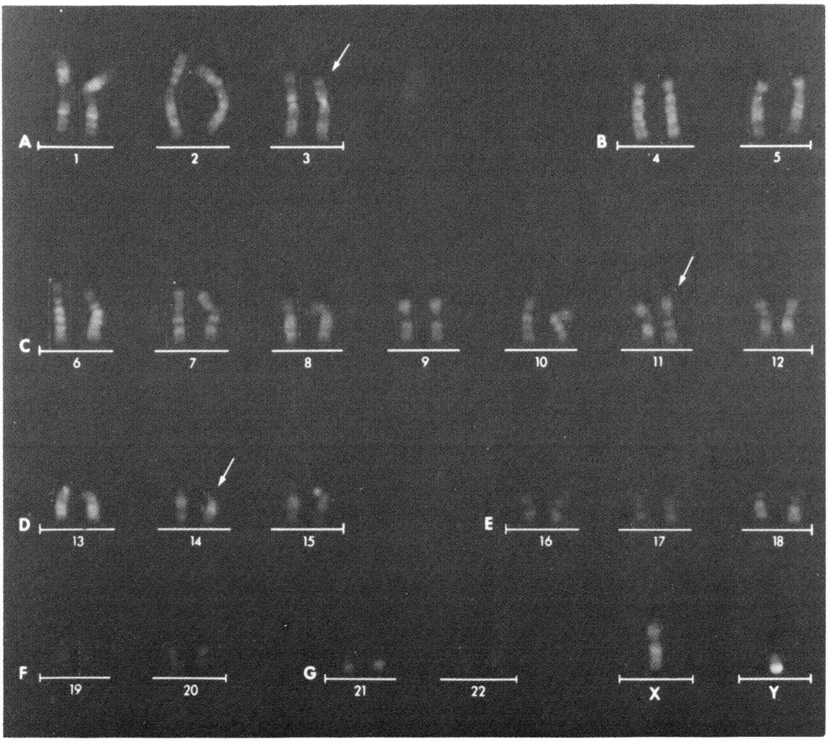

A
1
2
3
B
4
5
C
6
7
8
9
10
11
12
D
13
14
15
E
16
17
18
F
19
20
G
21
22
X
Y

2

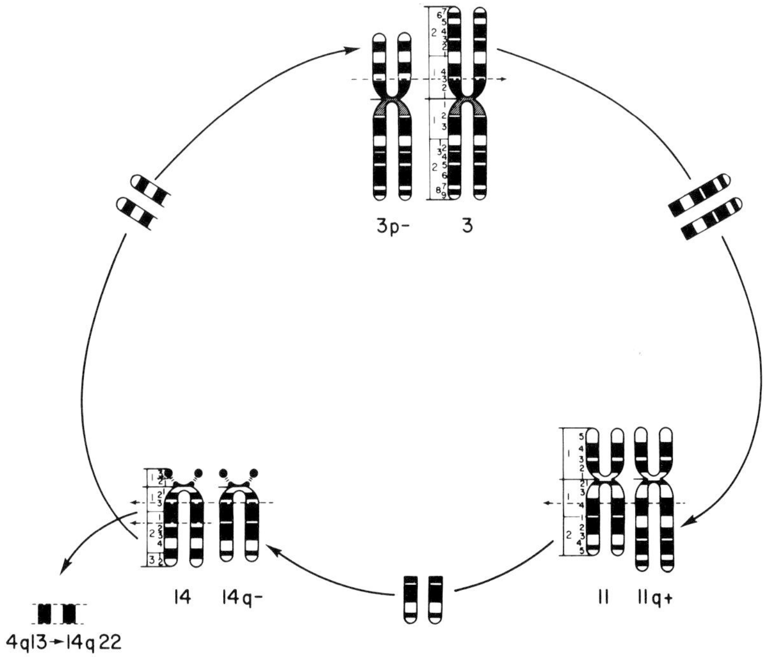

3p-
3
14
14q-
11
11q+
4q13→14q22

3

DISCUSSION

This abnormal karyotype arose de novo by one of two possible translocations involving break points 3q13, 11q14, 14q13, and 14q22. The first possibility involves the translocation of chromosome material between chromosomes 3, 11, and 14 with an interstitial deletion occurring in the long arm of chromosome 14, del(14) (q13→q22). The second possibility is a complex translocation that involves a fourth chromosome, another 14. A single break in each of two homologous chromosomes 14 could have produced this interstitial deletion.

Four breakpoints in the three chromosomes related to this complex translocation and an interstititial deletion of a small segment of the 14 long arm have been identified. The correlation of the clinical symptoms present in this patient with the interstitial deletion provides an opportunity for deletion mapping. Genes related to the development of glaucoma and cleft lip and palate may either be located in the proximal long arm of chromosome 14 between the breakpoints we have identified or possibly associated with position effects in the other translocation chromosomes.

The three chromosomes involved in the translocation in our patient have been previously reported to be involved in other complex translocations. Certain of the 23 pairs of human chromosomes have been found in complex rearrangements more frequently than others. Chromosomes 3 and 5 have been found in 5 out of 19 cases; and 4, 9, 11, 13, and 14 have been found in 4 out of 19 cases, including

TABLE 1. Complex Chromosomal Rearrangements

Chromosomes	Reference
6, 14, 20	Allderdice et al (1971) [1]
5, 7, 17	Berger et al (1974) [2]
4, E	Bobrow et al (1971) [3]
5, 15	Catti and Schmid (1971) [4]
4, 7, 15	Chandley at al (1975) [5]
3, 4, 14	Creasy et al (1974) [6]
2, 21, 21	Creasy et al (1976) [7]
9, 21, 22	Dallapiccola et al (1976) [8]
1, 2, D, 16	de Grouchy and Lautmann (1968) [9]
2, 3, (5 breaks)	Fitzgerald (1974) [10]
5, 9, 13	Fredga and Hall (1970) [11]
9, 14, 13	Fujita et al (1974) [12]
1, 5, 8, 9	Martinetti and Noel (1973) [13]
3, 15, C	Nuzzo et al (1968) [14]
3, 11, 20	Palmer et al (1976) [15]
11, 12, 13	Sanchez et al (1974) [16]
7, 11, 21	Schwinger et al (1975) [17]
5, 13, 4	Taillemete et al (1973) [18]
3, 11, 14	This patient

ours. This may be a coincidence or there may be an increased propensity for these chromosomes to be involved in such complex translocations [1–18] (Table 1).

In summary, we are reporting a child with a unique interstitial deletion derived during a complex translocation involving three chromosomes. The clinical features of this child, including congenital glaucoma and cleft lip and palate, may be associated with the deletion of the 14q13 to 14q22 chromosome segment.

REFERENCES

1. Allderdice PW, Miller OJ, Miller DA, Breg WR, Gendel E, Zelson C: Familial translocation involving chromosomes 6, 14, and 20, identified by quinacrine fluorescence. Humangenetik 13:205–209, 1971.
2. Berger R, Derre J, Ortiz MA: Les trisomies partielles du bras long du chromosome 7. Nouv Presse Med 3:1801–1804, 1974.
3. Bobrow M, Jones LF, Clarke G: A complex chromosomal rearrangement with formation of a ring 4. J Med Genet 8:235–239, 1971.
4. Catti A, Schmid W: A pericentric inversion, 5p–q+, and additional complex rearrangements in a case of cri-du-chat syndrome. Cytogenetics 10:50–60, 1971.
5. Chandley AC, Edmond P, Christie S, Gowans L, Fletcher J, Frackiewicz A, Newton M: Cytogenetics and infertility in man. Ann Hum Genet 39:231–254, 1975.
6. Creasy MR, Crolla JA, Daker MG: A familial reciprocal translocation between three chromosomes. Humangenetik 24:303–308, 1974.
7. Creasy MR, Crolla JA, Polani PE, Elliott MH: A three-chromosome translocation involving two homologous chromosomes in a trisomic abortus. Clin Genet 9:15–19, 1976.
8. Dallapiccola B, Bollea G, Mazzilli C, Gandini E: Complex translocation t (9;21) (9;22) (q12p13) (q12q11) in the family with a child with 9p trisomy syndrome. Humangenetik 33:73–76, 1976.
9. de Grouchy J, Lautmann F: Caryotype 46,XX,1q–.2q–.Dq+.16q chez une infant polymalformee. Ann Genet 11:129–131, 1968.
10. Fitzgerald MG: Complex five break rearrangement. Clin Genet 5:62–67, 1974.
11. Fredga K, Hall B: A complex familial translocation involving chromosomes 5, 19, 13. Cytogenetics 9:294–306, 1970.
12. Fujita H, Abe T, Yamamoto K, Furuyama J: Possible complex translocation (9; 14; 13) (q12;pl?;q31) in mother of a child with 9p– trisomy syndrome. Humangenetik 25: 83–92, 1974.
13. Martinetti J, Noel B: Complex de-novo rearrangement involving 4 chromosomes in a newborn. Ann Genet 16:285–288, 1973.
14. Nuzzo F, Marini A, Baglioni C, Ford CE, deCarli L, Sereni LP: A case of multiple chromosomal rearrangements with persistence of foetal hemoglobin. Cytogenetics 7:169–182, 1968.
15. Palmer CG, Poland C, Reed T, Kojetin J: Partial trisomy 11, 46,XX,–3,–20,+der3, +der20,t(3:11:20), resulting from a complex maternal rearrangement of chromosomes 3, 11, 20. Humangenetik 31:219–225, 1976.
16. Sanchez O, Yunis JJ, Escobar JI: Partial trisomy 11 in a child resulting from a complex maternal rearrangement of chromosomes 11, 12 and 13. Humangenetik 22:59–65, 1974.
17. Schwinger E, Mikkelsen M, Niesen M: Familial balanced (7;11;21) translocation and Down's syndrome in two siblings. Clin Genet 7:304, 1975.

18. Taillemete JL, Baheux-Morlier G, Cathelineau L, Roux C: Cri-du-chat syndrome and complex chromosome rearrangement in a dizygotic twin. Ann Genet 16:127–130, 1973.
19. Bergsma D (ed): "Paris Conference: Standardization in Human Cytogenetics." White Plains, NY: The National Foundation–March of Dimes, BD:OAS 8(7), 1972.

Trisomy 21 With Average Intelligence?!

Boris G. Kousseff, MD

In 1866 Dr. Down described the syndrome that now carries his name. A century later, the karyotype of the patients with Down syndrome was deciphered. Trisomy 21, either as the only cell line or as mosaic trisomy 21, emerged as by far the predominant chromosome abnormality, translocations being responsible for a few cases. In all three cytogenetic types mental retardation is considered a defining characteristic of the syndrome, but among patients with mosaicism, several have been reported to be of average intelligence [1, 2]. In those cases a low percentage of cells with the extra chromosome 21 has been found [2, 3]. It is well known that the degree of severity of mental retardation varies considerably from one patient to another even within the same cytogenetic type. Therefore, to predict accurately the intellectual potential of a newborn with the syndrome is difficult, if not impossible. Intensive stimulation since infancy is considered possibly to be instrumental in enhancing ultimate intellectual function and has become a favored trend in the management of the condition. Solid evidence to substantiate this trend, however, is still to be reported.

CASE REPORT

A white female was diagnosed at birth as having Down syndrome. The antenatal, perinatal, and neonatal periods were uneventful. Chromosome analysis of cultured peripheral lymphocytes showed an extra G chromosome in all 62 mitoses counted. The Trypsin-Giemsa banding technique identified the accessory chromosome as number 21. The karyotypes of both parents were normal. The mother was 45 years of age at the time of the proband's delivery. She has had 8 healthy children, and 4 first-trimester miscarriages. Following specific instructions

Birth Defects: Original Article Series, Volume XIV, Number 6C, pages 323–325

from her early infancy, the parents, both college graduates, embarked on intensive stimulation at home. All sibs participated as well, and the infant was literally bombarded with a variety of visual and auditory stimuli all day long. At the age of 9 months she was already in a program for children with psychomotor delay. As she grew older, consequent formal testing, always done by the same psychologist, showed remarkable progress in psychomotor development (Table 1). At age 3 years, 3 months mental age on the Stanford Binet Test of Intellectual Functioning was 2 years, 3 months, yielding an IQ of 69 – borderline retardation range. On follow-up at age 4 years, 2 months, using the same testing tool, the social age was 3 years, 3 months and the mental age was 3 years, 4 months, yielding an IQ of 68 – mild mental retardation. At age 5 years, 1 month, the social age was 5 years, 2 months, and on the Wechsler Preschool and Primary Scale of Intelligence (WPPSI), the patient showed a Verbal IQ of 79, a Performance IQ of 74, and a Full Scale IQ of 74 – borderline intelligence. Eight months later, at the age of 5 years, 9 months, the social age was recorded as 5 years, 4 months, and on the WPPSI the Verbal IQ was`82, the Performance IQ 85, and Full Scale IQ 83 – low average intelligence. Of even greater surprise was the average performance on the three subtests considered predictors of academic potential: Information, vocabulary, and arithmetic. The child performed slightly below average on only the three subtests measuring social judgment, visual discrimination, and visual motor skills. The speech evaluations done with every psychologic testing showed results corresponding to the respective intellectual performance.

TABLE 1. Psychometric Results on Consequent Testing of Down Syndrome Patient

Chronologic in years, months	Social age	Mental age	IQ Score		Range
3, 3		on SB1[b] 2, 3		69	Borderline
4, 2	on VSMS[a] 3, 3	on SBT 3,4		68	Mild mental retardation
5, 1	on VSMS 5, 2	on WPPSI[c]	Verbal	79	
			Performance	74	
			Full Scale	74	Borderline
5, 9	on VSMS 5, 4	on WPPSI	Verbal	82	
			Performance	85	
			Full Scale	83	Low average

[a]VSMS – Vineland Social Maturity Scale
[b]SBT – Stanford Binet Test of Intellectual Functioning
[c]WPPSI – Wechsler Preschool and Primary Scale of Intelligence

DISCUSSION

We can offer only speculations in an attempt to explain this unusually high intellectual performance for Trisomy 21. Certainly our patient has been growing in an intact intelligent middle-class family. The parents took the birth of their special child as a challenge and provided intensive intellectual stimulation since early infancy. Also, the child has always been in a program for children with psychomotor delay. Although it has been suggested by Golden and Pashayan [4] that the higher the education of the parents, the better the intellectual performance of the children with the syndrome, it is impossible to prove that this alone, or together with the intensive stimulation program is the cause of the remarkable performance. Our patient just might be a child with Trisomy 21 who has a priori unique potential. Whatever the explanation, however (or lack of it, to be more precise), we ought to keep in mind that the intellectual stimulation was provided and we would hope that the current trend in management of Down syndrome will result in similar reports. Another thing in the case that deserves mentioning is the constructive surgery (not plastic or cosmetic in such cases) that has been scheduled in the near future. The goal is to make the child more acceptable among her peers. The surgery will consist of removal of the epicanthal folds, correction of the eyelids, lifting of the bridge of the nose, and reconstruction of the chin with an inlay. Once this is accomplished, if the current intellectual performance is maintained throughout the years to come, we might be allowed to feel that nowadays we are doing a little bit more for our patients with the syndrome than Dr. Down was able to do for his patients a century ago.

REFERENCES

1. Clarke CM, Edwards GH, Smallpiece V: #21 Trisomy/normal mosaicism in an intelligent child with some mongoloid characters. Lancet 1:1028, 1961.
2. Fisher K, Koch R, Donell GN: Comparison of mental development in individuals with mosaicism and trisomy 21 Down's syndrome. Pediatrics 58:744–749, 1976.
3. Zellweger H, Abbo G: Chromosomal mosaicism and mongolism. Lancet 1:827, 1963.
4. Golden W, Pashayan HM: The effects of parental education on the eventual mental development of non-institutionalized children with Down's syndrome. J Pediatr 89:603–608, 1976.

A Comparison Between a Trisomy 21 Child (Probably Mosaic) With Normal Intelligence and a Mosaic Down Syndrome Population*

Mary Esther Carlin, MD, S. Leon, PhD, and J.D. Gilbert, MS

INTRODUCTION

Seguin in 1846 described the clinical presentation of a particular form of mental retardation that he called "furfuraceous idiocy" [1]. This syndrome was later studied by Langdon Down, who in 1866 introduced the name "mongolian idiocy," because he mistakenly believed the retardation and physical features resulted from a resurgence of traits found in the Mongolian race [2]. In 1909, Shuttleworth recognized the increased frequency of these findings in offspring born to mothers of advanced age [3]. Somewhat later, the name Down syndrome was proposed to forever eradicate the racial etiologic connotation [4]. In 1952, Benda suggested the term "congenital acromicria" to accent the growth disturbances seen, but it was not generally accepted [5, 6]. Finally an etiologic name, trisomy 21, emerged in 1959, when Lejeune and his co-workers noted the presence of an extra autosomal chromosome now identified as number 21 in individuals with the Down syndrome [7–12].

Despite the nosologic evolution there has never been any doubt that mental retardation is a hallmark feature. Usually it is moderate to severe, but occasionally may be only mild. Numerous investigators have studied the mental and motor development in individuals with trisomy 21 [13–25]. The typical IQ has been reported to be 30–50 [50, 70]. Because of this poor level of functioning many children were placed in institutions following the diagnosis.

This trend was gradually reversed as further studies clearly demonstrated that those children reared in a home environment with parental figures and sib

*This work was supported in part by HEW grants MCT–000-288-11-0 and MCT–000903.

Birth Defects: Original Article Series, Volume XIV, Number 6C, **pages 327–341**

stimulation, at least during the early years, developed faster and functioned on a higher level than matched individuals reared in institutions [13, 14, 23–32]. Dicks–Mireaux found the IQ range for those who were home-reared to be 27.4–62.4 compared to a range of 17.4–37.4 in institutionalized individuals [24]. In addition, developmental milestones such as walking and talking were achieved much earlier in the children reared at home [25–32].

This more rapid rate of reaching early milestones allows for increased potential and slightly increased adult functioning, but does not insure continued IQ ascent. In fact, the pattern of mental development in trisomy 21 usually demonstrates an early rise in IQ with a plateau from approximately 2–5 years, followed by a gradual decline [23, 25, 50, 51]. This reflects the limited communication abilities of these individuals, especially spontaneous language [33–40]. It should be noted, however, that many older individuals who have continued in an educational environment may add IQ points after the age of 20 [13].

Normal intellectual functioning in patients with trisomy 21 was not noted until the 1960s, when sporadic reports of individuals with features of the Down syndrome but with normal intellectual functioning began to appear. Most were identified as mosaic for trisomy 21 and normal cell lines [41–46]. Continued investigation into the performance of mosaic individuals has shown that their mean performance level is above that of nonmosaic individuals. Some do test in the normal range. The data suggest that the level of functioning tends to be higher in those individuals possessing higher percentages of normal cells [47–54].

This paper will present a child with trisomy 21 who tests and functions in the normal range. Her performance will be contrasted with that of 18 children identified at the Mailman Center for Child Development as having mosaic Down syndrome. The mosaic child with the lowest percentage trisomic and the highest percentage normal cells will also be discussed.

CASE REPORTS

Case 1

(See Figs. 1 and 2.) The proband is a white female, 7 years 2 months old. She is the youngest of six children. Her mother was 35 and her father 38 at the time of her conception. The obstetric history included one previous spontaneous abortion in the first trimester between the fourth and fifth child. This last pregnancy was complicated by first trimester bleeding treated with Provera injections and rest. The proband was born at term following a rapid labor and weighed 4,100 gm, measured 55.8 cm long and had a head circumference of 34.3 cm. Her newborn physical examination revealed facies typical of the Down syndrome

Fig 1. Proband, *Case 1* at 6½ years of age, showing how a smile and a hair style can modify the typical facial features.

and x-ray films of the hip demonstrated an iliac index of 27 (below 60 is presumptive of the Down syndrome). Her neonatal course was complicated by bilirubin elevations in each of the three days prior to discharge and by failure to emit any cries for the first 9 days of life.

At 15 days of age the infant came to our center for chromosome studies. The physical examination at that time revealed upward slanting of the palpebral fissures, bilateral epicanthal folds, a small, flat nose, an open mouth with a protruding tongue, a short, broad neck, mild generalized skin mottling, and very slight hypotonia. Dermatoglyphic findings included bilateral single palmar flexion creases, bilaterally elevated atd angles, 10 ulnar loops and an arch-tibial pattern on both soles [55–58].

At 7 months the infant's development was complicated by meningitis associated with seizures and several days' obtundation. No organism was identified, but a bacterial origin was suspected. Other medical therapy consisted of a brief trial of thyroid extract during the first year and several courses of antibiotics and finally bilateral polyethylene tube insertions for otitis media. Her developmental milestones include sitting at 7 months, crawling at 9 months, standing

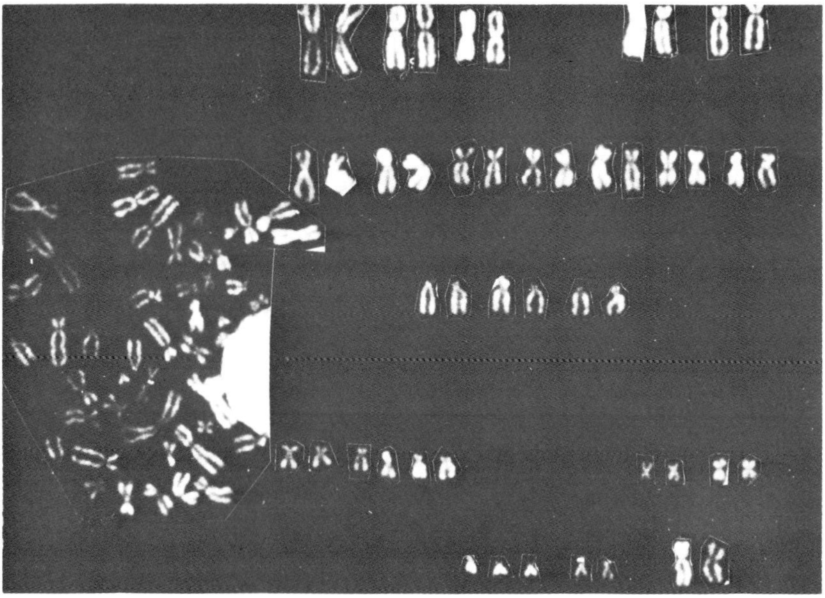

Fig. 2. Fluorescent 47,XX,+21 karyotype from *Case* 1.

alone at 10 months, walking at 16 months, speaking words at 17 months and sentences at 3 years, and being completely toilet-trained at 3½ years. In all these areas her rates of achievement lagged behind those of her sibs.

The proband's parents have completed high school. Her mother has attended college and her oldest sister has a degree in special education. All of the family members, especially the oldest sister, have been the proband's most consistent educators. Each one spends some time daily working with her. In addition, she was enrolled in a preschool before age 3 and a Montessori school by 4 years of age. At 6 years of age she began regular classes in a public elementary school as well. She also takes acrobatic dancing lessons to improve her muscle tone and coordination and receives speech therapy on a regular basis [33, 59].

Table 1 shows the results of the child's annual psychologic tests. The mental age and resultant IQ in 1976 are probably excessively high, as a jump in her mental age of 20 months in 10 months time would not seem likely. Nonetheless, all the scores are in the normal range. Her recent educational achievement tests have likewise reflected normal performance, with an oral reading word recognition at the 1.7 grade level, reading comprehension at the 1.3 grade level, spelling at the 1.2 grade level, and arithmetic at the 1.4 grade level.

Laboratory Results. Table 2 outlines the results obtained from the numerous chromosome studies performed on the patient. When she was first studied in

Table 1. Annual Psychologic Test Scores on *Case 1.*

Year	Chronologic age	Mental age	Test given	IQ score
1972	23 mo	20.4 mo	Cattell	88
1973	34 mo	29.6 mo	Stanford-Binet	87
1974	4 yr 1 mo	4 yr 1 mo	Stanford-Binet	88
1975	5 yr 1 mo	5 yr	Stanford-Binet	89
1976	6 yr 1 mo	6 yr 8 mo	Stanford-Binet	104
1977	7 yr 2 mo	6 yr 10 mo	Stanford-Binet	90

Table 2. Results of Numerous Chromosome Studies Performed on *Case 1.*

Date	No. cells counted	No. with 5Gs	%Trisomic	No. with 4 Gs	%Normal
4/70	30	27	90	3[a]	10
3/74	30	28	93.3	2	6.6
8/74[b]	27	21	77.7	6	22.2
1/76	30	30	100	0	0
2/76	26	23	88.4	3	11.5
4/77	9	9	100	0	0
7/76[c]	25	25	100	0	0
3/77[c]	31	31	100	0	0
4/77[c]	19	19	100	0	0

[a]No photographs available; modal number listed as 46 with no indication of which chromosome is missing.
[b]Lymphocyte culture analyzed at Children's Hospital of Los Angeles.
[c]Analyses made on skin fibroblast metaphases.

1970, 30 metaphases were examined and her karyotype was reported as 47,XX,+G (see Fig. 3). Three cells were identified as having a modal number of 46. No drawings or pictures of these particular cells are available. No notation was made to indicate whether these were normal cells with a 46,XX karyotype or cells that had 5 G group chromosomes and some other random chromosome missing to account for the 46 modal number.

The patient was studied at Los Angeles Children's Hospital by Alfi, Koch, and Fishler in 1974. Their chromosomal analysis yielded 6 normal cells out of the 27 cells counted, indicating mosaicism for trisomy 21. It is noteworthy that the results of their developmental testing paralleled ours with an IQ of 91 on the Stanford-Binet, 92 on the Peabody, 92 on the Leiter, and a social quotient of 111 on the Vineland.

In an effort to resolve the question of mosaicism, the patient was restudied on several occasions. In addition to examining 125 lymphocyte metaphases, 75 fibroblast metaphases from a skin culture were also studied. No normal cells were identified in the skin, but five normal cells with a 46,XX karyotype were

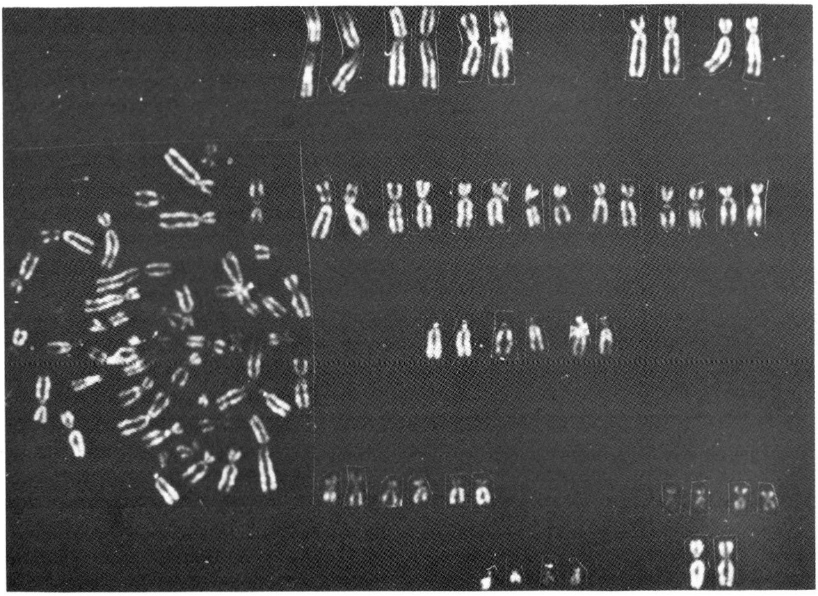

Fig. 3. Rare fluorescent 46,XX karyotype from *Case* 1.

found in the peripheral blood (see Figs. 3 and 4). A few cells in the blood demonstrated random loss of some autosome other than a G group chromosome.

Table 3 summarizes the analyses. It combines our analyses and those performed in Los Angeles, and if one assumes that the original 3 metaphases containing 4 G chromosomes were normal cells, then the percentage of trisomy 21 cells is 91% in the blood and 100% in the skin. If all the metaphases are totaled, regardless of origin, the percentage of normal cells is 6%. It should also be noted that blood group analysis failed to reveal the presence of chimerism [52–54].

The Mosaic Population. As well as reexamining the proband's chromosomes, it seemed appropriate to evaluate our mosaic Down syndrome population and attempt to correlate their performance levels with their percentage mosaicism. Table 4 lists these children in order of increasing percentage of normal cells. The IQ scores range from 30 to 73. Those scoring above 60 are either below 3 years of age or have a greater percentage of normal cells than of trisomic cells. The person with the lowest percentage of trisomic and the highest percentage of normal cells (9% 47,XX,+G/91% 46,XX) is patient 18. Her developmental history follows.

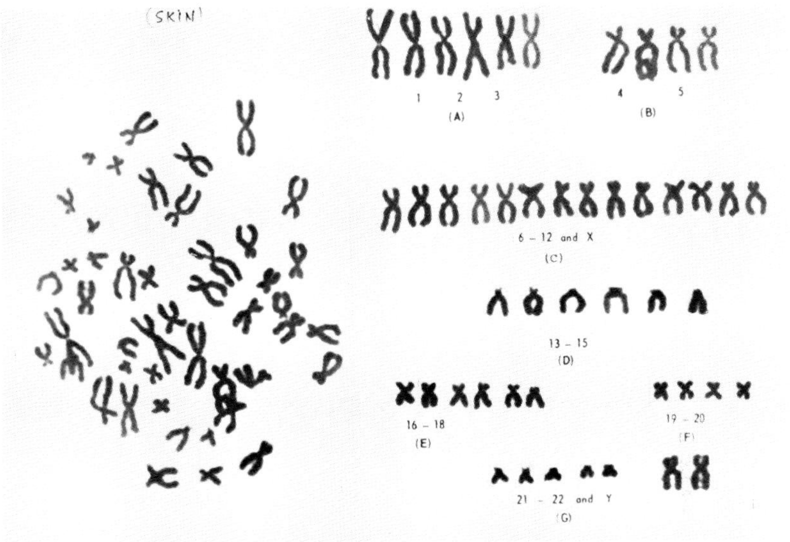

Fig. 4. 47,XX,+G karyotype from skin fibroblasts of *Case* 1.

Case 2

The proband is a white female, 15 years 2 months old (see Figs. 5 and 6). She is the older of two daughters born to a mother who was 16 and a father who was 17 at the time of conception. The baby was born at term following 12 hours of labor, weighing 3,940 gm, and measuring 50.8 cm in length. Her total bilirubin was 18 mg % on the third day of life. No exchange transfusion was deemed necessary but she remained jaundiced for two weeks.

The proband has always been healthy. Her developmental milestones were slightly delayed. She walked at 17 months and was completely toilet-trained at 2½ years. She began saying words at 1 year, but did not speak in sentences until 3½ years. Her physical examination reveals a slight upward slant to the palpebral fissures, small epicanthal folds, Brushfield spots, a somewhat wide nasal bridge, short 5th fingers with clinodactyly, and an increased interdigital space between the 1st and 2nd toes bilaterally. Her dermatoglyphics demonstrate elevated atd angles with 6 ulnar loops and 4 whorls. Both index fingers have ulnar loops and both soles have an arch-tibial pattern. She does not have a simian crease [55–58].

Fig. 5. Proband, *Case 2,* at 10 years of age.

Table 3. Summary of Chromosome Results on *Case 1.*

Culture	No. cells counted	No. with 5 Gs	% Trisomic	No. with 4 Gs	% Normal
Blood[a]	125	117-120	94-96	5-8	4-6
Blood[b]	152	138-141	91-93	11-14	7-9
Skin[a]	75	100	100	0	0

[a]Summary of studies from our center.
[b]Summary of studies from our center and from Los Angeles.

Both parents have completed high school; her mother is currently a medical student and her father, a professional artist. The proband attended a Montessori preschool. Her parents have likewise spent extra time teaching her at home. At the present time she is placed in a educable mentally retarded class at her local junior high school. She also has received speech therapy and takes ballet lessons. On the Wechsler Intelligence Scale for Children she achieved a mental age of 9 years with a verbal score of 66, a performance score of 61, and a full scale score of 60.

Fig. 6. Proband, *Case 2,* at 15 years of age.

DISCUSSION

Many abnormalities and central nervous system deficits have been hypothesized to account for the retardation seen in trisomy 21. Fetal brain growth is clearly delayed, so that these infants are frequently microcephalic from birth [5, 13]. Autopsies have demonstrated the cerebellum and certain nuclei in the brain stem to be smaller than normal [60]. Specific deficits in areas such as auditory sequencing [37, 40, 61, 62], color retention [63], short-term memory [62, 64], articulation [34, 36, 38, 40], visual-motor tasks [39, 59, 61, 62, 64], ability to differentiate between symbols [61, 62, 64], and language development have all been documented. Emotional disturbance and frank psychoses have been seen as well [65].

Despite these built-in obstacles some children with trisomy 21 function considerably better than others [16, 17]. As was mentioned before, development is faster and reaches higher levels in those children reared at home rather than in institutions, even in those centers that have an enriched environment [13, 14,

Table 4. Mosaic Trisomy 21 Population With Percentage Mosaicism and IQ Scores

Patient number	% Trisomy 21/% normal cells	Chronologic age when tested	Mental age	IQ score
1	97/3	Not tested	Neonate	--
2	94/6	26 mo	16.8 mo	64
3	90/10	13 mo	8 mo	61
4	90/10	10 yr 1 mo	4 yr 9 mo	48
5	90/10	30 mo	22 mo	73
6	90/10	Not tested	Neonate	--
7	85/15	2 mo	a	--
8	85/15	8 yr 9 mo	4 yr 1 mo	46
9	80/20	15 mo	8.3 mo	55
10	80/20	25 mo	12.6 mo	50
11	80/20	4.5 mo	3 mo	66
12	80/20	27 mo	19.2 mo	71
13	80/20	3 yr 4 mo	1 yr	30
14	76/24	3 yr 8 mo	21.8 mo	50
15	25/75	3.8 mo	2.5 mo	69
16	12/88[b]	6 yr 2 mo	4 yr 5 mo	72-verbal 77-performance 72-full scale
17	10/90	Not tested	Neonate	--
18	9/91[b]	15 yr 2 mo	9 yr	66-verbal 61-performance 60-full scale

[a]No delays evident on Denver Developmental Screening Test.
[b]Percentages are derived from lymphocytes cultured on two separate occasions. (See Figs. 7 and 8).

23–32]. The data available from studies of those reared at home also suggest that the IQs of the trisomy 21 population are roughly 2SD below that of the normal population [24]. Some feel, therefore, that the development potential of offspring with trisomy 21 is higher in those families where the parents have higher IQs [24, 48, 66]. Parental education is also a factor. In the study by Golden and Pashayan, the mean IQ of children with trisomy 21 whose parents had graduated from high school was 50 (range 32–74). If the parents had completed only grade school, the mean IQ was 34.6 (range 0–45) [67]. The degree of hypotonia also seems important in that it affects not only the motor milestones, but langauage as well [23, 33, 36, 39, 59].

Mosaicism for trisomic and normal cells has also been observed to increase the intellectual functioning of the individual [47–49, 50–53, 68]. Mosaic subjects are frequently reported to have less visual-perceptual difficulties, to develop clearer speech, and to demonstrate increased vocabulary, comprehension, and ability to deal with numerical concepts [50, 51]. In addition, they are often incidentally noted to have fewer of the physical stigmata or to manifest milder

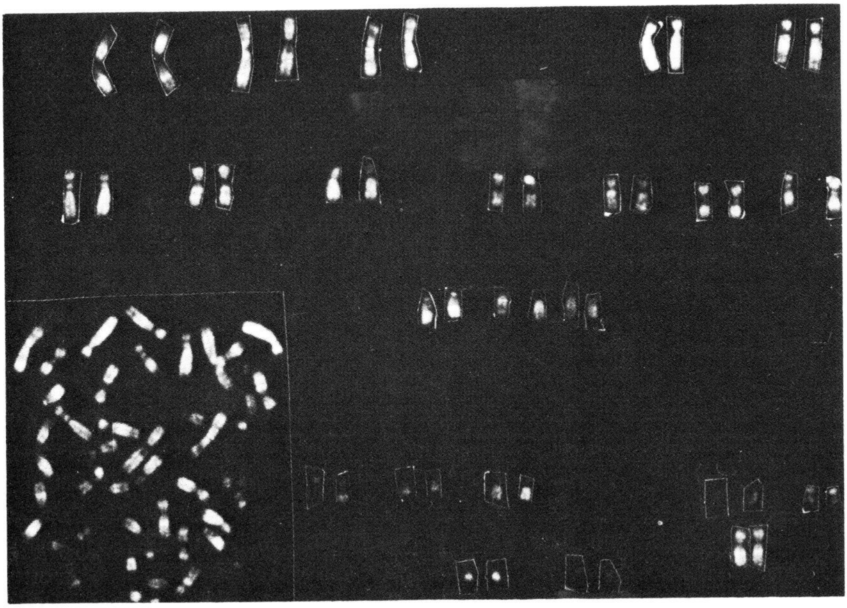

Fig. 7. Fluorescent 46, XX karyotype from *Case 2* (91%).

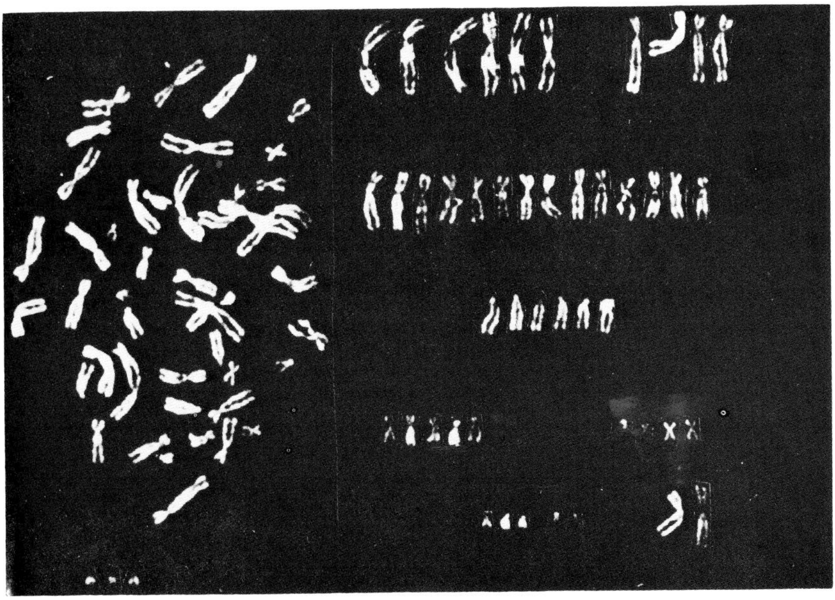

Fig. 8. Fluorescent 47,XX,+21 karyotype from *Case* 2 (9%).

presentations of these features than seen in the nonmosaic trisomy 21 population [41, 43, 48, 50, 51, 69].

In spite of the added strengths possessed by many mosaic subjects, most still do not test in the normal range [45, 47–51]. The 15 children Fishler followed had a mean IQ score of 67 [50, 51], while the 20 children Rosecrans studied had a mean score of 65 [47]. The mean score for our sample is 58.2, but it should be noted that 4 of the children have not yet been tested and many are still quite young, when test scores are not as reliable. Patients 15, 16, and 18 might be expected to have higher scores. However, while they test above the mean, their scores are still in the retarded range. They function lower than might be anticipated even in the face of special educational programs to enhance their developmental potential and, at least in the one case parental IQ and education that might predict a better outcome.

The remainder of our mosaic population contrasts sharply with the proband, *Case 1.* Whether one feels that she is a child with nonmosaic trisomy 21 and that the normal cells found are spurious, or whether one considers her to be a mosaic child with a low proportion of available tissues comprised of 46,XX cells, her performance is truly outstanding. Of course, since she is still young and has only just begun her academic endeavors, she may not continue to function so well. Of some concern is the fact that her emotional maturity has not kept pace with her cognitive abilities. Nonetheless, she has performed better for longer than any of our mosaic Down syndrome children. While a few scattered individuals have been reported to test in the normal range [41, 43, 45, 47, 50, 51], indications as to their percentage of normal cells have been infrequent. When listed, it is usually not as low as the number found in *Case 1.* Rather, the proband's achievement in this case demonstrates how a child with trisomy 21 (even if mosaic) from an intelligent and highly motivated family can occasionally exceed expectations if maximally stimulated.

SUMMARY

A 7-year-old girl with (probably mosaic) trisomy 21 who tests and functions in the normal range of intelligence has been presented. Her parents and 5 sibs have worked intensely in an attempt to maximize her learning. She has also taken acrobatics and speech therapy to refine her motor skills and coordination. Her performance is compared with that of 18 other mosaic trisomy 21 children. None of these children tests in the normal range. The mosaic girl with the lowest trisomic cell percentage in our series (9%) is also described. Despite a favorable cytogenetic constitution and similar family background and involvement, schooling, and speech therapy, she has an IQ of 60. The performance of the proband in *Case 1* is felt to be unusual and may reflect the premise that correlations between trisomic cell percentages and the degree of mental retardation in mosaic trisomy 21 should be drawn with extreme caution [51, 71, 72]. It may also lend support to the gathering body of evidence that early intervention and education may be fruitful in children with Down syndrome.

REFERENCES

1. Séquin E: "Traitement moral, hygiene et education des idiots et des autres enfants arrieres ou retardes dans leur developpement." Paris: JB Bailliere, 1846.
2. Down JL: Ethnic classification of idiots. Lond Hosp Rep 3:259–262, 1866.
3. Shuttleworth GE: Mongolian imbecility. Br Med J 2:661–669, 1909.
4. Wolstenholme GEW, Porter R (eds): "Mongolism." Ciba Foundation Study Group No. 25. Boston: Little, Brown, 1967.
5. Benda C: "Mongolism and Cretinism." New York: Grune and Stratton, 1949.
6. Benda CE: Research in congenital acromicria (mongolism) and its treatment. Q Rev Pediatrics 8:79–96, 1953.
7. Lejeune J, Gautier M, Turpin R: Les chromosomes humains en culture de tissus. C R Acad Sci (Paris) 248:602–603, 1959.
8. Lejeune J, Turpin R, Gautier M: Le mongolisme, premier exemple d'aberration autosomique humaine. Ann Genet (Paris) 2:41–49, 1959.
9. Lejeune J, Turpin R, Gautier M: [Chromosome diagnosis of mongolism] (Fr). Arch Fr Pediatr 16:962–963, 1959.
10. Lejeune J, Turpin R: Somatic chromosomes in mongolism. Res Publ Assoc Res Nerv Ment Dis 39:67–77, 1962.
11. Lejeune J: Autosomal disorders. Pediatrics 32:326–337, 1963.
12. Lejeune J: The 21 Trisomy – current stage of chromosomal research. Prog Med Genet 3:144–177, 1964.
13. Durling D, Benda CE: Mental growth curves in untreated institutionalized mongoloid patients. Am J Ment Defic 56:578–588, 1952.
14. Quaytman W: The psychological capacities of mongoloid children in a community clinic. Q Rev Pediatrics 8:255–267, 1953.
15. Wolman IJ (ed): "Symposium on Mongolism." Q Rev Pediatrics 8:76–96, 126–138, 253–291, 1953.
16. Levinson A, Friedman A, Stamps F: Variability of mongolism. Pediatrics 16:43–53, 1955.
17. Kugel RB, Reque D: A comparison of mongoloid children. JAMA 175:959–961, 1961.
18. Nakamura H: Nature of institutionalized adult mongoloid intelligence. Am J Ment Defic 66:456–458, 1961.
19. Nakamura H: An inquiry into systemic differences in the abilities of institutionalized adult mongoloids. Am J Ment Defic 69:661–665, 1965.
20. Dameron LE, Lawrence E: Development of intelligence in infants with mongolism. Child Dev 34:733–738, 1963.
21. Share J, Koch R, Webb A, Graliker B: The longitudinal development of infants and young children with Down's syndrome (mongolism). Am J Ment Defic 68:685–692, 1964.
22. Silverstein AB: Mental growth in mongolism. Child Dev 37:725–729, 1966.
23. Carr J: Mental and motor development in young mongol children. J Ment Defic Res 14:205–220, 1970.

24. Dicks-Mireaux MJ: Mental development of infants with Down's syndrome. Am J Ment Defic 77:26–32, 1972.
25. Melyn M, White D: Mental and developmental milestones of non-institutionalized Down's syndrome children. Pediatrics 52:542–545, 1973.
26. Farrell MJ: Adverse effects of early institutionalization of mentally subnormal children. Am J Dis Child 91:278–281, 1956.
27. Tizard J: Residential care of mentally handicapped children. Br Med J 1:1041–1046, 1960.
28. Centerwall SA, Centerwall WR: A study of children with mongolism reared in the home compared to those reared away from home. Pediatrics 25:678–685, 1960.
29. Shotwell AM, Shipe D: Effect of out-of-home care on intellectual and social development of mongoloid children. Am J Ment Defic 68:693–699, 1964.
30. Stedman DJ, Eichorn DH: A comparison of the growth and development of institutionalized and home-reared mongoloids during infancy and early childhood. Am J Ment Defic 69:391–401, 1964.
31. Shipe D, Shotwell AM: Effect of out-of-home care on mongoloid children: A continuation study. Am J Ment Defic 69:649–652, 1965.
32. Cicchetti D, Sroufe LA: The relationship between affective and cognitive development in Down's syndrome infants. Child Dev 47:920–929, 1976.
33. Strazzula M: Speech problems of the mongoloid child. Q Rev Pediatrics 8:268–273, 1953.
34. Kolstoe OP: Language training of low-grade mongoloid children. Am J Ment Defic 63:17–30, 1958.
35. Lenneberg EH, Nicholas IA, Rosenberger EF: Primitive stages of language development in mongolism. Br J Discord Commun VVL:119–137, 1962.
36. Blanchard I: Speech pattern and etiology in mental retardation. Am J Ment Defic 68:612–617, 1964.
37. Bilovsky D, Share J: The ITPA and Down's syndrome: An explanatory study. Am J Ment Defic 70:78–82, 1965.
38. Evans D, Hampson M: The language of mongols. Br J Disord Commun 3:171–181, 1969.
39. Cornwall AC: Development of language, abstraction and numerical concept formation in Down's syndrome children. Am J Ment Defic 79:179–190, 1974.
40. Dodd B: Recognition and reproduction of words by Down's syndrome and non-Down's syndrome retarded children. Am J Ment Defic 80:306–311, 1975.
41. Clarke CM, Edwards JH, Smallpiece V: 21-Trisomy/normal mosaicism in an intelligent child with some mongoloid features. Lancet 1:1028–1030, 1961.
42. Hayashi T, Hsu TC, Chao D: A case of mosaicism in mongolism. Lancet 1:218–219, 1962.
43. Richards BW, Stewart A: Mosaicism in a mongol. Lancet 1:275–276, 1962.
44. Zellweger H, Abbo G: Chromosomal mosaicism and mongolism. Lancet 1:827, 1963.
45. Finley WH, Finley SC, Rosecrans CJ, Tucker CC: Normal/21-trisomy mosaicism. Am J Dis Child 112:444–447, 1966.
46. Reisman L, Shipe D, Williams RDB: Mosaicism in Down's syndrome: Studies in a child with an unusual chromosome constitution. Am J Ment Defic 70:855–859, 1966.
47. Rosecrans CJ: The relationship of normal/21 trisomy mosaicism and intellectual development. Am J Ment Defic 72:562–566, 1968.
48. Richards BW: Mosaic mongolism. J Ment Defic Res 13:66–83, 1969.
49. Richards BW: Investigation of 142 mosaic mongols and mosaic parents of mongols; cytogenetic analysis and maternal age at birth. J Ment Defic Res 18:199–208, 1974.

50. Fishler K: Mental development in mosaic Down's syndrome as compared with trisomy 21. In Koch R, de la Cruz FF (eds): "Down's Syndrome (Mongolism): Research, Prevention and Management." New York: Brunner/Mazel, 1975.
51. Fishler K, Koch R, Donnell GN: Comparison of mental development in individuals with mosaic and trisomy 21 Down's syndrome. Pediatrics 58:744–748, 1976.
52. Ford CE: "Mongolism." CIba Foundation Study Group No. 25. Wolstenholme GEW and Porter R (eds). London: J & A Churchill, 1967, p 71.
53. Penrose LS: Studies on mosaicism in Down's analomies. In Jervis GA (ed): "Mental Retardation." Illinois: CC Thomas, 1967.
54. Ford CE: Mosaics and chimaeras. Br Med Bull 25:104–109, 1969.
55. Penrose LS: Dermatoglyphics in mosaic mongolism and allied conditions. Genetics Today 3:973, 1964.
56. Holt SB: "The Genetics of Dermal Ridges." Springfield, Charles C Thomas,1968.
57. Loesch D: Dermatoglyphic characteristics of 21 trisomy mosaicism in relation to the fully developed syndrome and normality. J Ment Defic Res 18:209–269, 1974.
58. Loesch D, Smith CAB: Discriminant diagnosis of 21 trisomy mosaicism. J Ment Defic Res 20:207–219, 1976.
59. Frith U, Frith CD: Specific motor disabilities in Down's syndrome. J Child Psychol Psychiatry 15:293–301, 1974.
60. Crome LC, Cowie V, Slater E: A statistical note on cerebellum and brain stem weight in mongolism. J Ment Defic Res 10:69–72, 1966.
61. Scheffelin M: A comparison of four stimulus response channels in paired-associate learning. Am J Ment Defic 73:303–307, 1968.
62. Ray AB, Shotick AL: Short-term and long-term recall of familiar objects by trainable and educable mentally retarded and normal individuals of comparable mental age. J Ment Defic Res 20:183–190, 1976.
63. Sinson JC, Wetherick NE: The nature of the colour retention deficit in Down's syndrome. J Ment Defic Res 19:97–100, 1975.
64. Machay DN, McDonald G: The effects of varying digit message structures on their recall by mongols and non-mongol subnormals. J Ment Defic Res 20:191–197, 1976.
65. Menolascino FJ: Psychiatric aspects of mongolism. Am J Ment Defic 69:653–660, 1965.
66. Fraser FC, Sadovnick AD: Correlation of IQ in subjects with Down's syndrome and their parents and siblings. J Ment Defic Res 20:179–182, 1976.
67. Golden W, Pashayan HM: The effect of parental education on the eventual mental development of non-institutionalized children with Down's syndrome. J Pediatr 89: 603–605, 1976.
68. Polani PE: Cytogenetics of Down's syndrome. In Gardner LL (ed): "Symposium on Genetics." Pediatr Clin North Am 10:423–448, 1963.
69. Gibson D, Gibbons RJ: The relation of mongolian stigmata to intellectual status. Am J Ment Defic 63–65, 1958.
70. Dunsdon MI, Carter CO, Herntley RMC: Upper end of range of intelligence in mongolism. Lancet 1:565–568, 1960.
71. Kohn G, Taysi K, Atkins TE, Mellman WJ: Mosaic mongolism. I. Clinical correlations. J Pediatr 76:874–879, 1970.
72. Taysi K, Kohn G, Mellman WJ: Mosaic mongolism. II. Cytogenetic studies. J Pediatr 76:880–885, 1970.

A Terminal Long Arm Deletion of Chromosome 16 in a Dysmorphic Infant: 46,XY,del(16) (q22)

Kutay Taysi, MD, Marvin Fishman, MD,* and Gurbax S. Sekhon, PhD

INTRODUCTION

Although terminal deletions of most human chromosomes have been reported in the literature [1], there has been no clinical case report of the long arm deletion of chromosome 16 in a liveborn infant. This communication describes a patient who has a terminal long arm deletion of chromosome 16 and congenital hydrocephalus, hypospadias, and dysmorphic facial features.

MATERIALS AND METHODS

The infant was hospitalized for the evaluation of the congenital hydrocephalus.

Chromosome analyses of the patient and his parents were made on 72-hour lymphocyte cultures which were harvested using potassium chloride (0.075 M) as hypotonic solution, fixed in methanol:acetic acid (3:1), spread on wet slides, and dried on a hot plate. Slides were kept at 60°C for 24 hours, incubated 20 minutes in 2 × SSC and stained with Giemsa R66 [2]. Skin was cultured in minimal essential medium containing 20% fetal calf serum. The fibroblast cultures were harvested according to the method of Harnden and Brunton [3]. The slides of fibroblast cultures were treated and stained as those of blood cultures.

*Dr. Fishman was supported in part by The Georgia Warm Springs Foundation, New York.

Birth Defects: Original Article Series, Volume XIV, Number 6C, pages 343–347

CLINICAL AND CYTOGENETIC FINDINGS

A 5-hour-old male infant was admitted to St. Louis Children's Hospital for the evaluation of a large head. The baby was born after an uncomplicated, full-term pregnancy to a 31-year-old Gravida 1 Para 0 healthy white female. Delivery was by Cesarian section because of a breech presentation and failure of engagement. The Apgar score was 4 at 1 minute and 8 at 5 minutes. The father was 35 years old and in good health. There was no family history of congenital malformation, including those of the central nervous system.

On examination the weight was 4.58 kg, length 55 cm, and head circumference 51.5 cm. Marked frontal bossing was present. The palpebral fissures had an antimongoloid slant, and the nasal bridge was depressed. The ears were low-set and posteriorly rotated, the palate had a moderately high arch, and the neck was short (Fig. 1). Examination of the genitalia disclosed first-degree hypospadias and a chordae on the dorsal side of the penis. Transillumination demonstrated marked increase in transmission of light, especially on the right. The falx cerebri was displaced to the left and the torcula appeared to lie low on the occipital region and was also displaced to the left. Funduscopic examination revealed small greyish optic disks. The remainder of the neurologic examination revealed no abnormality except for slightly increased deep tendon reflexes and

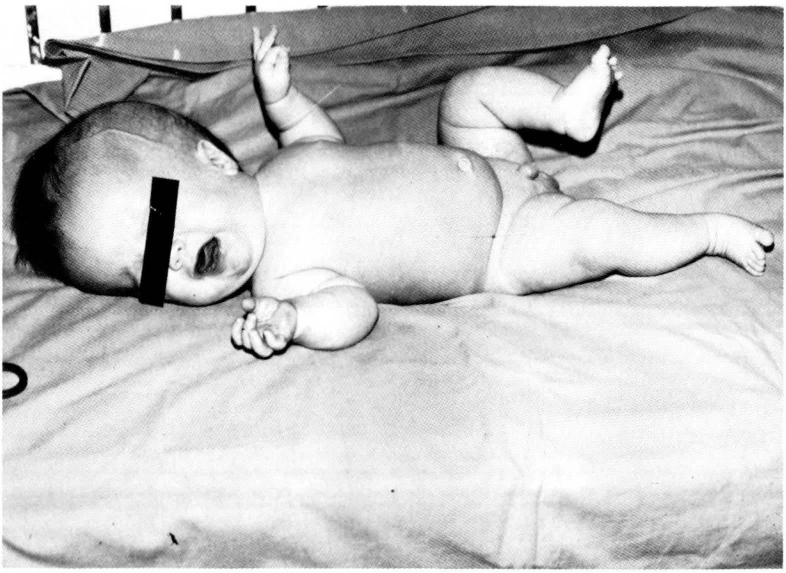

Fig. 1. The proband. Note frontal bossing, low-set ears, short neck, and increased head circumference.

poor head control. The dermatoglyphic examination revealed 10 ulnar loops, normally placed axial triradii, and bilateral tibial arch pattern on the hallucal area. There was a partial simian line on the right.

Pertinent diagnostic investigation included negative toxoplasmosis fluorescent antibody titers in the mother and baby. Viral antibody titers in the mother were: cytomegalovirus less than 1:4, herpes simplex 1:16, and rubella hemagglutination inhibition 1:40. The corresponding antibody titers in the infant were less than 1:4, 1:8, and 1:80. Computerized tomography of the cranium revealed massive hydrocephalus with the falx displaced to the left in its posterior course.

The chromosome analyses from peripheral lymphocyte and skin fibroblast cultures revealed a terminal long arm deletion of chromosome 16 (Fig. 2). The missing part involved a segment of the long arm distal to q22. The karyotypes of the parents were normal. The partial karyotypes from different cells of the mother (A), the father (B), and the proband (C) are shown in Figure 3.

DISCUSSION

A terminal deletion of the human chromosomes is not a rare event. However, there has been no clinical case report with the terminal long arm deletion of

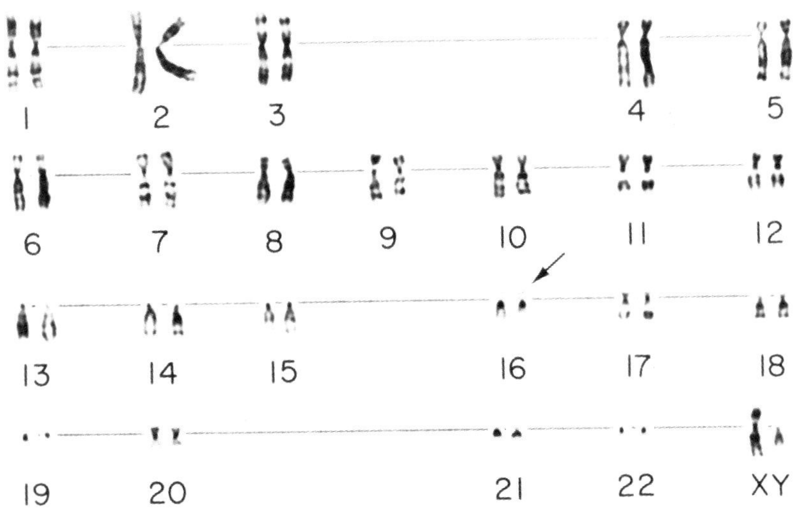

Fig. 2. The karyotype of the proband showing the terminal long arm deletion of chromosome 16.

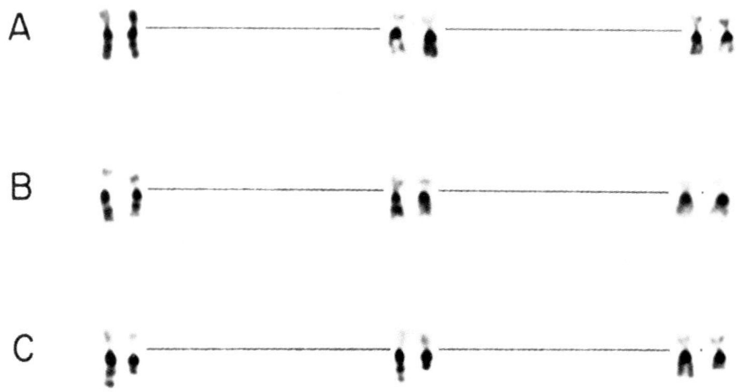

Fig. 3. The partial karyotypes showing normal chromosome 16 in the mother (A), the father (B), and the terminal long arm deletion in the proband (C).

chromosome 16 in the literature. Only 5 cases of 16q– were cited in the third listing of the "Repository of Chromosomal Variants and Anomalies" [4] without a description of the clinical features of the patients. Thus, the long arm deletion of chromosome 16 must be a rare event in a liveborn infant, presumably because of failure of embryos with such a chromosome abnormality to survive. However, the difficulty of the cytogenetic diagnosis of a single-band deletion of a chromosome in a routine preparation may be another reason for the rarity of this cytogenetic abnormality.

Most of the frequent causes of congenital hydrocephalus were excluded by appropriate laboratory tests in this case. It is tempting to attribute the congenital hydrocephalus, hypospadias, and dysmorphic facial features, such as the antimongoloid slant of the eyes, depressed nasal bridge, short neck, high-arched palate, and low-set ears, of the present case, to the terminal deletion of the long arm of chromosome 16. However, clinical and cytogenetic reports of further cases with long arm deletion of chromosome 16 are needed to confirm the existence of a 16q– syndrome.

SUMMARY

A terminal deletion of the long arm of chromosome 16 was found in a newborn infant who had congenital hydrocephalus, hypospadias, and dysmorphic facial features. The chromosome complement was 46,XY,del(16)(q22). To the best of our knowledge this is the first reported case of partial long arm deletion of chromosome 16 in a liveborn infant with multiple congenital malformations.

REFERENCES

1. Borgaonkar DS: "Chromosomal Variation in Man. A Catalog of Chromosomal Variants and Anomalies." 2nd ed. New York: Alan R. Liss, 1977.
2. Sumner AT, Evans HJ, Buckland RA: A new technique for distinguishing between human chromosomes. Nature (New Biol) 232:821–822, 1971.
3. Harnden DG, Brunton S: The skin culture technique. In Yunis JJ (ed): "Human Chromosome Methodology." New York: Academic Press, 1965, pp 57–73.
4. "Repository of Chromosomal Variants and Anomalies in Man. Third Listing." Baltimore: The Johns Hopkins University Press, 1976, pp 94, 95, 97.

Asymmetric Clinical and Cytogenetic Findings in a 4-Year-Old Girl With Trisomy 18 Mosaicism*

Kathleen W. Rao, BS, Philip D. Buchanan, PhD, and Arthur S. Aylsworth, MD

Characteristic clinical phenotypes have been delineated for the more common autosomal trisomies. There is more variability, however, in the clinical features associated with trisomy mosaicism. It is reasonable to assume that out current concept of the clinical presentation of patients with trisomy mosaicism is biased by the more severe cases because they are more likely to come to our attention. We are presenting a case of trisomy 18 in which the asymmetry of clinical features is correlated with an asymmetry in the distribution of two cell lines: one trisomic for chromosome 18 and the other cytogenetically normal.

CASE HISTORY

The patient is a 4-year-old white female who was referred to the Genetics Clinic by our Pediatric Cardiology Division for evaluation of multiple anomalies and dysmorphic features. She was the 4.3 kg product of a 42-week pregnancy, born to a 23-year-old primiparous mother who gained 5.4 kg during the pregnancy. Fetal activity was normal and there was no history of serious maternal illness during the pregnancy. The patient's early measurements probably indicate a mild but significant degree of intrauterine growth retardation for a term infant. She developed congestive heart failure and was found to have a ventricular septal defect, an atrial septal defect, and a patent ductus arteriosus, as well as dislocation of the right hip and a right metatarsus adductus. She remained in the newborn nursery for 28 days.

*Supported in part by NIGMS GM07092, and Genetic Counseling Center grants from The National Foundation–March of Dimes and the North Carolina Department of Human Resources.

Birth Defects: Original Article Series, Volume XIV, Number 6C, pages 349–354

Her height and weight have paralleled the 3rd percentile since early in the first year of life and are currently at the 3rd percentile. Her early developmental milestones were delayed; she sat at 10 months, said her first words at 18 months, and walked at 20 months. At the age of 3 years, 8 months, her receptive language ability was at the 2 year, 3 month level, while her expressive language ability was estimated to be at approximately the 1 year, 6 month level. She is currently functioning in the moderately retarded range.

There is no family history of infertility, increased fetal wastage, or birth defects.

At 4 years of age, her head circumference is below average but within 2 standard deviations of the mean for her chronologic age. She has moderate facial asymmetry with the right side being smaller than the left (Fig. 1). On her right hand, which is smaller than her left hand, she has a proximal t triradius, a distally placed t′ triradius, and arches on all 5 fingertips. On her left hand she has a normal proximal t triradius, arches on her first 3 fingers, and ulnar loops on her 4th and 5th fingers. She has slight asymmetry of the thorax and a grade 3/6

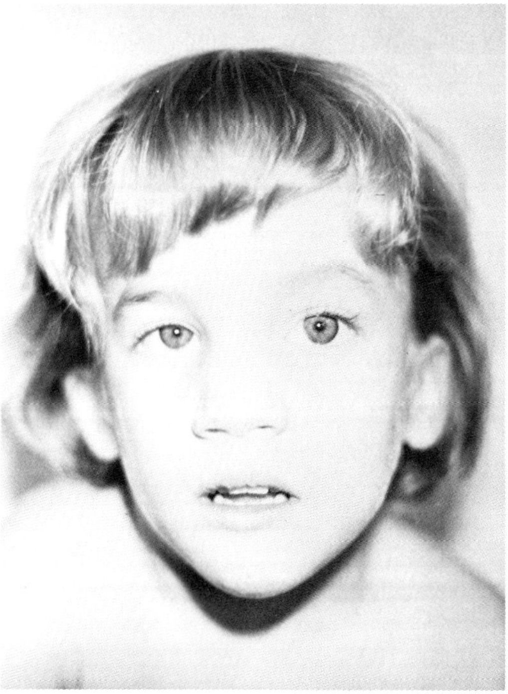

Fig. 1. The proband: moderate facial asymmetry.

systolic murmur. There is no abdominal organomegaly. She has normal prepubertal genitalia.

CYTOGENETIC STUDIES

Peripheral blood lymphocytes were cultured and chromosome preparations were made in a conventional manner. Two separate blood samples were obtained from the patient (2.5 months apart) and 50 metaphases were studied from each sample. An extra chromosome 18 was observed in every cell. G-banding by a modified trypsin-Giemsa technique [1] and Q-banding with quinacrine mustard (QM) [2] revealed 47 structurally normal chromosomes (47,XX,+18) in each cell, with the extra chromosome identified as a chromosome 18 (Fig. 2). Lymphocyte chromosome studies with Q-banding done on both parents were normal.

Because of the poor phenotype-karyotype correlation in this patient and the observed morphologic asymmetry, skin biopsies were obtained from the inner

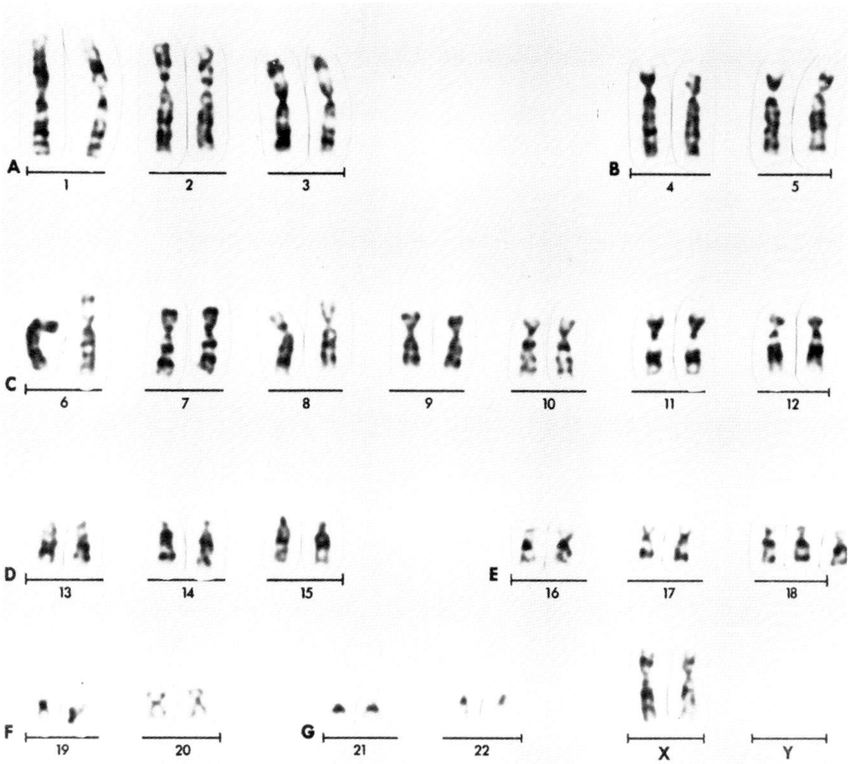

Fig. 2. Lymphocyte karyotype of trisomy 18 cell with G-banding.

surfaces of both forearms. These biopsies were cultured separately in RPMI 1640 with 10% fetal calf serum and antibiotics, and chromosome preparations were made when the cultures were at the 3rd passage. Fifty metaphase cells were counted from each biopsy; the results are presented in Table 1. The right arm biopsy, from the smaller side of the patient, contained an extra chromsome 18 in each cell counted. The left biopsy was made up of two cell populations; 14% of the cells contained an extra chromosome 18 and 86% of the cells had a normal 46,XX karyotype.

DISCUSSION

A similar patient with more severe malformations was described by Schuler et al [3]. The child had severe deformities on the left side of his body, while his right side was more nearly normal. He had a normal lymphocyte karyotype and trisomy 18 mosaicism in skin cells. Fibroblasts grown from the left side of the body contained 82% trisomy 18 cells while those cultured from the right side had only 12% trisomy 18. Another patient, reported by Brøgger et al [4] also displayed asymmetry, the left side being larger than the right. This patient had two abnormal cell populations interpreted as an unbalanced B-G translocation on her left side and mosaicism on her right side. The cell lines on the right side included a minor line containing the same B-G translocation and a predominant line containing the B-G translocation with trisomy 2 and monosomy of a C-group chromosome. In these two cases, as in ours, the observation of asymmetry or unilateral malformations led to the discovery of chromosomal mosaicism.

Hamerton [5] reported that the mean survival of 153 patients with trisomy 18 was 71 days. Our patient is slightly over 4 years old. She does not fit the clinical picture of typical trisomy 18, although two blood karyotypes revealed trisomy 18 in every cell. Survival beyond the age of 1 year is rare in children with pure trisomy 18 and only a few cases have been reported. In two of these [6, 7] blood and skin karyotyping failed to show a normal cell line, but in four others [8–11] only peripheral blood karyotypes were done. While these last four patients may indeed

TABLE 1. Chromosome Findings in Blood Lymphocytes and Skin Fibroblasts of Patient

Tissue	Number of metaphases examined	46,XX cells (%)	47,XX +18 cells (%)
1st blood	50	0	100
2nd blood	50	0	100
Forearm skins			
Left	50	86	14
Right	50	0	100

have pure trisomy 18, the present case and that of Schuler et al [3] illustrate that the lymphocyte karyotype can fail to disclose mosaicism which may exist in other tissues.

There are 3 possible origins of trisomy mosaicism: chimerism, mitotic nondisjunction or anaphase lag in a trisomic embryo, and mitotic nondisjunction in a cytogenetically normal embryo. We used QM intensity variants to address the question of whether this patient could be a chimera and found that her parents have a total of 5 independently segregating fluorescent variants. The patient displays concordance for all 5 markers in the 46,XX and 47,XX,+18 cell lines, which significantly reduces the probability of chimerism. We can only speculate about the relative probabilities that mitotic nondisjunction may have occurred in a trisomy 18 or in a normal fetus. The patient's relatively mild clinical features lend weight to the argument that the trisomy 18 cell line arose relatively late in the development of a cytogenetically normal embryo. The cytogenetic studies, however, indicate that in the tissues studied the trisomy 18 cell line predominates. In either case, we cannot assume that the percentage of normal and abnormal cells found in the tissues that we have examined accurately reflects the cell distribution throughout the patient's body.

SUMMARY

In summary, this case report illustrates several points.

1) Peripheral blood karyotyping may fail to detect mosaicism. The frequency of this occurrence is unknown.

2) A poor phenotype-karyotype correlation (including unusually long survival in a patient with an autosomal trisomy other than trisomy 21) is an indication that additional tissues should be karyotyped.

3) Asymmetric development can be associated with chromosomal mosaicism.

4) In the case of an individual with unilateral malformations, cytogenetic studies of both the abnormal and normal tissues may be necessary to accurately assess the patient's genetic status.

REFERENCES

1. Seabright M: A rapid banding technique for human chromosomes. Lancet 2:971–972, 1971.
2. Caspersson T, Zech L, Johansson C, Modest EJ: Identification of human chromosomes by DNA-binding fluorescent agents. Chromosoma 30:215–227, 1970.
3. Schuler D, Ferenczi I, Görgényi A, Dubos M, Fekete G, Ruzicska R: Unilateral congenital malformations with chromosome aberration. Hum Hered 22:198–203, 1972.
4. Brøgger A, van der Hagen CB, Støren A: Mosaicism involving two karyotypes with 45 chromosomes and a new autosomal translocation in a girl with mental defect, asymmetry, congenital heart disease and other malformations. Acta Paediatr Scand [Suppl] 159: 44, 1965.

5. Hamerton JL: "Human Cytogenetics." New York: Academic Press, 1971, vol 2.
6. Weber WW, Mamunes P, Day R, Miller P: Trisomy 16-18(E): Studies in long-term survival with report of two autopsied cases. Pediatrics 34:533–540, 1964.
7. Surana RB, Bain HW, Conen PE: 18-Trisomy in a 15-year-old girl. Am J Dis Child 123:75–77, 1972.
8. Hook EB, Lehrke R, Roesner A, Yunis JJ: Trisomy-18 in a 15-year-old female. Lancet 2:910–911, 1965.
9. Geiser CF, Schindler AM: Long survival in a male with 18-trisomy syndrome and Wilms' tumor. Pediatrics 44:111–116, 1969.
10. Bargman GJ, Museles M: 18-Trisomy syndrome in a girl of eight. Lancet 1:143, 1970.
11. Stoll C, Levy J-M, Terrade E: Les survies prolongées dans la trisomie 18. Ann Pédiatr 21 (2): 185–190, 1974.

Deletion of the Long Arm of Chromosome 4: A Clinically Identifiable Syndrome?

Jaime L. Frias, MD, Robert M. Nelson, MD, and Shari L. Ray, CT

A deletion of the long arm of chromosome 4 was demonstrated in a 3-month-old male infant with multiple congenital malformations and marked developmental deficiency. The infant – born to a 16-year-old woman whose previous pregnancy had resulted in an early spontaneous abortion – was delivered by C section at 38 weeks gestation weighing 2,600 gm.

At birth the infant was noted to have multiple congenital malformations, marked cyanosis, and hypotonia. He was resuscitated and transferred to a secondary care hospital where he stayed for approximately 3 months. During that time the infant experienced choking and cyanosis with feedings and frequent regurgitation.

At 3 months of age, when admitted to the University of Florida Shands Teaching Hospital, his weight (3.5 kg), length (54 cm), and head circumference (34.5 cm) were all below the 3rd% and his psychomotor development was markedly retarded. Physical examination demonstrated the following craniofacial anomalies: abnormally shaped head with prominent occiput; asymmetric forehead and shallow supraorbital ridges; upward slanting palpebral fissures, more marked at left, epicanthic folds with dysplasia of the inner canthi; posteriorly rotated and abnormal ears, each with a pointed helix, hypoplastic antihelix and narrow, slit-like, external auditory meatus; depressed nasal bridge and anteverted nostrils; cleft lip; cleft palate and micrognathia (Fig. 1). In addition, he had proximal implantation of the thumbs, camptodactyly of the 2nd through 4th fingers bilaterally, clinodactyly of the 5th fingers, bilateral dysplasia of the hip, irregular implantation of the toes and cutaneous syndactyly between the 2nd and 3rd toes. Dermatoglyphic analysis demonstrated 8 ulnar loops, 1 radial loop, 1 whorl, and normally placed axial triradii on both palms. The hands also showed bilateral simian creases and a single flexion crease on the left 5th finger.

Birth Defects: Original Article Series, Volume XIV, Number 6C, pages 355–358

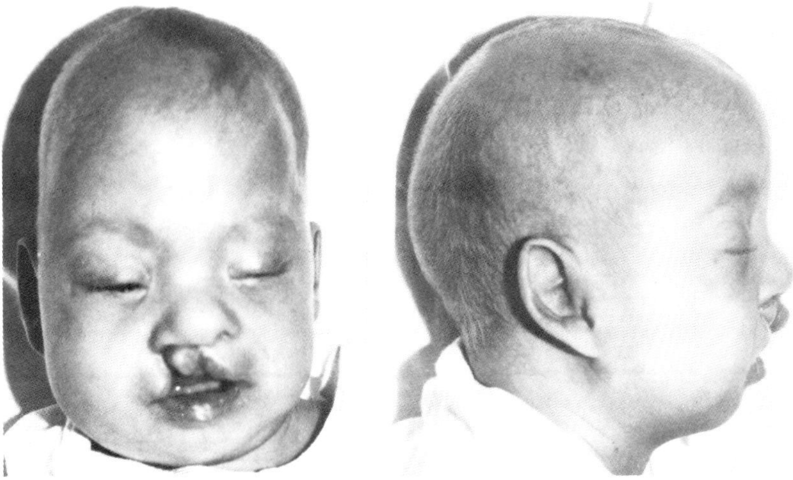

Fig. 1. Infant at 3 months old. Note head shape, shallow supraorbital ridges, upward slanting palpebral fissures, epicanthic folds, abnormal position and shape of auricle, depressed nasal bridge, cleft lip, and micrognathia.

During his hospitalization the infant had recurrent episodes of aspiration, and roentgenograms of the chest showed confluent pulmonary infiltrates suggestive of chronic aspiration. A barium swallow failed to demonstrate any anatomic deformity but confirmed massive gastroesophageal reflux. A feeding gastrostomy was performed to alleviate this problem and he was transferred back to the referring hospital, where he died at 6 months of age. Postmortem examination showed a patent foramen ovale, and interstitial infiltration of lymphocytes and numerous histiocytes in the alveoli of both lungs.

Chromosome analyses in peripheral blood lymphocyte and cultured skin fibroblasts showed a deletion of the distal one-fifth of the long arm of chromosome 4. Trypsin-Giemsa banding demonstrated that the break occurred at band q31 (Fig. 2). No evidence of translocation of the deleted segment to another chromosome was found. Thus far we have been unable to locate the parents to investigate the possibility of a balanced translocation in one of them.

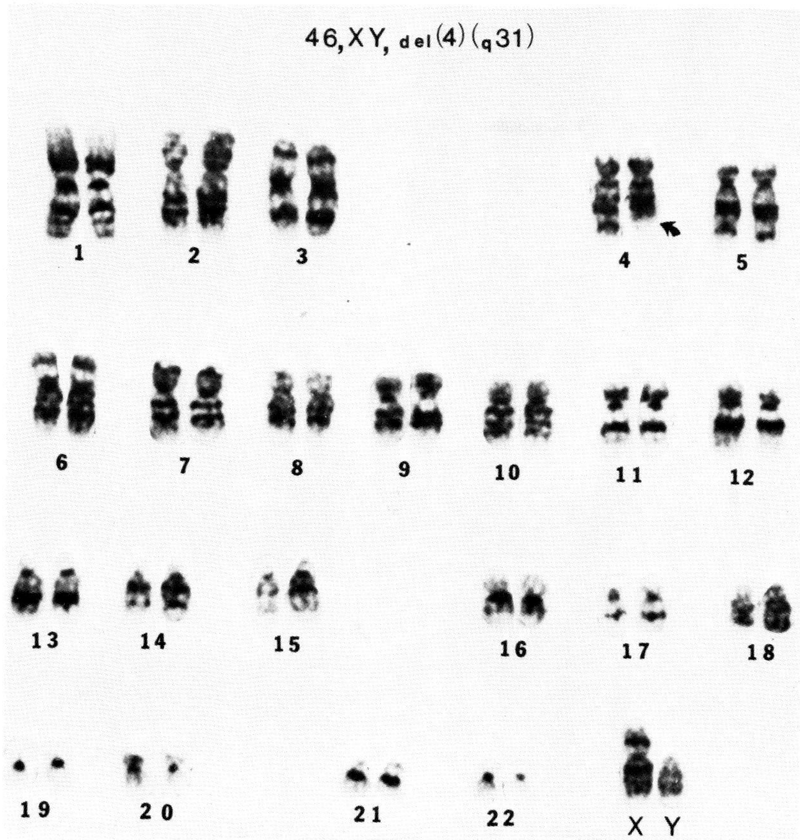

Fig. 2. Trypsin-Giemsa banding karyotype of infant showing the deletion of the long arm of chromosome 4.

DISCUSSION

Three other infants with multiple congenital malformations and deletion of the long arm of chromosome 4 have been reported in the literature [1–3]. Findings common to these children, including our patient, are indicated in Table 1.

Although the study of additional cases will be necessary to more completely define the phenotype, the similarity of the pattern of malformation observed in these patients suggests that the del(4q) syndrome constitutes a clinically identifiable disorder.

TABLE 1. Findings in Reported Cases of del(4q) Syndrome

	Ockey [1]	Golbus [2]	Ferrier [3]	Our patient
General				
normal birthweight	+	+	+	+
failure to thrive	?	+	–	+
regurgitation	?	+	–	+
cyanosis with feeding	+	+	–	+
Craniofacial				
prominent occiput	?	+	+	+
asymmetric forehead	?	?	+	+
narrow face	(+)	?	+	+
oblique palpebral fissures	+	?	+	+
dysplastic inner canthi	(+)	?	+	+
epicanthic folds	(+)	?	+	+
depressed nasal bridge	+	+	+	+
anteverted nares	+	+	+	+
low-set, posteriorly rotated ears	+	+	+	+
abnormally shaped ears	+	+	+	+
cleft lip and/or cleft palate	?	+	+	+
micrognathia	(+)	+	+	+
Other				
congenital heart disease	+	+	+	+
prox. implant. thumbs	p	?	+	+
clinodactyly 5th finger	–	?	+	+
cong. hip dysplasia	+	?	–	+
abn. impl. toes	+	+	+	+

+, present at examination; (+), present by picture; –, absent; ?, unrecorded; p, phocomelia

REFERENCES

1. Ockey CH, Feldman GV, Macaulay ME, Delaney MJ: A large deletion of the long arm of chromosome no. 4 in a child with limb abnormalities. Arch Dis Child 42:428–434, 1967.
2. Golbus MS, Conte FA, Daentl DL: Deletion from the long arm of chromosome 4 (46,XX,4q–) associated with congenital anomalies. J Med Genet 10:83–85, 1973.
3. Ferrier S, Freund M: A propos d'un cas de délétion du bras long du chromosome B4 (B4q–). Arch Genet (Zur) 47:16–26, 1974.

Autosomal Imbalance With a Near-Normal Phenotype: The Small Effect of Trisomy for the Short Arm of Chromosome 18

R. J. M. Gardner, MB, ChB, MSc, N. L. Rudd, MD, FRCP(C),
L. J. Stevens, PhD, and R. G. Worton, PhD

INTRODUCTION

It is generally held that excess autosomal material (other than heterochromatin) that is sufficiently substantial to be detected by current cytogenetic methods has an important distorting effect upon the phenotype. The usual clinical picture associated with partial autosomal trisomy is, at the very least, one of considerable mental retardation and minor physical anomalies. We here present evidence that trisomy for the short arm of chromosome 18 may be an exception to this rule.

CASE REPORT

The family was ascertained through an infant girl presenting with multiple malformations. On examination at 2 months of age her weight, length, and head circumference were well below the 3rd percentile. She had a curious facial appearance, with minor anomalies of the palate, auricles, and fingers. The neck and sternum were short; she had generalized hypotonia; and there was early evidence of developmental delay. Chromosome analysis showed a 46,XX,11p+ karyotype. The nature of the abnormality was revealed when the phenotypically normal mother was karyotyped: In addition to the 11p+ chromosome, virtually the entire long arm was deleted from one chromosome 18. The maternal karyotype is described as 46,XXt(11;18) (p15;q11). The proband is therefore 46,XX der (11), t(11;18) (p15;q11) mat.

Birth Defects: Original Article Series, Volume XIV, Number 6C, pages 359–363

Other family members were then examined. The maternal grandmother carried the same translocation as the mother (Fig. 1). A maternal aunt of the proposita carried the der(18) as well as 2 normal 18s, and 2 normal 11s (Fig. 1). Her karyotype is described as 47,XX,+der(18),t(11;18) (p15;q11) mat. She is thus effectively trisomic for all of the short arm of 18 and, presumably, for a very small piece of the proximal long arm (Fig. 2). If the translocation is reciprocal, she is also trisomic for a small segment of band 11p15. This karyotype was found in all of the 17 cells examined from a lymphocyte culture as well as in all of 31 cells examined from a skin fibroblast culture.

This aunt, who carries the extra der(18) chromosome, was examined at the age of 20 years. She was a quiet, slight woman. Her height (160 cm), weight (52 kg), and head circumference (55 cm) were well within the normal range. She had minor anomalies of the face, skin, and digits. Her ears were small, with an unusual helical fold, particularly on the right. There was nothing unusual about her facial appearance, with the exception of a deviation of the mandible approximately half an inch to the right, rendering the mouth asymmetric, with the appearance of a right facial palsy. However, full facial movements were present. She had some dental crowding, and malocclusion secondary to the abnormal position of the mandible. Her hands were unusual, with ulnar deviation of the fingers. The fingers were long, thin, and stiff, with decreased abduction of the thumb. Examination of the skin revealed generalized roughness over the entire trunk, especially the anterior abdominal wall, with pinpoint pigmented eleva-

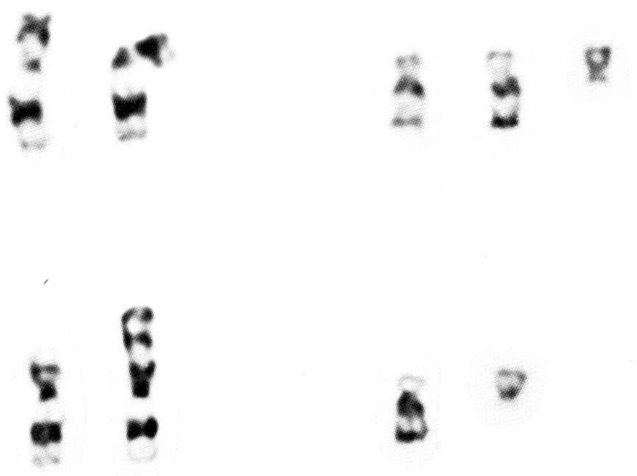

Fig. 1. Partial karyotypes of the maternal aunt of the proposita (top row) and her mother (bottom row), showing chromosomes 11 and 18 and their derivatives. The aunt of the proposita carries 47 chromosomes with 2 normal 11s and 2 normal 18s plus the der(18), t(11;18) (p15;q11). Her mother is a balanced carrier of the translocation t(11;18) (p15;q11).

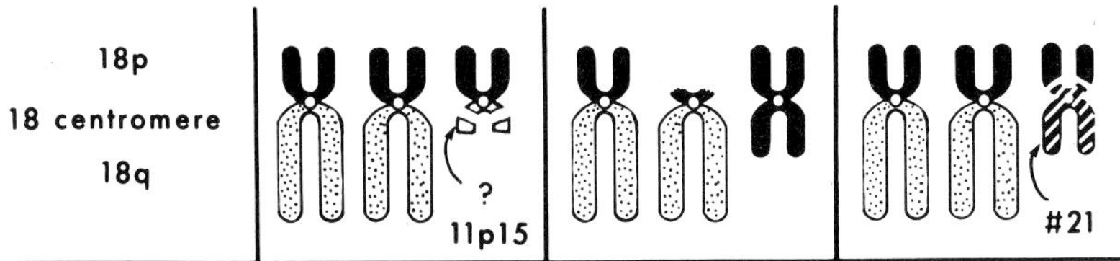

Most of 18p	tri	tri	tri
Proximal 18p	tri	tetra	di
18 centromere	tri	tri	di
Proximal 18q	tri	di	di
Region 11p15	?tri	–	–

Fig. 2. Diagrammatic representation of partial karyotypes of the cases of trisomy 18p in the present report (first column), the families of Taylor et al (second column) and of Jacobsen and Mikkelsen (third column). The short arm of 18 is indicated by the blacked-in portions. In the columns is indicated whether the chromosome segment concerned is present in disomic (di), trisomic (tri), or tetrasomic (tetra) state.

tions over the entire abdomen. There was bilateral hallux valgus. Dermatoglyphic analysis revealed a low ridge count (64), a radial loop on the right ring finger, and an arch fibular in each hallucal area. The remainder of the physical examination was entirely within normal limits. She completed a high school education, gaining a diploma in business and commerce, and is presently employed as a full-time clerk in a factory. Talking with her, one had an impression of a person of average intelligence. We did not feel it was justified to suggest that she undergo formal psychologic assessment.

DISCUSSION

Our patient with 18p trisomy has only subtle physical malformations and, of particular importance, seems to be of average intelligence. However, her phenotype is, perhaps, less normal when she is compared to her immediate family. She is smaller and shorter than her two sisters (one a balanced carrier, one whose chromosomes have not been analyzed). Her mother was a teacher, one sister was a kindergarten teacher, and the other has a university degree in home science; these facts, and the impressions gained on talking with the mother and one of the sisters, indicate that they are of above average intelligence.

We could find only one report of a case of undoubted 18p trisomy in the literature. Taylor et al [1] described a family in which the mother carries an 18p isochromosome and an 18p– chromosome, making her effectively trisomic for

18p. She had been ascertained through her two retarded daughters, one of whom was tetrasomic for 18p (carrying the i(18p) chromosome), the other monosomic for 18p (carrying the 18p–). Taylor et al stated that "parents (of these two girls) are normal, though the mother is small: 152.4 cm and 34.0 kg." and the authors later speak of trisomy for 18p producing "little or no deleterious effect in the mother." Apparently, then, it may be inferred that she was of normal intelligence.

Taylor's case is not absolutely comparable to ours. The chromosome constitution resulting in the effective 18p trisomy arose in a different way, and is not exactly the same with respect to the autosomal imbalance. In addition to trisomy 18p, our case is trisomic for the centromere and a very small piece of proximal 18q (and possibly region 11p15); Taylor's case, as well as being trisomic for 18p and the centromere, is tetrasomic for a very small piece of proximal 18p (see Fig. 2).

From the prebanding days, Jacobsen and Mikkelsen [2] describe a family ascertained through a retarded child with a deletion of 18p. The mother also carried the 18p–chromosome along with a Gp+. Her karyotype is interpreted as 46,XX,t(18;G) (p1;p1). Her two other children and two of her brothers carried the Gp+ chromosome in unbalanced state, and were interpreted as being effectively trisomic for 18p. Three were assessed as mildly mentally retarded; the youngest, assessed at 6 months of age, had a developmental quotient of 106. Some had minor physical anomalies (see Table 1). These authors raised the question of whether the Gp+ may simply have been a G with large satellites. They concluded this to be unlikely in view of the fact that the 3 persons with the Gp+ and the 18p– were normal, while the 4 with the Gp+ and a normal pair of 18s had some signs of neurologic impairment.

Again (assuming the interpretation of the karyotype to be correct), the family of Jacobsen and Mikkelson is not exactly comparable to ours. The 4 cases of supposed trisomy 18p are, in fact, only disomic for the centromere and, probably, a very small segment of proximal 18p (Fig. 2). If anything, one would have expected these cases, on theoretic grounds, to have been less severely affected than our patient and the patient of Taylor et al. The situation is a little confused by the fact that the parents of two of the 18p trisomic individuals were first cousins.

The clinical findings in the three families are summarized in Table 1. The modes of chromosomal arrangements leading to the partial trisomy 18p are set out diagrammatically in Figure 2, along with an interpretation of the theoretic extent of the chromosomal imbalance.

We conclude that trisomy for 18p, while not being completely innocuous, has a less harmful effect upon the phenotype than a triple dose of a substantial amount of other parts of the autosomal chromatin.

TABLE 1. Summary of Clinical Findings in Six Cases of Trisomy of 18 Short Arm

Jacobsen and Mikkelsen [2]	
Case 3[a]	Born 36 weeks gestation, weight 2.1 kg. Anxiety and depression on entering school. IQ (WISC) = 75. Tests interpreted as "organic cerebral affection." Did well in school for mentally retarded. Examined at approx. 9 years: No neurologic abnormalities, normal EEG.
Case 4[a]	Birthweight 2.74 kg. Examined at 6 months: Appearance and development normal, DQ = 106, slightly abnormal EEG with "occasional 2–3 Hz trains with sharp-wave predominance in right temporal region."
Case 5	Normal appearance except for left congenital facial palsy. Physical health normal. Estimated to be mentally subnormal on leaving school at age 14. Industrial worker. Marriage is childless.
Case 6	Estimated to be mentally subnormal on leaving school at age 14. Operations for severe pes splanovalgus and metatarsus varus at 16. Examined at age 24: Appearance "perfectly normal." Brickyard laborer.
Taylor et al [1]	
Mother of index case	Normal, although small (152.4 cm, 34.0 kg). Inferentially of normal intelligence.
Present authors	
	Rather small and short. Minor anomalies of auricles, hands; mandibular asymmetry and dental crowding; rough skin. Appears of average IQ. Works as a clerk.

[a]Parents are first cousins.

SUMMARY

A woman with trisomy for the short arm of chromosome 18 appeared to be of normal intelligence, and had only minor physical anomalies. One other adult case of trisomy 18p described in the literature was short of stature but, by inference, of normal intelligence. Three other adult cases of probable trisomy 18p were of "subnormal" intelligence, and two had a minor physical malformation. We conclude that trisomy for 18p is a relatively innocuous partial autosomal trisomy.

REFERENCES

1. Taylor KM, Wolfinger ML, Brown MG, Chadwick DL: Origin of a small metacentric chromosome. Familial and cytogenetic evidence. Clin Genet 8:364–369, 1975.
2. Jacobsen P, Mikkelsen M: Chromosome 18 abnormalities in a family with a translocation t(18p–,21p+). J Ment Defic Res 12:144–161, 1968.

Follow-Up on a Human X-Autosome Translocation First Studied in 1963 and 1964*

John M. Opitz, MD, Robert I. DeMars, MD, Stanley L. Inhorn, MD, and B. Rafael Elejalde, MD

INTRODUCTION

The patient presented below was referred to the University of Wisconsin in May 1963. Her gonadal dysgenesis and minor somatic malformations were then interpreted as a form of the Ullrich-Turner syndrome which suggested that the C chromosome involved in her apparently balanced C-F translocation was an X. However, because of the fact that X-chromatin mass and autoradiographic studies could not resolve the question of X- versus C-autosome involvement, the case remained unpublished. It is presented because it raised some cytologic and clinical questions still unanswered, and because it is an historic footnote on the subject of human X-autosome translocations in connection with the review by Summitt et al [1] and the report on the KOP translocation [2].

CLINICAL REPORT

A 21-year-old woman (MP170642) was referred for secondary amenorrhea. Her 32-year-old normal brother had 4 healthy children. A miscarriage occurred between the birth of this brother and the patient. The 56-year-old mother was living and well; she had a younger sister with 3 normal sons. The 63-year-old father, an only child, was living and healthy. Thus, the patient was born when

*Studies of Malformation Syndromes of Man LII; Paper No. 2188 from the Genetics Laboratory of the University of Wisconsin. These studies were supported by DHEW/PHS grants GM 18982 and GM 20,130 from the National Institute of General Medical Sciences.

Birth Defects: Original Article Series, Volume XIV, Number 6C, pages 365–375

her mother and father were 35 and 42 years old, respectively. She was born overdue after an induced labor and after a pregnancy complicated by bleeding at 2 months, with a birthweight of 2,268 gm and a length of 48.25 cm. She had much lanugo at birth and mild ptosis of the left upper eyelid; the small size of her hands was remarked upon since early childhood.

She had been healthy all her life and her growth and maturation were considered normal. Her menarche occurred at 13 years and was associated with normal maturation of secondary sexual characteristics. She had 5 irregular periods during her 13th year and 3 the following year; secondary amenorrhea ensued at the age of 15.

Examination (Fig.1): height 160 cm (slightly less than average for her family), weight 51.67 kg, occipitofrontal head circumference 54.6 cm. She was considered a "handsome, happy, and good natured" young woman of average intelligence. She had slight ptosis of the left upper lid, long, narrow, anteverted and poorly differentiated ears, many nevi over face, neck and upper trunk, bilateral cubitus valgus and strikingly small and delicate hands with clinodactyly of the 5th digits. Her fingers were relatively long and tapered, and she had narrow and transversely hyperconvex fingernails. The palmar axial triradii were proximally placed, and her digital dermal patterns consisted of 10 arch-like ulnar loops. Her breasts were small and showed signs of secondary involution; she had normal amounts of axillary and pubic hair, and normal external genitalia. Pelvic examination showed normal vagina, cervix, and uterus, but adnexa could not be palpated. Intravenous pyelograms showed normal kidneys but spina bifida occulta of the 1st sacral vertebra.

On June 24, 1964 Dr. Gloria E. Sarto performed an exploratory laparotomy on the patient and removed dysgenetic gonadal streaks which did not contain a single detectable follicle (Fig.2).

CYTOGENETIC AND CYTOLOGIC STUDIES

In 1963–1964, on the basis of the physical examination, gonadal dysgenesis on a chromosomal basis was suspected. In a Feulgen-stained buccal smear 63% of mucosal cells were X-chromatin negative; 37% "showed an X-chromatin body distinctly smaller than normal . . . tending to support the initial impression of cytogenetic gonadal dysgenesis, possibly on the basis of a long arm X deletion." Drs. R.R. Race and Ruth Sanger of the MRC Blood Group Research Laboratory, the Lister Institute, London, found the patient and all relatives to be Xg(a–).

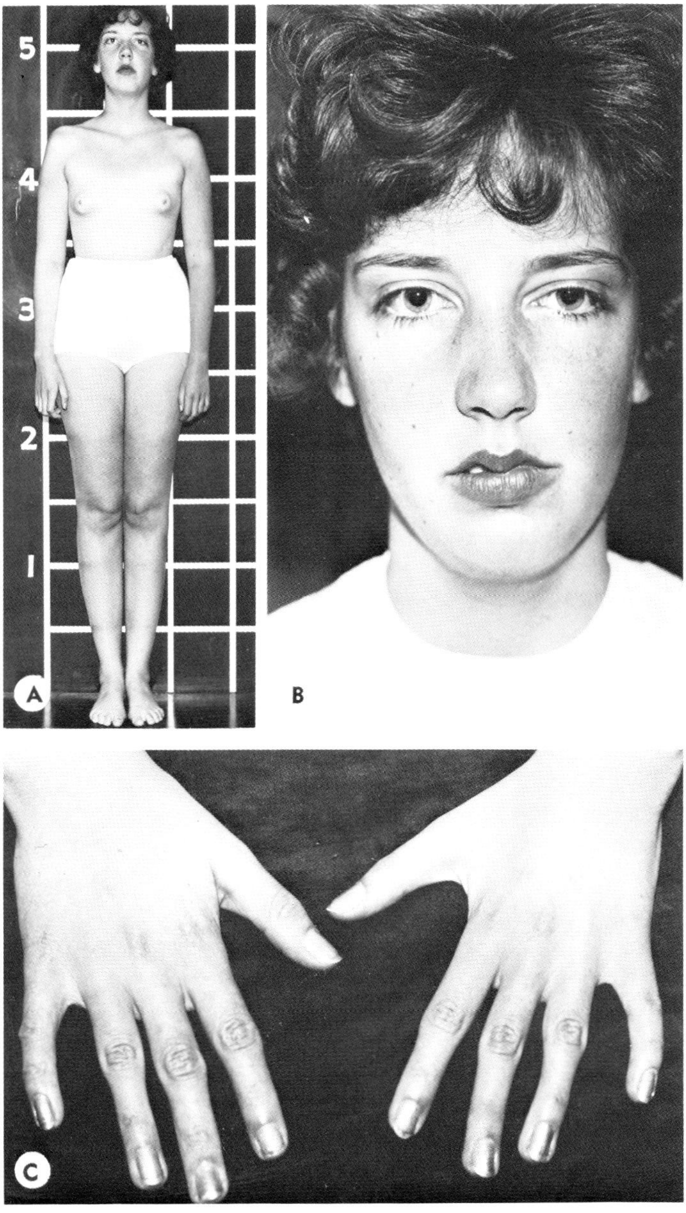

Fig 1. Patient at 21 years; see text.

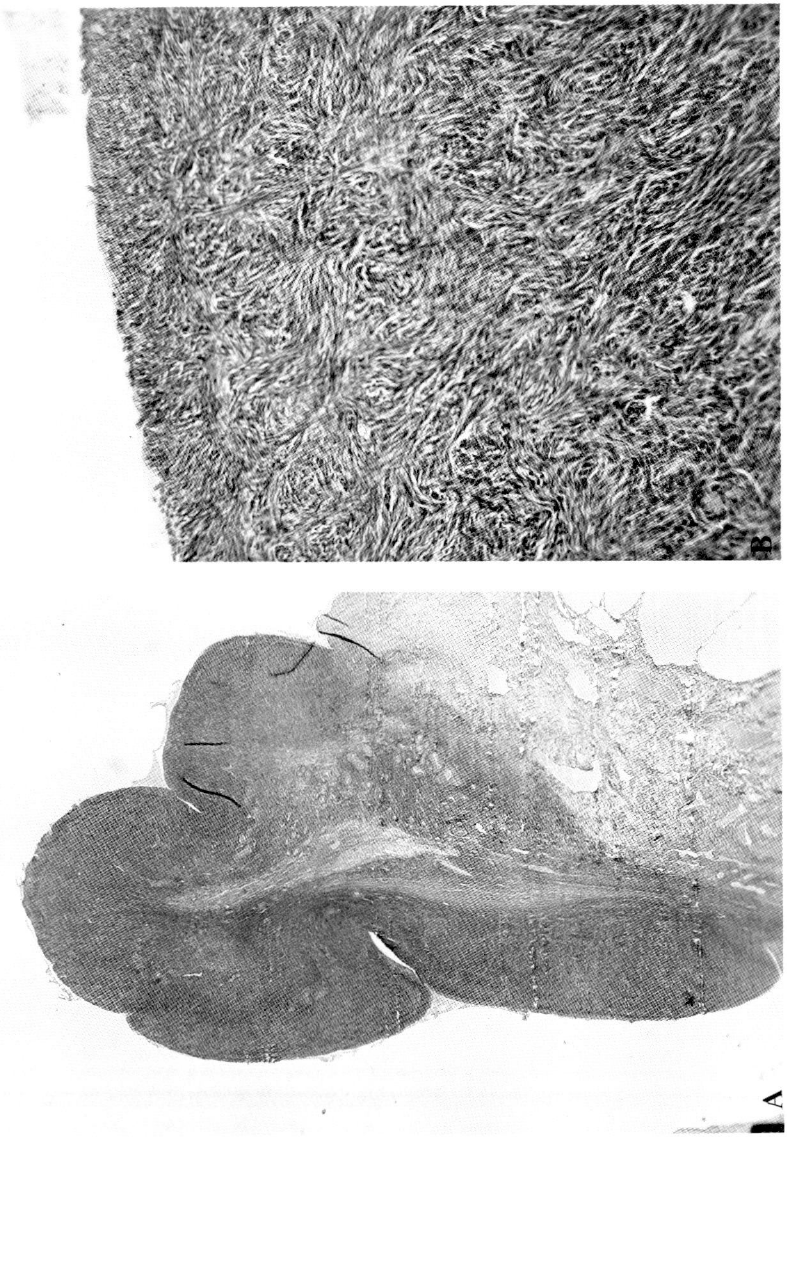

Fig 2. Gonadal streak, A) at low and B) higher magnification showing only stroma and no follicle cells.

Analysis of lymphocyte and fibroblast chromosomes showed a modal number of 46 (Fig. 3, Table 1). One F chromosome was absent and an extra small submetacentric, E-like chromosome ($\bar{E}$) was found. This chromosome was sometimes difficult to identify with certainty; it most resembled a 17. She had 16 C chromosomes, but one of the smallest was a clearly abnormal chromosome ($\bar{C}$) which could be identified consistently in all cells by its very high arm ratio.

The two abnormal chromosomes ($\bar{C}$ and $\bar{E}$) were considered translocation chromosomes, and in view of the patient's phenotype, and the (apparent) small size of the buccal smear X-chromatin mass it was postulated that the translocation involved one of the X chromosomes. The two interpretations of the chromosome breaks are shown in Figure 4; in 1963 we favored alternative A, since the short arm of $\bar{C}$ matched the short arm of an F better than did the short arm of $\bar{E}$. With either interpretation the sum of $\bar{C}$ and $\bar{E}$ seemed to roughly equal the total mass of a normal X and a normal F, and the chromosome rearrangement suggested that this could be a balanced reciprocal translocation.

Studies of the incorporation of tritiated thymidine into lymphocyte chromosomes during DNA replication (Fig. 5) consistently showed the *normal* X to be late labeling.

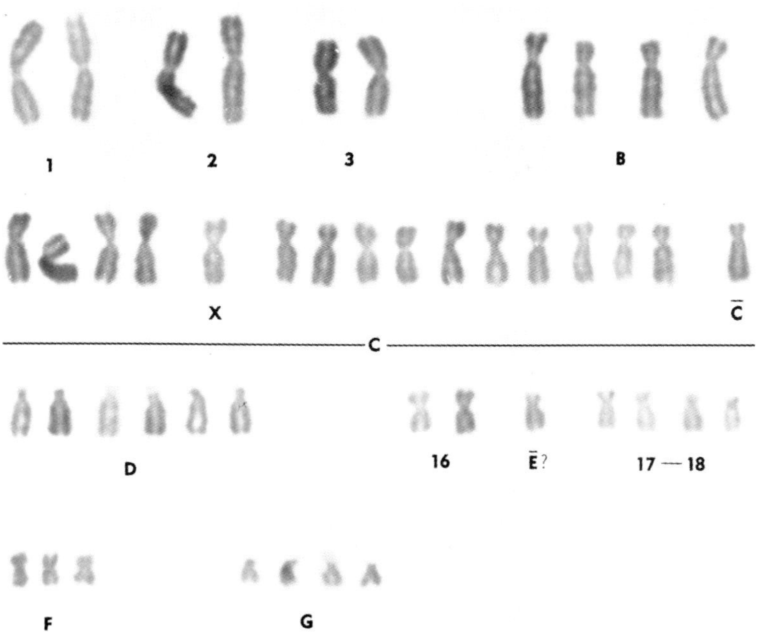

Fig 3. Karyotype prepared in 1963; the X chromosome was identified autoradiographically. Note 3F, and $\bar{C}$ and $\bar{E}$ translocation chromosomes.

Table 1. Chromosome Analysis

Tissue	No. of chromosomes	No. of cells	Remarks
Blood	44	1	F, C, missing
	45	8	1 cell difficult to analyze, others lacking 1 D, E, F, B, and G; 2 cells each lacked 1 C
	46	71	All had $\bar{C}$ and $\bar{E}$ chromosomes
	47	1	Extra C
Skin	45	1	Lacking G
	46	17	All had $\bar{C}$ and $\bar{E}$ chromosomes

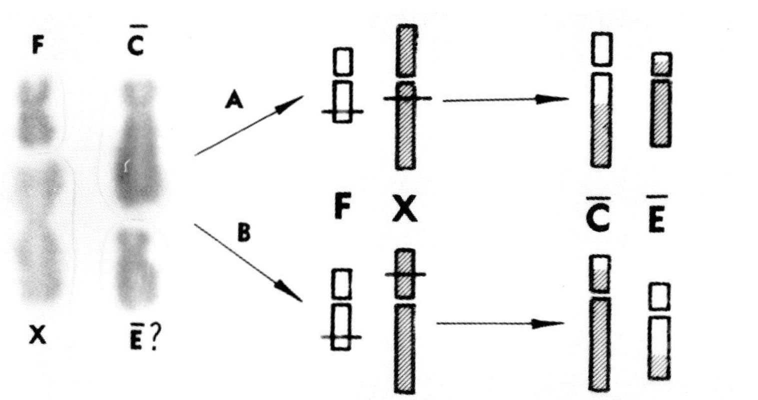

Fig 4. Two hypotheses of origin of $\bar{C}$ and $\bar{E}$ chromosomes; A was favored and confirmed later.

The X-chromatin interpretations in 1963 and 1964 were based on blind studies (supervised by K. Patau) of buccal smear cells by Dr. John M. Opitz, Dr. Santo and Dr. R. L. Summitt who consistently ($p = 0.023$) picked out the patient's slides from the controls on the basis of a smaller than normal X-chromatin mass.

X-Chromatin mass size in fibroblasts was studied by DeMars who cultured fibroblasts from a skin biopsy. The cell populations seemed to have a higher than usual proportion of cells with small X-chromatin masses. This suggested the possibility that the X-chromatin of many cells derived from skin was formed by a small X (possibly the $\bar{E}$) chromosome. Many cells with X-chromatin masses of normal size also occurred in the skin-derived cell populations indicating the existence of a second subpopulation of cells in which the normal X formed X-chromatin.

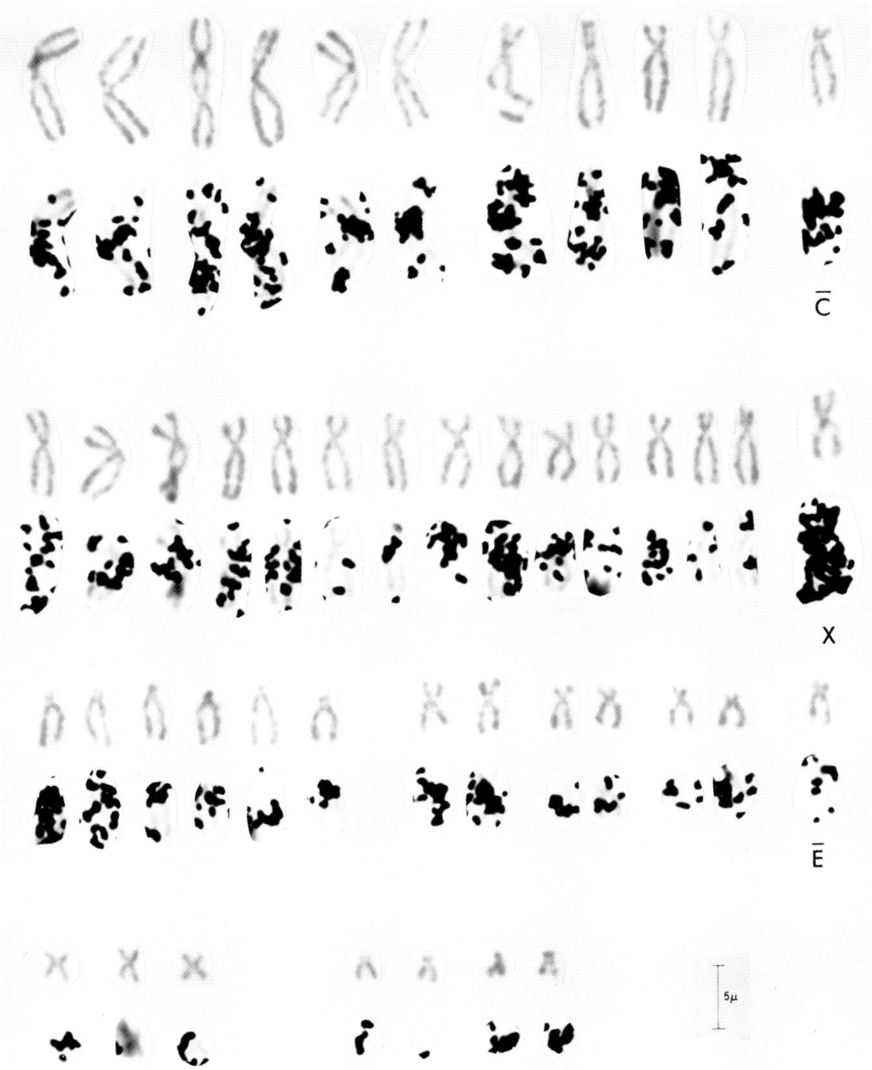

Fig 5. The same karyotype before and after labeling with tritiated thymidine: the X chromosome is at the right-hand end of the row of chromosomes; the $\bar{C}$ at the right-hand end of the B group; and the $\bar{E}$ at the right-hand end of the E group.

Attempts to test this hypothesis were made by isolating clonal populations. The cells in such populations should form only small X-chromatin masses or large ones. Presumably the $\bar{E}$ chromosomes should be late replicating in populations of the former type.

In Figure 6 the X-chromatin masses of one such clonal population (A) are compared with those of cells cultivated from a normal female (B). Four cylindrical stainless steel wells (5 mm diameter) were placed on a coverslip. Each well

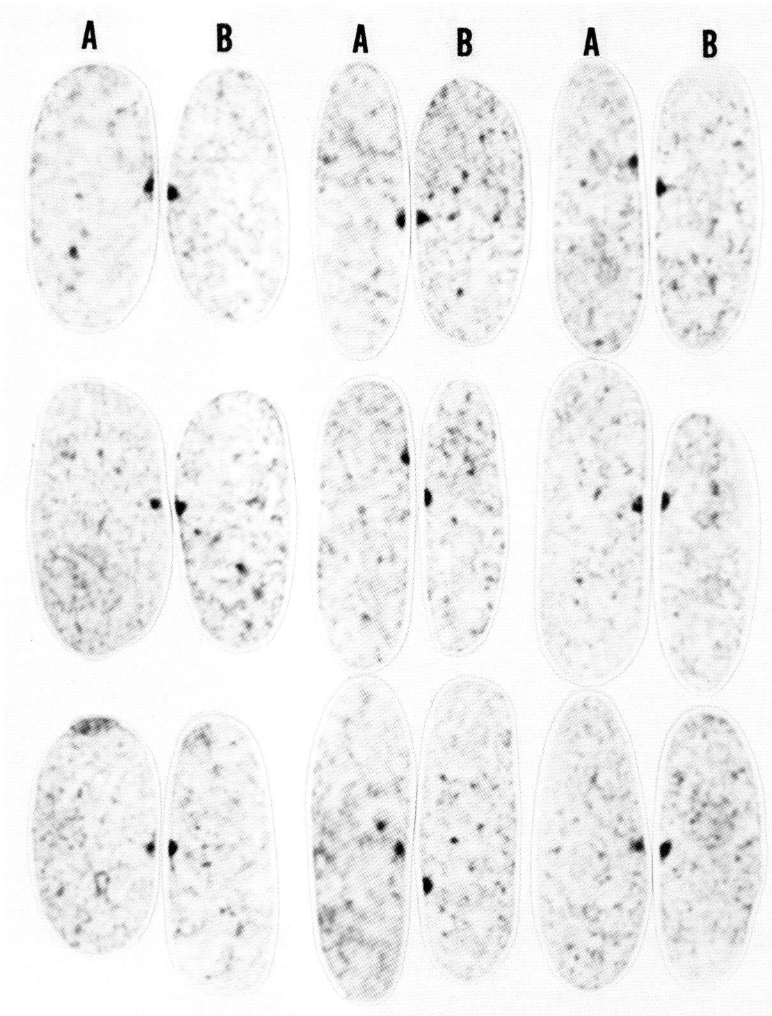

Fig 6. Fibroblast nuclei from patient (A) and normal 46,XX control female (B) to demonstrate apparent size difference of X-chromatin mass; see text for details.

was inoculated with about 1,000 cells either of strain No. 78 or of the control female. The nutrient medium in the wells was changed every 2nd day for 8 days by which time the cells had formed a continuous monolayer. The wells were removed on the 10th day and were fixed on the coverslip with methanol (3 vol): acetic acid (1 vol) and stained with the Feulgen reaction. The cells were dehydrated with increasing concentrations of methanol, then xylol and finally mounted with permount. Undeformed nuclei were selected for having a well-defined X-chromatin mass near the middle of the long margin. These were photographed with Kodak high contrast film using a 100X apochromat Zeiss objective. The negatives obtained were of similar contrast and prints were made under as nearly identical conditions as possible. The nuclei in Figure 6 are mounted in pairs consisting of one nucleus of the patient (A) and one nucleus of the normal control female (B). The nuclei were paired as much as possible on the basis of having similar size and having X-chromatin masses that could be reasonably closely apposed. Figure 6 depicts the first 9 nuclei of each strain that were found to satisfy the criteria of selection defined above. The comparison indicated that the X-chromatin masses of the clone isolated from (A) appeared smaller than those of the control female.

By Q-band staining Dr. DeMars first confirmed th X-autosomal nature of the translocation and found that the 1963 interpretation (A) of the karyotype (Fig. 4) was correct, ie that the breaks involved the proximal portion of Xq and the tip of a chromosome 19.

Recent studies by Dr. B.R. Elejalde, using Q, G, and BSG stains confirmed this and placed the breaks at Xq13 and 19q13 (Figs. 7 and 8).

COMMENT

The presence, if verified, of a cell line in which one or the other of the two translocation chromosomes is inactivated would violate the Lyon hypothesis, but might explain the patient's clinical symptoms as partial monosomy of autosomal material. However, her clinical manifestations then and now impress more as a mild form of the Ullrich-Turner syndrome. Phenotypic abnormality is not unprecedented in apparently balanced X-autosome translocations in which the normal X is inactivated, but remains as unexplained here as in the other cases: X deletion and position effect have been invoked as possible explanations. The q13 location of the X breakpoint in this case is in keeping with the findings in previously reported cases. In the review by Summitt [1] an Xq13 breakpoint was noted in two patients with gonadal failure and in one patient with normal gonadal function. On the basis of these cases, Summitt suggested that gonadal failure will result if the break is within a region which extends proximally to a point within band Xq13. Thus, balanced X-autosome translocations with q13

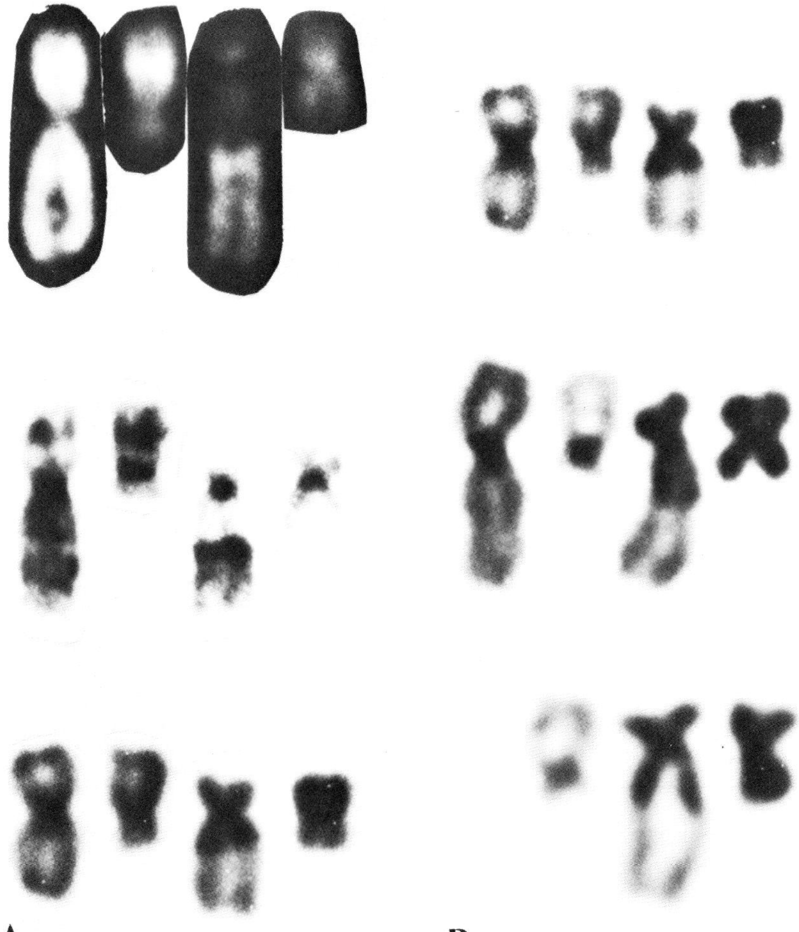

A B

Fig 7. Q-, G-, and BSG-stained X, E, C, and 19 chromosomes from 3 cells A): BSG-stained chromosomes from 3 cells B) – the bottom row lacking an X. BSG stains chromosome 19 especially well.

X breakpoints may or may not result in gonadal failure. If, in the future, techniques allow more specific localization of breakpoints within a band, the suggestion of Summitt leads to the proposal that those balanced X-autosome translocations with proximal q13 breakpoints will be associated with normal gonadal function; those with distal q13 breakpoints with gonadal failure.

The only other apparently balanced X-19 translocations known at this time in man were studied by Gerald et al [3, 4]. One of these was a retarded girl from an institution in Wrentham, Massachusetts. To our knowledge clinical details of

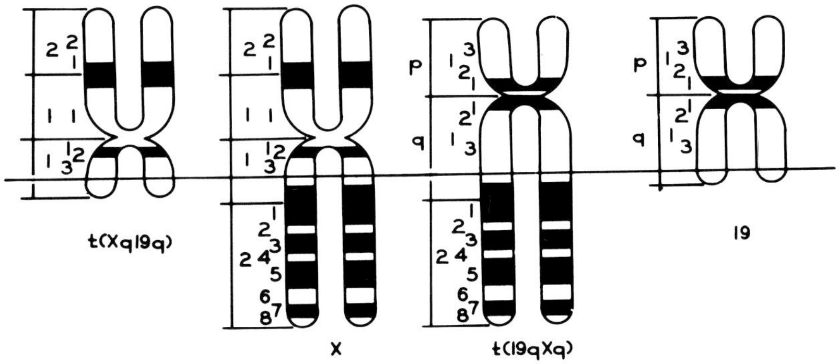

Fig 8. Diagrammatic representation of postulated break and reunion points in X and 19, and in the $\bar{C}$ [t(19qXq)] and $\bar{E}$ [t(Xq19q)] translocation chromosomes; the exact size of the 19q13 piece is unknown, but the break at Xq13 seems quite firm.

these 2 cases have never been published. In one of these (the retarded girl) the *normal* X was inactivated. However, in both cases the breaks in one or both chromosomes were in different locations than in our case; in the first the breaks were located in the middle of the 19q and of the Xq; in the other they were in 19p and around Xq13.

Cells from this patient have been deposited at the Camden (New Jersey) Human Genetic Mutant Cell Repository.

REFERENCES

1. Summitt RL, Tipton RE, Wilroy RS Jr, Phelan JP, Martens PR: X-autosome translocations: A review. This volume pp 219–248.
2. Pallister PD, Opitz JM: Studies of malformation syndromes of man LI: The KOP translocation. This volume pp 133–146.
3. Gerald PS: X-autosome translocations in hybrid cells. In Davidson RL, de la Cruz F (eds): "Somatic Cell Hybridization." New York: Raven Press, 1974, pp 49–51.
4. Gerald PS, Burns G, Monedjikova V: Localization of genes on the X-chromosome by somatic cell hybridization. Pediatr Res 7:344, 1973.

Trisomy 18 (29 Cases) and Trisomy 13 (19 Cases): A Summary*

Marion E. Hodes, June Cole, Catherine G. Palmer, and Terry Reed

During the period of 1963–1976, 29 cases of trisomy 18 (T18) and 19 cases of trisomy 13 (T13) were seen in the Department of Medical Genetics. Chromosomes were studied in all and 30 were banded. One patient had tertiary T18 and 8 had translocations of chromosome 13. These were 46,XY,–14,t(13;14)(13qter→cen→14qter)mat; 46,XX,–Dt (D;D); 46,XX,–13,t (13;13) (13qter→cen→13qter); 46,XY,–13,t(13;13) (13qter→cen→13qter); 46,XY,–14, t(13, 14) (13qter→cen→14qter); 46,XX,–22t (13;22) (13qter→cen→22qter)mat; 46,XX, –14,t(13;14) (13qter→cen→14qter); and 46,XY,der(8),t(8;13) (p23;q14)mat. The T18 translocation was 46,XX,–13,t(13;18) (p11;q11). Postmortem examinations were performed on 8 of the T18 cases and 5 of the T13. Clinical, cytogenetic, and dermatoglyphic observations were also summarized.

For the purpose of this report, the clinical features were grouped by those common to both trisomies (ear and cardiac abnormalities, micrognathia, overlapping fingers, microcephaly, prominent heels, high-arched palate, microphthalmia, hypertonia, umbilical hernia, and cryptorchidism), those common in T18 (prominent occipit, dislocated hips, hypoplastic nails, clubfoot deformities, wide-spaced nipples, hypertrophic clitoris, hammer toe, narrow palpebral fissures, short sternum, small mouth, excess skin of the neck, hypotonia, duplicate urinary system, and short hallux), and those common in T13 (cleft lip and palate, short neck, polydactyly, capillary hemangiomas, dextrocardia, hypotelorism, coloboma of the iris, scalp defects, and apneic episodes). Our findings, summarized in Table 1, were generally consonant with the broad ranges reported by

*This is publication No. 77-26 from the Department of Medical Genetics and was supported in part by the Indiana University Human Genetics Center, PHS GM 21054, and Dental Genetics Training Grant PHS DE 00007.

Birth Defects: Original Article Series, Volume XIV, Number 6C, pages 377–382

TABLE 1. Features of the Trisomies 18 and 13*

Feature	T18			T13		
	This series (present/mentioned)	Magenis and Hecht [2]	Smith [1]	This series (present/mentioned)	Hecht [3]	Smith [1]
Abnormal ears	29/29	>80	>50	14/16	>80	>50
Abnormal heart	27/29	>95	>50	16/17	50–80	80
Micrognathia	25/26	>80	>50	2/3	50–80	<50
Overlapping or flexed fingers	26/26	>80	>50	11/15	50–80	>50
Microcephaly	19/27	50–80	10–50	12/14	50–80	>50
Prominent heels	14/16	50–80	10–50	5/8	50–80	>50
High- or arched-palate	14/16	>80	50	5/7		<50
Microphthalmia	9/11	10–50	<10	15/17	50–80	>50
Hypertonia	6/10	50–80	>50	4/6	20–30	<50
Umbilical hernia	4/6	50–80[b]	>50	5/6	10–50[b]	>50
Cryptorchidism	3/7	100	>50	7/7	100	>50
Prominent occiput	20/22	>80	>50			
Abnormal hips	18/22	>80	>50	1/–		
Hypoplastic nails	14/14	10–50	50		10–50	<50
Clubfoot	8/9	40–60	10–50		10–20	<50
Wide-spaced nipples	9/10	10–50	10–50			
Hypertrophic clitoris	8/9		10–50			
Hammer toe	8/9	50–80[a]	>50[a]		10–50[a]	
Narrow palpebral fissures	8/10		>50			
Short sternum	8/8	>80	>50			
Small mouth	6/7		>50			
Excess skin of neck	6/7	40–60	>50		50–80	>50
Seizures	5/8				20–30	>50

Hypoplastic genital labia	5/5		10–50			
Hypotonia	4/10	50–80			40–50	<50
Duplicate urinary system	4/4	50–80	10–50			<50
Epicanthal folds	4/5	10–50	10–50		50–80	
Syndactyly	4/5	10–50	10–50			<50
Short hallux	3/3	50–80[a]	>50[a]		10–50[a]	
Cleft palate	1/–	10–20[c]	10–50	12/19	50–80	>50
Polydactyly	2/–		<10	11/14	50–80	>50
Scalp defect	1/–			9/12	10–50	>50
Cleft lip	1/–	10–20[c]	10–50	9/18	50–80	60–80
Apneic episodes	4/4		±	8/8	50–80	>50
Capillary hemangioma			<10	7/8	50–80	>50
Dextrocardia				7/19	20–50	>50
Hypo-/hypertelorism			<10	6/6	>80	<50
Coloboma of iris			<10	6/9	10–50	>50
Sloping forehead	1/–			5/5		>50
Flat head	1/–			3/4		
Hypoplastic nipples				3/3		
Prominent nasal bridge				3/3		
Short neck	2/–	50–80	<10	3/3	50–80	

*Horizontal lines denote conditions common to both T18 and T13 (upper portion), or to T18 (middle portion) or T13 (lower portion) separately.

[a] Same entry

[b] Or inguinal

[c] Same entry

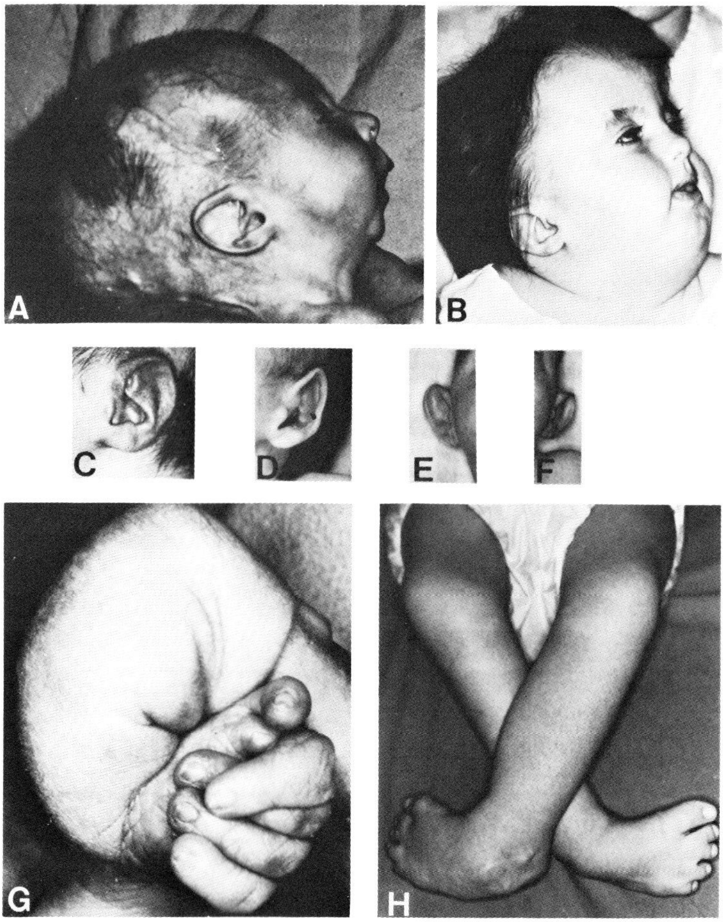

Fig. 1. Some abnormalities of trisomy 18. A) Profile showing prominent occiput, Grecian nose, micrognathia, and simple ears. B) Unusual face; low-set, rotated simple ear; abnormal eyebrows; hirsutism; long eyelashes; obesity. C–F) Various ear types. G) Preaxial polydactyly and radial aplasia. H) Scissored legs, clubfeet, and fleshy syndactyly.

Smith [1], Magenis and Hecht [2], and Hecht [3], when allowance is made for the errors introduced by physicians who failed to note the absence of a particular finding and our refusal to accept "normal" as a sufficient description. One notable difference was that only 43% of the T18 males had cryptorchidism (versus 100% in the literature). The most characteristic dermatoglyphic finding in T18 was simple arches on fingers and toes, and in T13, hallucal fibular arches. The absence of arches on the big toe and presence of non-arch patterns on all fingers is strong evidence against the diagnosis of T18. At autopsy, patent ductus arteriosus and some form of intraventricular septal defect were almost universal in both trisomies. The chief postmortem findings were brain abnormalities in 86% of the T18 and 100% of the T13 bodies examined, and cardiac anomalies of some sort in all. The major defects are summarized in Table 2.

The overall appearance of these children generally aroused suspicion of a chromosomal disorder. Typical facies of T18 (Fig. 1A) and T13 (Fig. 2A) were frequently seen. Occasionally the face was unusual (Fig. 1B), although low-set, rotated, or malformed ears and other features led to a correct presumptive diagnosis. Ear deformities of many kinds were seen and examples are shown in Figures 1A–F and 2A. Polydactyly was a common feature of T13 (Fig. 2D) but was occasionally seen in T18 (Fig. 1G). Scalp defects were usually small, but could be massive (Fig. 2B). An unusual finding in 1 case of T13, genu recurvatum, is shown in Figure 2C.

Primary and tertiary T13 patients did not differ in mean birthweight or mean maternal or paternal age. Of the T18 babies, 36% were born from February through April. For T13, 44% of births were from June to November (versus Taylor's 79% [4]). Mean maternal age in translocation T13 was 24.6 years (0.6 year less than nontranslocation). There was some evidence of bimodality in maternal age distribution. The sex ratio for T18 was 22 females, 7 males; and for T13, 11 females, 8 males.

TABLE 2. Postmortem Findings in Trisomy 18 (11 Cases) and Trisomy 13 (5 Cases)

	T18 (%)	T13 (%)
Patent foramen ovale	27	60
Patent ductus arteriosus	82	60
Interventricular septal defect	82	80
Abnormal aortic valve	45	
Accessory spleen		60
Meckel diverticulum	9	40
Horseshoe kidney	36	
Double uterus		60

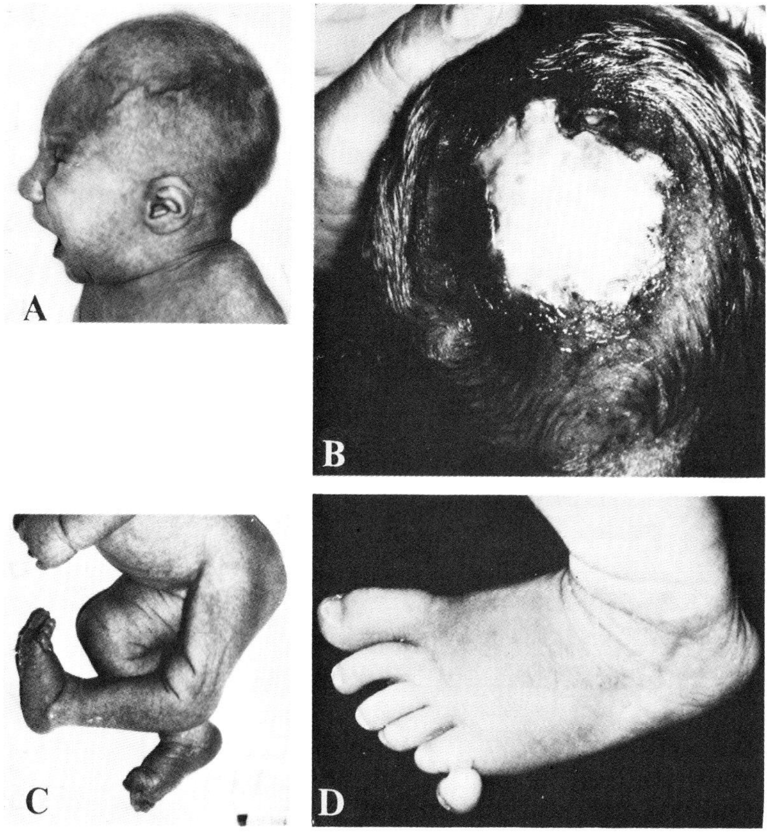

Fig. 2. Some abnormalities of trisomy 13 and trisomy 18. A) Profile showing prominent nose and nasal bridge and small, low-set ears in T13. B) Scalp defect, T13. C) Genu recurvatum, T18. D) polydactyly, T13.

REFERENCES

1. Smith DW: "Recognizable Patterns of Human Malformations: Genetic, Embryologic, and Clinical Aspects." 2nd Ed. Philadelphia: W.B. Saunders, 1976.
2. Magenis E, Hecht F: Chromosome thirteen trisomy syndrome. In Bergsma D (ed): "Birth Defects Atlas and Compendium." Baltimore: Williams & Wilkins for The National Foundation – March of Dimes, 1973, pp 251–252.
3. Hecht F: Chromosome eighteen trisomy syndrome. Ibid pp 246–247.
4. Taylor AI: Autosomal trisomy syndromes: A detailed study of 27 cases of Edward's syndrome and 27 cases of Patau's syndrome. J Med Genet 5:227–252, 1968.

Trisomy 11q(q21→qter)

Patricia I. Bader, MD, MS, Marjorie Jansch, MD, Donald Hoffman, MD, Catherine G. Palmer, PhD, Helen Gerber, MT (ASCP), and Gail Taylor, MS, MT (ASCP)

INTRODUCTION

This report concerns a patient with an unbalanced chromosome translocation resulting in a partial trisomy for the long arm (q) of chromosome 11 (duplication of 11q21→11qter). Segregation of a balanced chromosome translocation (3;11)(p26;q21) was observed in four generations of the patient's family. The patient's dysmorphology and clinical course are discussed and a review of published cases of authenticated 11q duplication suggests that this specific anomaly leads to a distinct clinical syndrome. In addition, relatives who have apparently balanced translocations showed maxillary prognathism and periodontitis.

CASE REPORT

The patient was born to a 22-year-old mother (para 2 gravida 2) after a 40-week pregnancy. She weighed 2,750 gm, was 48.25 cm long, and was noted to be hypotonic and apneic, and to cry weakly. No Moro reflex could be elicited.

Upper respiratory tract infections with choking and apnea necessitated several hospital admissions. A cardiac murmur was detected and a description of the murmur, chest x ray, electrocardiogram, and cardiac flow studies were consistent with the diagnosis of ventricular septal defect. The patient had eczema of the scalp, face, and intercrural areas. The patient had a delayed motor and social development, and at 7 months her development was estimated to be that of a 16-week-old child.

Birth Defects: Original Article Series, Volume XIV, Number 6C, pages 383–392

On physical examination at age 7 months, she weighed 6,524 gm, and was 63.5 cm long (both measurements were between the 3rd and 10th %). Occipitofrontal head circumference was 39 cm (slightly below 3rd %). There were epicanthal folds, upward slanting palpebral fissures, and alternating esotropia (Figs. 1 and 2). The inner canthal distance was 2.25 cm. The outer canthal distance was 9 cm. The nose was short and saddle-shaped, with upturned and flared nostrils. The external nares were characterized by unusual membranous projections in the vestibulas of the nasal passages (Fig. 3). There was an increased nasolabial distance, a small mouth, glossoptosis, pronounced micrognathia and a high-arched palate. The ears were cartilaginous and somewhat low-set, with a prominant and horizontally placed anthelix. The thorax was small and the nipples widely spaced (12 cm). A grade 1–2/6 systolic ejection murmur was heard best along the third left intercostal space. The digits were long and tapering and the nails rather hyperconvex. Muscle stretch reflexes were extremely brisk in the

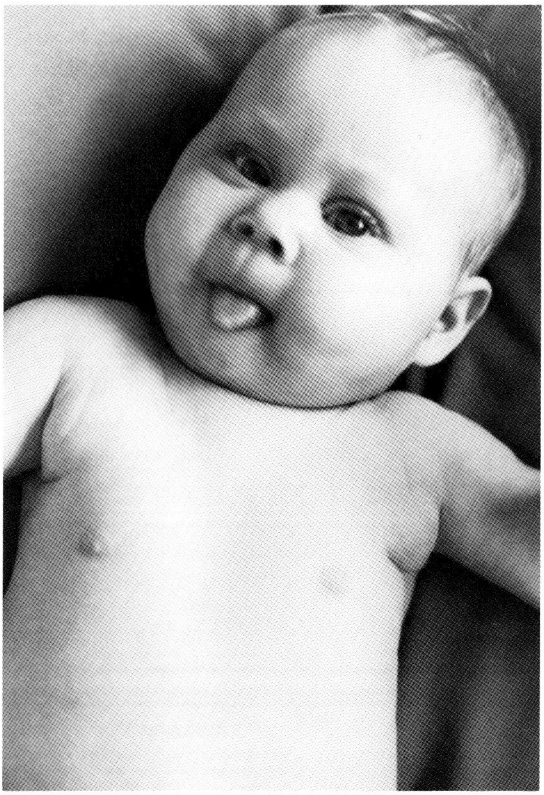

Fig. 1. Facies of the proband at 7 months.

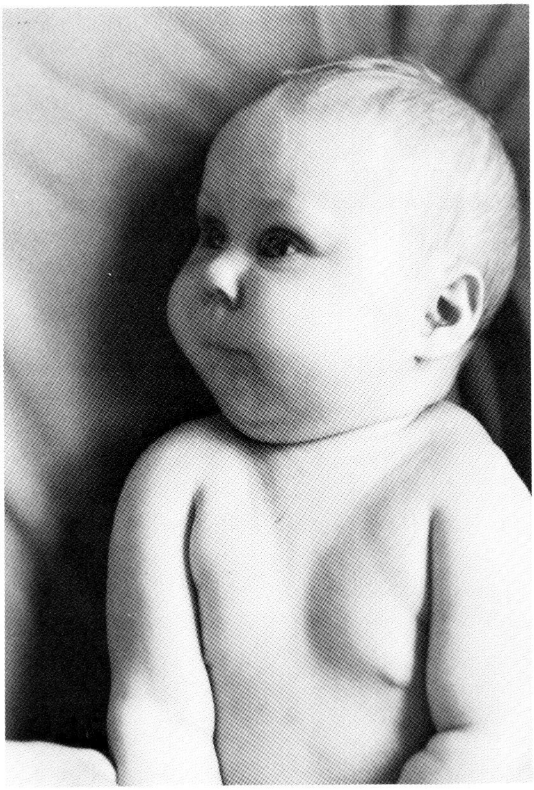

Fig. 2. Oblique view of the proband's facies.

lower limbs and spasticity of all limbs was noted. The bony structure of the thorax, vertebras, and clavicles was interpreted as normal on radiologic examination. Quantitative study of immunoglobulins was normal.

At age 8 months the patient developed a seizure disorder characterized by generalized convulsive activity. An electroencephalogram showed diffuse slowing of electrical activity and was interpreted as compatible with diffuse encephalopathy. The patient developed status epilepticus and temperature elevation of 105°F. The white blood count was elevated (20,500/cu mm), but the cerebrospinal fluid showed no white blood cells and normal levels of protein and glucose. The patient expired after continuing seizure activity. The CSF culture eventually showed growth of Hemophilus influenzae. The parents did not permit an autopsy examination.

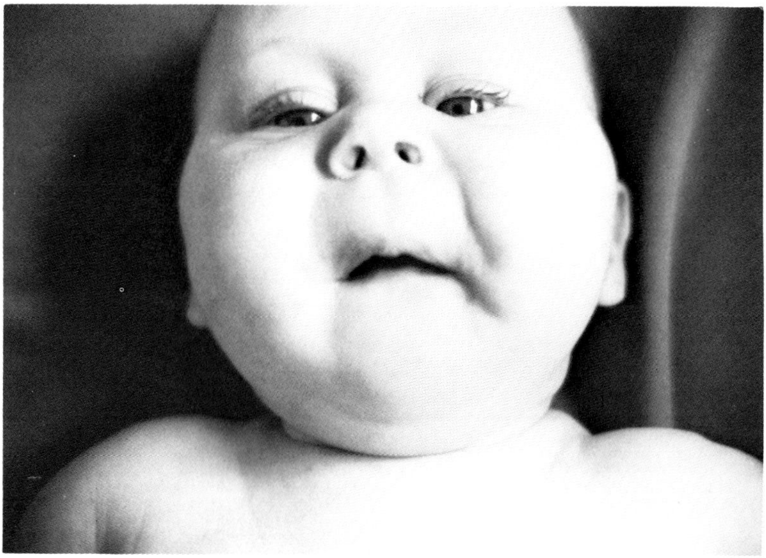

Fig. 3. Membranous projection into nasal vestibulas.

CYTOGENETIC STUDIES

Leukocyte cultures were stained by Giemsa and Giemsa banded with trypsin (GTG). An additional band of chromosome material on the short arm of chromosome 3(3p+) was seen in all metaphases (Fig. 4). A study of the mother's lymphocytes revealed a modal chromosome number of 46 with a balanced translocation which could be designated 46,XX,t(3;11)(3qter→3p26::11q21→11qter; 11pter→11q21::3p26→3pter) (Fig. 5). A few slides were studied by R-banding techniques following heat denaturation in order to confirm this designation (Fig. 6). The chromosome imbalance observed in the patient consists of duplication of the distal end of the long arm of chromosome 11 (region 11q21→11qter), and assuming a reciprocal translocation deficiency of a distal portion of 3p(3pter→3p26).

FAMILY STUDIES

Dermatoglyphic studies of the patient, her mother, father, and sib were conducted and the observations made are listed in Table 1. Twenty-three relatives of the patient were studied with GTG (Fig. 7). In seven family members a balanced translocation, t(3,11), was noted. Twelve relatives had normal karyotypes.

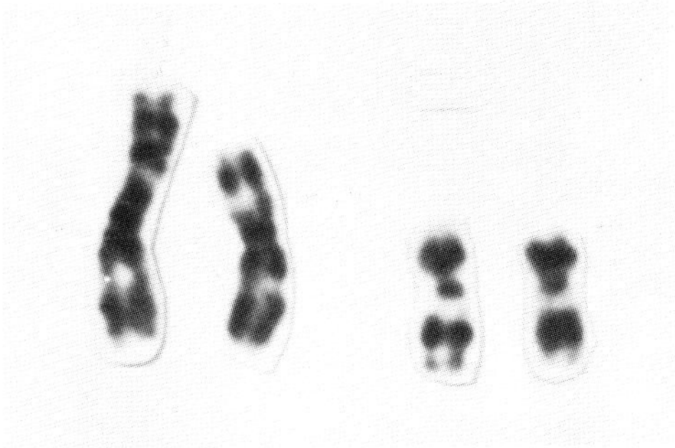

Fig. 4. Partial karyotype of the proband, showing GTG banding patterns of chromosomes 3p+, 3, 11, 11 (left to right).

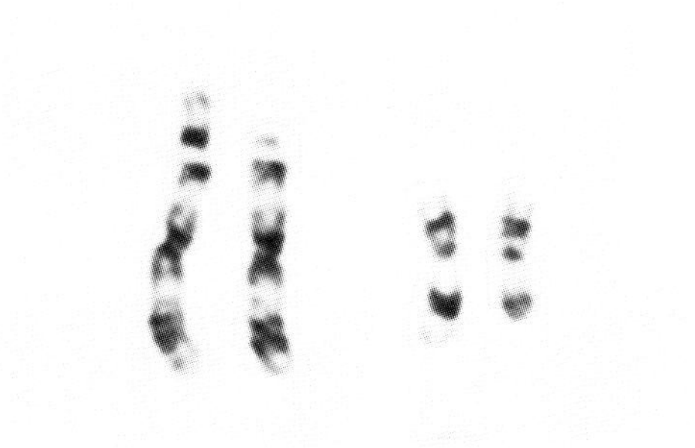

Fig. 5. Partial karyotype of the proband's mother, showing GTG banding patterns of chromosomes 3p+, 3, 11, 11q− (left to right).

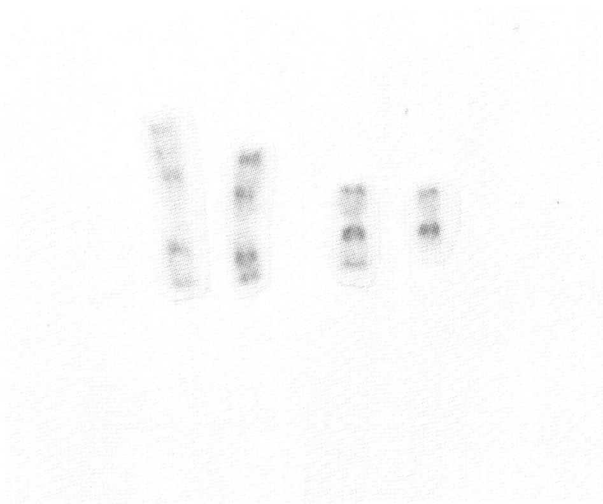

Fig. 6. Partial karyotype of the proband's mother showing R banding patterns of chromosomes 3p+, 3, 11, 11q− (left to right).

TABLE 1. Dermatoglyphic Patterns

Subject	Right					Left					*atd* angle		A–B ridge count		
	1	2	3	4	5	1	2	3	4	5	R	L	R	L	T
Proposita	U	W	U	U	U	U	W	U	U	U	52°	47°	60	57	117
Father	U	U	U	W	U	U	W	A	U	U	38°	38°	38	41	79
Mother	U	R	A	U	U	U	A	U	U	U	42°	37°	38	39	77
Sib	U	A	U	U	U	U	A	U	U	U	43°	43°	39	37	76

Note: W = whorl, U = ulnar loop, R = radial loop, A = arch, A–B ridge count = no. of ridges between triradii A and B.

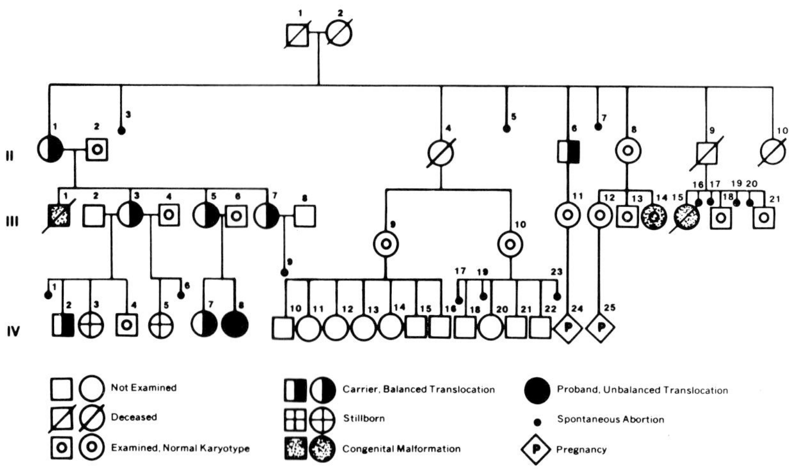

Fig. 7. Pedigree of the proband (*IV,8*).

CLINICAL INVESTIGATION

Carriers of the balanced translocation are the patient's mother, maternal grandmother, maternal aunts, maternal great uncle, sister (age 3), and male cousin (age 7). Each of these individuals was noted to have a prognathic maxilla (class II malocclusion) (Fig. 8). The adults uniformly showed severe gingivitis with moderate to severe periodontitis and inflammatory gingival hyperplasia secondary to chronic infection (Fig. 9). Each was noted to be a mouth breather, although no nasal obstruction could be demonstrated. Prognathism and periodontitis were not noted in family members with normal karyotypes.

Fifteen of 51 descendants of *I, 1* and *I, 2* were lost as a result of spontaneous abortion or stillbirth. One child (*III, 15*) died during the neonatal period and had

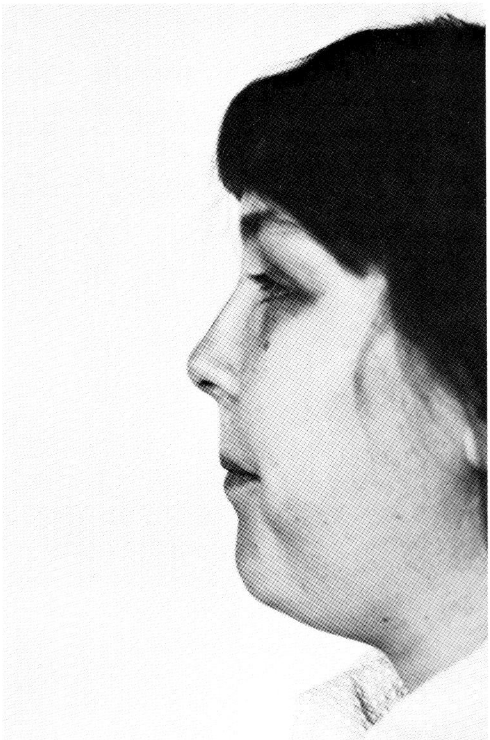

Fig. 8. Profile of the proband's mother, showing maxillary prognathism.

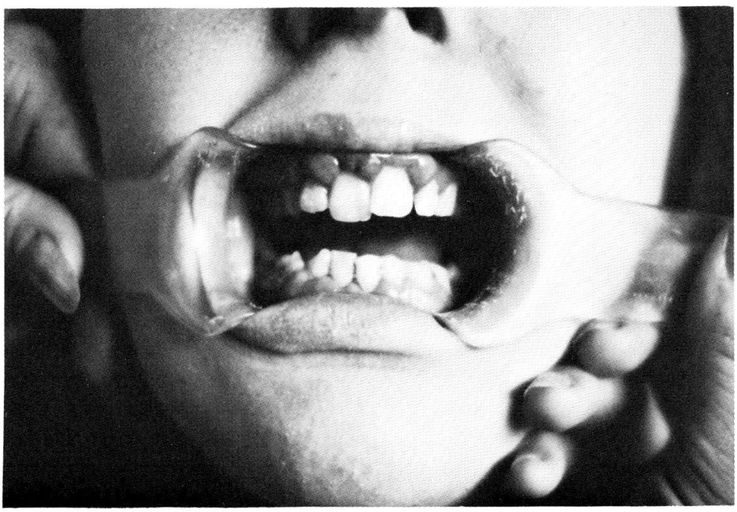

Fig. 9. Periodontal disease as seen in the proband's mother.

undescribed multiple congenital anomalies. Two children died before reaching 2 years of age (*III, 1* and *II,10*). The former (*III, 1*) was known to have hydrocephalus and failure to thrive and to have died as a result of infection. The latter (*II, 10*) was not known to have congenital abnormalities, but her death was secondary to infection. A cousin (*III, 14*) has been institutionalized since she was a child. She has psychomotor retardation and is congenitally blind. Her karyotype was normal, however.

DISCUSSION

Aurias and Laurent [1] summarized common features of nine patients [2–7] with 11q duplication and suggested that these features comprised a syndrome. These nine patients, three additional reports in the literature [8–10], and the above-reported patient combine to provide 13 cases of 11q duplication demonstrated by R, Q, or G banding and inherited from a parent with a balanced translocation (paternal in 4 cases [3, 5, 6] and maternal in 9 cases [2, 4, 7–10]. In eight patients, the modal chromosome number is 46 with variable segments of distal 11q duplicated, and there is a possibility of a partial monosomy (3p [this case], 4q, 6p, 10q, 13q [2 cases], 17p, 21q). The modal chromosome number is 47 in all other case reports [4, 7, 8, 10] and a small extra chromosome is composed of the short arm and centromere of chromosome 22 (pter→q111) and the distal portion of the long arm of chromosome 11(q231→qter).

The principal features reported are low birthweight (average is 2,290 gm), cardiovascular and genitourinary malformations, CNS dysfunction, and craniofacial abnormalities (Table 2), including a broad, flat nose with prominent alae, upturned nostrils, increased nasolabial distance, prominent philtrum, micrognathia, cleft or high-arched palate, glossoptosis, prominent and low-set ears, prominent and horizontally placed anthelix, epicanthal folds, and often oblique

TABLE 2. Dysmorphology of 11q Duplication

Broad, flat nose with upturned nostrils	10/13
Micrognathia	12/13
Cleft or high-arched palate	10/13
Low implantation of ears	10/13
Prominence of the horizontal anthelix	6/13
Excessive intermammillary space	6/13
Cardiopathy	11/13
Micropenis	5/7
Urinary tract abnormalities	6/13
Clubfoot	6/13
Dermatologic abnormalities	8/13
Developmental retardation or perinatal death	13/13

palpebral fissures. The neck is short and nipples are widely spaced [1, 8–10].

Cardiac malformations are frequent (11 of 13 cases). Truncus arteriosus [8], coarctation of the aorta [8], patent ductus arteriosus [5], patent foramen ovale [5], auricular septal defect [2, 3, 5, 8], ventricular septal defect [5, 8, 10], pulmonary artery atresia [3], pulmonary stenosis [10], aortic stenosis [4], mitral stenosis [5], stenosis of the aortic isthmus [5], and right ventricular hypoplasia [3, 5] have been reported.

Hypoplasia of the penis is reported in 5 of 7 males [4, 5, 7, 10]. Urinary tract abnormalities include hypoplasia or complete agenesis of a single kidney [2, 8, 10], unilateral cystic dysplasia of the kidney [4], high implantation of the ureters [5], and vesical ureteral reflux [7].

In addition to axial hypotonia and hypertonia of the limbs [1], abnormalities of the central nervous system include agenesis of the corpus callosum reported in two cases [2, 5], and micropolygyrae noted in another case [8]. Craniorachischisis with spina bifida and myelomeningocele was reported in one patient [6]. In addition a child (with many of the features of 11q trisomy, myelomeningocele, and spina bifida) was born to a translocation carrier [5]. Four patients [4, 5, 8, 10] were reported to have cranial deformities.

The most common orthopedic anomalies include clubfeet [1, 3, 4, 7], dysplastic hip [2, 3, 7], malformations of the scapula and the clavicle [1, 5], and synostoses of the radiocubital bones [3].

Skin abnormalities are frequent and cutis laxa has been reported in the most severely affected infants [1, 3, 5]. However, the oldest survivor [9] has been reported to have prurigo. The patient discussed in this report had severe eczema.

Dysplasia of the skin in some instances has made dermatoglyphic studies impossible to interpret [5]. We and other investigators noted an increased distance between the palmar triradii A and B in this patient and in two previously reported cases [3, 4]. We verified this impression by recording an elevated ridge count between triradii A and B. A long palmar flexion crease which ended between the 2nd and the 3rd digits was also seen in all three patients. Unusual palmar creases were noted in case reports by Tusques et al, Aurias et al, and Giraud et al [3, 4, 8].

Other reported anomalies include agenesis of the left hemidiaphragm in one case [4], hypoplasia of the gall bladder in one patient [5], and anomalous vascularization and position of the spleen in one case [3].

There were several family members who had apparently balanced translocations and had, in common, severe periodontal disease as well as maxillary prognathism. There have been previous reports of abnormalities in apparently balanced carriers [12, 13]. These abnormalities have been attributed to duplications or deletions near the translocated segment, mutation, or position effect. It is of interest that severe periodontal disease has been observed in kindreds in the past and attributed to an autosomal recessive [11] or X-linked dominant gene with decreased penetrance [14].

ACKNOWLEDGMENTS

We would like to thank Robley Evans, DDS, Bluffton, Indiana, for his assistance in the medical evaluation of the patient and her family. We would also like to thank Mrs Nancy Campbell, Bluffton, Indiana, for secretarial assistance.

REFERENCES

1. Aurias A, Laurent C: Trisomie 11q. Individualisation d'un nouveau syndrome. Ann Génét 18:189, 1975.
2. Francke U: Quinacrine mustard fluorescence of human chromosomes: Characterization of unusual translocations. Am J Hum Genet 24:189, 1972.
3. Tusques J, Grislain JR, Andre MJ et al: Trisomie partielle 11q identifiée grace à l'étude en "dénaturation ménagée" par la chaleur, de la translocation équilibrée paternelle. Ann Génét 14:167, 1972.
4. Aurias A, Turc C, Michiels Y et al: Deux cas de trisomie 11q (q231→qter) par translocation t(11;22)(q231;q111) dans deux familles différentes. Ann Génét 18:185, 1975.
5. Rott HD, Schwanitz G, Grosse KP, Alexandrow G: C 11/D 13-translocation in four generations. Humangenetik 14:300, 1972.
6. Wright YM, Clark WE, Breg WR: Craniorachischisis in a partially trisomic 11 fetus in a family with reproductive failure and a reciprocal translocation, t(6p+;11q−). J Med Genet 11:69, 1974.
7. Laurent C, Biemont MC, Bethenod M et al: Deux observations de trisomie 11q(q23.1→qter) avec la même anomalie des organes genitaux externes. Ann Génét 18:179, 1975.
8. Giraud F, Mattei JF, Mattei MG, Bernard R: Trisomie partielle 11q et translocation familiale 11-22. Humangenetik 28:343, 1975.
9. Jacobsen P, Hauge M, Henningsen K et al: An (11;21) translocation in four generations with chromosome 11 abnormalities in the offspring. Hum Hered 23:568, 1973.
10. Ayraud N, Galiana A, Llyod M, Deswarte M: Trisomie 11q(q23.1→qter) par translocation maternelle t(11;22)(q23.1;q11.1). Une nouvelle observation. Ann Génét 19:65, 1976.
11. Jacobs PA: Correlation between euploid structural chromosome rearrangements and mental subnormality in humans. Nature 249:164, 1974.
12. Monteleone PL, Volk LR, Sekhon G et al: Rob (14q15q) Translocation in several members of a family. This volume.
13. Jorgenson RJ, Levin LS, Hutcherson ST, Salinas CF: Periodontitis in sibs. Oral Surg 39:396, 1975.
14. Melnick M, Shields ED, Bixler D: Periodontosis: A phenotypic and genetic analysis. Oral Surg 42:32–41, 1976.

Interchromosomal Effect in Balanced Translocations

Claude G. Stoll, MD, Elisabeth Flori, MD, and Denys Beshara, MD

Numerous chromosomal studies of children with Down syndrome have been done since Lejeune [1], Gianelli, and Carter demonstrated that this disease is directly related to the presence of three chromosomes 21. In approximately 95% of the cases [2] the trisomy is free and homogeneous, but in 5% of the cases, and in 10% of observations when the mother is less than 30 years of age [3], the trisomy is caused by a translocation. While we understand perfectly well how a parent carrier of a Robertsonian translocation (D/21 or 21/22) can have a child with trisomy 21, we know much less about a parent carrier of a translocation of a chromosome other than 21 who produces a child with trisomy 21. This latter case, however, is not as uncommon as one usually thinks, and this has been shown by three recent family studies. One must hold open the possibility of the phenomenon of interchromosomal interference; ie an incidence at meiosis of a carrier of a structural chromosomal anomaly producing an anomaly involving another pair of chromosomes. This study proposes to demonstrate that genetic counseling given to parents with translocations should take into account this interchromosomal effect.

MATERIALS AND METHODS

Our two unrelated families both had one child with trisomy 21, which permitted us to uncover in one of the parents a reciprocal translocation involving two other chromosomes (already published: a father with 46,XY,t(13;17)(q14;q22) producing a 47,XX,+21 child [4] and a mother with 46,XX,t(2;15)(q14;q25) producing a 47,XX,+21,t(2;15)(q14;q25) child [5]. We saw another family in which a child with typical trisomy 18 was born. The mother and the maternal

Birth Defects: Original Article Series, Volume XIV, Number 6C, pages 393–398

grandmother were carriers of a balanced translocation 46,XX,t(16;19)(q24;p11). No other child was born in this family. Our observations concerning children with trisomy 21 born of parents who were carriers of balanced translocations or pericentric inversions of other chromosomes are to be found in Table 1. Other similar published cases are also included.

DISCUSSION

When the birth of a trisomic child permits us to discover a balanced translocation or a pericentric inversion of other chromosomes of one of the parents, the first problem that poses itself is that of the further risks directly attributed to the chromosomal rearrangement. These risks, well-documented for a long time now, have been well established for certain translocations and less well established for others, whose frequency one can consider with Lejeune (personal communication, 1974), to be about 20%.

A second problem that should be considered is the risk of a recurrence of trisomy. This risk is not directly related to chromosomal rearrangement but may be due to an interchromosomal effect.

The notion of an interchromosomal effect was introduced by Sturtevant, in 1919 [23], who noticed that in Drosophila melanogaster a heterozygous inversion of chromosome 3 increases the frequency of chromosomal exchanges in chromosome 2. Following that, it was shown that heterozygous inversion of any pair of the large chromosomes increases the frequency of exchanges in the other pairs [24]. The same effect due to pairing of the nonhomologous chromosomes has been shown in translocations in Drosophila [25, 26]. To explain the relationship between chromosomal exchanges and the nondisjunction of certain chromosomes, Grell [27] proposed the hypothesis of distributive pairing. According to this hypothesis, chromosome pairing is a stepwise phenomenon. An exchange pairing, involving only the homologous chromosomes, occurs before exchanges take place. Distributive pairing follows these exchanges and concerns all chromosomes, homologous or not, which for several reasons (translocation, inversion, isochromosome, etc) may not have exchanges. Novitski [28] has suggested another hypothesis, in which only one pairing occurs. The centromeric regions of all chromosomes are fused in the chromocenter during interphase and prophase of the first meiotic division. In this manner, homologous chromosomes synapse and the chromosomes which do not exchange can pair with other chromosomes, homologous or not. White and Morley [29], during cytologic studies on the frequency of chiasmata in grasshopper carriers of inversion, arrived at the same conclusions as those arrived at using Drosophila. The interchromosomal effect then is a more general phenomenon, not reserved exclusively to Drosophila. Lejeune [30], in 1963, was the first to envisage the existence of an interchromosomal effect in man. A few years later a study of a lineage of individual

TABLE 1. Cases of Trisomy 21 Children Produced by a Parent Carrier of a Translocation or an Inversion of a Chromosome Other Than the Chromosome Causing the Trisomy

Authors	Karyotype of parent carrier of balanced translocation or inversion	Karyotype of trisomic 21 child
Robson [6]	46,XY,t(2p–;16q+)	47,XX,+G,t(2p–;16q+)
Subrt [7]	46,XY,t(3p–;17q+)	47,XY,+G,t(3p–;17q+)
Kahn [8]	46,XX,t(Bp+;Cp–)	47,XX,+G,t(Bp+;Cp–)
Yanagisawa [9]	46,XX,t(5q–;?7p+)	47,XY,+21,t(5q–;?7p+)
Lundsteen [10]	46,XX,t(2p–;5q+)	47,XY,+21,t(2p–;5q+)
Seabright		
(personal communication, 1975)	46,XX,t(1;20)(p36;q13)	47,XY,+21,t(1;20)(p36;q13)
Tenconi [11]	46,XX,t(6;18)(q16;q23)	47,XX,+21,t(6;8)(q13;q23)
Michalova [12]	46,XY,t(8;10)(p23;q11)	47,XY,+21
Ruffié [13]	46,XY,t(Cp+;Dq–)	47,XX,t(Cp+;Dq–),+Gqi
Mattei [14]	46,XY or XX,t(7;14)	47,XY or XX,+21
Stoll [4]	46,XY,t(13;17)(q14;q22)	47,XX,+21
Stoll [5]	46,XX,t(2;15)(q14;q25)	47,XX,+21,t(2;15)(q14;q25)
Stoll [in press]	46,XX,t(16;19)(q24;p11)	47,XX,+18
Catti [15]	46,XY,inv(11)	47,XX,+21
Hamerton [16]	46,XX,inv(19)	47,XX,+21,inv(19)
Wolf [17]	46,XX,inv(D)	47,XY,+21,inv(D)
Berger [18]		
Sparkes [19]		
Bayeux [20]	46,XY,inv(Y)	47,XX,+21,or 47,XY,inv(Y),+21
Leisti [21]		
Stoeckenius [22]		

carriers of D/D translocations [31] confirmed Lejeune's hypothesis. In the lineage of 45,XX,–D,–D, +t (DqDq) women, among live-born infants the respective frequencies of 13 and 21 trisomies were on the order of 2% [31]. The existence of an interchromosomal effect in these women is very possible. In the lineage of male carriers of t (DqDq) translocation, the frequency of anomalies does not appear to be significantly increased [31]. The interchromosomal effect, however, is not limited to t(DqDq) women. It is not uncommon to see trisomy 21 children born of mothers or fathers who have sex chromosome anomalies. This could be the case notably for the triple X syndrome [32, 33], for Klinefelter syndrome [34, 35], or for pericentric inversions of Y chromosomes [18, 19–22]. De la Chapelle [36] noticed a possible inversion interchromosomal effect since his patient, a carrier of a t(4;11) translocation, also had "4'X's" and "Y". In these aberrations of the sex chromosomes the existence of an interchromosomal effect cannot be demonstrated with certainty but seems probable. Aside from our three patients there are only 13 published cases in which a trisomy 21 child was born of a parent who was a carrier of a non-21-chromosome translocation or who himself had a pericentric inversion of a chromosome other than Y or 21. However, the karyotypes of the parents of a trisomy 21 child are not always studied. Trisomy 21 is very widespread. Most observers generally agree that 1 out of 650 newborns is affected by trisomy 21 [37, 38]. Balanced translocations are equally frequent, affecting 1 subject out of 700 [39]. The birth of a trisomy 21 child by a parent carrier of a non-"G" chromosome translocation may be the result of chance. The possibility of the occurrence of such an event would be $1/600 \times 1/700 = 1/420{,}000$.

An interchromosomal effect which may increase the possibility of such an event may intervene. As evidenced in favor of the existence of such an interchromosomal effect, apart from studies done with animals and in particular with Drosophila, we can adduce its intervention in DqDq translocations, and in certain gonosomal anomalies (XXX or XXY syndromes), and also our own three personal observations along with those already found in the literature. Such an effect only rarely manifests itself in balanced translocations. Dutrillaux (personal communication, 1974), studying 911 balanced translocations, finds this interchromosomal effect in only four of these cases. Even if this interchromosomal effect intervenes rarely, it would be difficult not to take it into account during genetic counseling of carriers of reciprocal translocations. The intervention of such a chromosomal influence usually results in the birth of trisomy 21 children or, less often, a child with another trisomy (our patient with a trisomy 18) or with gonosomal anomalies [36]. In man there are too few studies as yet to know whether this is more predominant in males or females. Further studies are necessary, and examination of the chromosomes of the parents of a child carrier of a chromosomal anomaly would be worthwhile.

REFERENCES

1. Lejeune J, Gauthier M, Turpin R: Etude des chromosomes somatiques de neuf enfants mongoliens. C R Acad Sci 248:1721, 1959.
2. Mikkelsen M: Down's syndrome at young maternal age: Cytogenetical and genealogical study of eighty-one families. Ann Hum Genet 31:51, 1967.
3. Gianelli G, Hamerton JL, Carter CO: Cytogenetics of Down's syndrome (mongolism). II. The frequency of interchange trisomy in patients born at a maternal age of less than 30 years. Cytogenetics 4:186, 1965.
4. Stoll C, Levy JM, Dörr R, Kapps R: Translocation t(13;17)(q14;q22) familiale diagnostiquée après la naissance d'un enfant trisomique 21. J Génét Hum 23:327, 1975.
5. Stoll C, Kuss JJ, Schneegans E: Inherited t2q–/15q+ translocation and Down's syndrome. Clin Genet 9:5, 1976.
6. Robson EB, Polani PE, Dart SJ, Jacobs PA, Renwick JH: Probable assignment of the alpha-locus of haptoglobin to chromosome 16 in man. Nature 223:1163, 1969.
7. Subrt I, Prchlikova H: Double chromosomal aberration. Trisomy G and the balanced translocation t(3p–;17q+). Humangenetik 8:111, 1969.
8. Kahn J, Abe K: Consistent and variable chromosome anomalies in parents of children with Down's syndrome. J Med Genet 6:137, 1969.
9. Yanagisawa S: A boy with trisomic Down's syndrome and a familial 5/?7 translocation 47,XY,+21,t(5q–;?7p+). Jpn J Hum Genet 17:38, 1972.
10. Lundsteen C, Vestermark S, Philip J: A familial balanced 2/5 translocation associated with trisomy 21 in one individual. Hum Hered 24:88, 1974.
11. Tenconi R, Baccichetti C, Dussini N, Panizza G: Translocation familiale t(6;18)(q16; q23) avec trisomie 21 libre. Ann Génét (Paris) 17:275, 1974.
12. Michalova K, Malkova J, Chrz R: Two cases of C-group balanced translocations. Humangenetik 27:157, 1975.
13. Ruffié J, Virenque J, Bardier A, Colmbies P: Remaniements chromosomiques complexes portant sur les autosomes, s'accompagnant d'anomalies crânio-faciales et d'un omphalocèle. C R Acad Sci 262:386, 1966.
14. Mattei JF. Etude génétique des parents d'enfants trisomiques 21. Thèse de Médecine, 1974, Marseille.
15. Catti A: Inversions familiales et autres aberrations chromosomiques concomittantes. Communication aux Troisièmes Journées Européennes de Conseil Génétique. Genève, October 25–27, 1974.
16. Hamerton JL, Gianelli F, Carter CO: A family showing transmisson of a D/D reciprocal translocation and a case of regular 21 trisomic Down's syndrome. Cytogenetics 2:194, 1963.
17. Wolf U, Baitsch H, Kunzer W, Reinwein H: Familiäres Auftreten eines anormalen D-chromosomes. Cytogenetics 3:112, 1964.
18. Berger R, Lejeune J, Roy J: Trisomie 21 et Y métacentrique. Ann Génét (Paris), 13:187, 1970.
19. Sparkes RS, Muller HM, Veomett IC: Inherited pericentric inversion of a human Y chromosome in trisomic Down's syndrome. J Med Genet 7:59, 1970.
20. Baheux G, Emerit I, Roux C: Trisomie 21 et anomalie de structure de l'Y. Ann Génét (Paris) 13:191, 1970.
21. Leisti J: Structural variation in human mitotic chromosomes. Ann Acad Sci Fenn (Med), Series A, IV, Biologica 179:1, 1971.

22. Stoeckenius M: A probable Y isochromosome in father and son of tall stature. Lancet 2:391, 1966.
23. Sturtevant AH: Contributions to the genetics of Drosophila melanogaster. III. Inherited linkage variations in the second chromosome. Carnegie Inst Wash Publ 278: 305, 1919.
24. Schultz J, Redfield H: Interchromosomal effect on crossing-over in Drosophila. Cold Spring Harbor Symp Quant Biol 16:175, 1951.
25. Grell RF: Non random assortment of non-homologous chromosomes. Genetics 42: 374, 1957.
26. Grell RF, Grell EH: The behaviour of non-homologous chromosomal elements involved in non-random assortment in Drosophila melanogaster. Proc Natl Acad Sci USA 46: 51, 1960.
27. Grell RF: A new hypothesis on the nature and sequence of meiotic events in the female of Drosophila melanogaster. Proc Natl Acad Sci USA 48:165, 1962.
28. Novitski E: An alternative to the distributive pairing in Drosophila. Genetics 50:1449, 1964.
29. White MJP, Morley FHW: Effects of pericentric rearrangements on recombination in grasshopper chromosomes. Genetics 40:604, 1955.
30. Lejeune J: Autosomal disorders. Pediatrics 32:326, 1963.
31. Dutrillaux B, Lejeune J: Etude de la descendance des individus porteurs d'une translocation t(DqDq). Ann Génét 13:11, 1970.
32. Breg WR, Cornwell JG, Miller OJ: The association of the triple-X syndrome and mongolism in two families. Am J Dis Child 104:534, 1962.
33. Kadotani F, Ohama K, Makino S: A case of 21 trisomic Down's syndrome from the triple-X mother. Proc Jpn Acad 46:709, 1970.
34. Ford CE, Jones KW, Miller OJ, Mittwoch V, Penrose LS, Ridler M, Shapiro A: The chromosomes in a patient showing both mongolism and the Klinefelter syndrome. Lancet 1:709, 1959.
35. Taylor AI, Moores EC: A sex chromatin survey of newborn children in two London hospitals. J Med Genet 4:258, 1967.
36. De la Chapelle A, Schroder J: Autoradiographically identified karyotype 49,XXXXY, t(4;11)(q35;q23) confirmed by banding. Hereditas 74:291, 1973.
37. Engel W, Reinwein H, Muller J, Kunze G: Chromosomenbefunde bei 365 Patienten mit Down-syndrom oder Verdacht auf Down-syndrom. Humangenetik 8:307, 1970.
38. Richards BW, Stewart A, Sylvester PE, Jasiewicz V: Cytogenetic survey of 225 patients diagnosed clinically as mongols. J Ment Defic Res 9:245, 1965.
39. Dutrillaux B: Les aberrations chromosomiques transmissibles. Journées Parisiennes de Pédiatrie. Paris: Flammarion, 1972, pp 13–21.

Case Report: Partial Trisomy 12p Associated With 4p Deletion Due to Paternal t(12p−;4p+) Translocation*

Mary Esther Carlin, MD and Charles Norman, CT

The proband is a white male who was first seen at 3 months of age. He was referred to our center because of an odd looking physiognomy and the presence of a questionable chromosomal abnormality on karyotypes done at a community hospital.

He is the fourth of four male offspring. The mother was 31 and the father 32 at the time of the proband's conception. The pregnancy was uneventful. He was born at term immediately following intravenous sedation. He was cyanotic and had poor muscle tone at birth. He weighed 2,875 gm, measured 49.5 cm in length, and had a head circumference of 30.5 cm and a chest circumference of 31.8 cm. He was noted at the time of delivery to appear immature and to have small, low-set ears, redundant sacral skin, and an undescended left testicle. He was admitted to the nursery in respiratory distress.

During the neonatal period he continued to have acrocyanosis and was a poor feeder. He had a gingival cyst which ruptured, revealing a tooth. This was removed. In addition, a tight tongue frenulum was clipped and bilateral preauricular skin tags were removed. X-ray films of the spine revealed spina bifida occulta underlying the excess sacral skin.

When seen at 3 months of age (Fig. 1), proband weighed 3,829 gm, measured 53.2 cm in length, and had a head circumference of 35.5 cm. His hands, feet, and forehead were cyanotic and the rest of his skin was pale. He had a weak cry and was in respiratory distress with stridor audible on both inspiration and expiration.

He had a number of dysmorphic features. His mouth was fish-shaped, with a high-arched palate. His nose was small, with a flat bridge and asymmetric nares.

*This study was supported in part by HEW grants MCT-000-288-11-0 and MCT-000903.

Birth Defects: Original Article Series, Volume XIV, Number 6C, pages 399–406

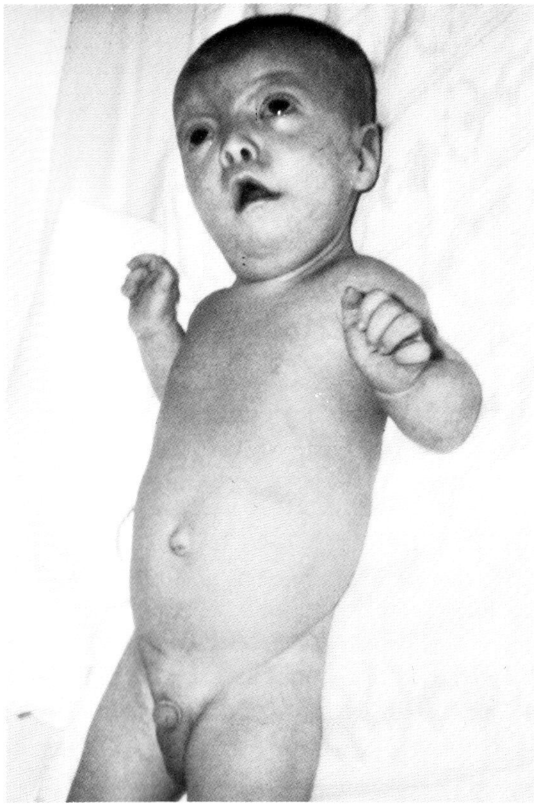

Fig. 1. The proband at 3 months of age.

The ears were low-set and posteriorly rotated, and they had large, flat pinnae and narrow external auditory canals. The eyes were prominent and the palpebral fissures slanted downward. He had bilateral epicanthal folds and an inferior coloboma on the right. Skin dimples were present over both elbows and a redundant skin fold was evident in the lumbosacral area. The thumbs were proximally placed. Hypotonia and hyperreflexia were noted. Dermatoglyphic analysis revealed a left atd angle of 56°, bilateral simian creases, and generalized hypoplasia of the ridges.

Peripheral blood lymphocytes yielded karyotypes (Fig. 2) demonstrating a partial trisomy: 46,XY,der(4)t(4;12)(4qter → 4p16::12p11→ 12pter; 12qter → 12 p11::4 p16 → 4 pter). The parents, who are both phenotypically normal, were also studied. The mother demonstrates a normal, 46,XX karyotype. The father's karyotype (Fig. 3) shows a balanced translocation: 46,XY,t(4:12)

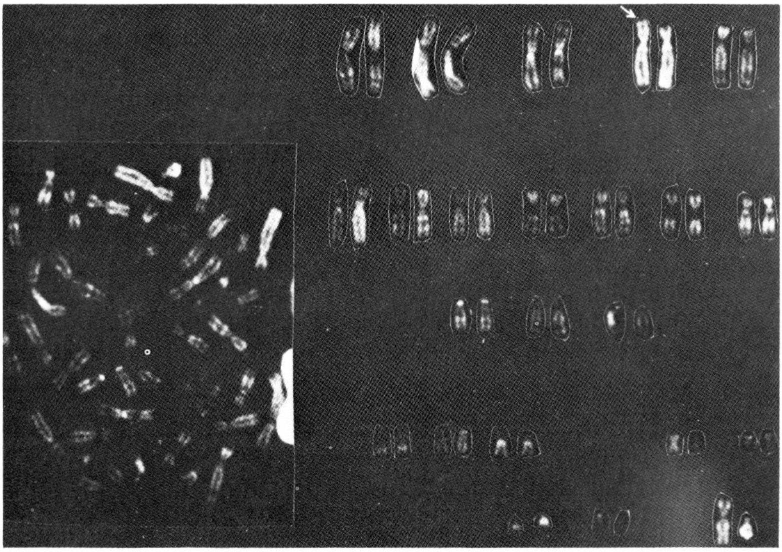

Fig. 2. 46,XY,der(4), t(4;12)(p16;p11) fluorescent karyotype from the proband.

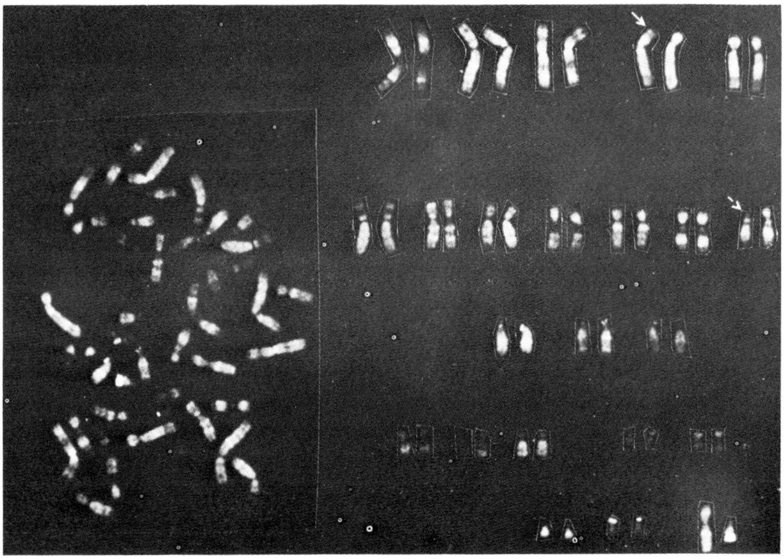

Fig. 3. 46,XY,t(4;12)(p16;p11) fluorescent karyotype from the proband's father, showing the balanced translocation.

(4qter → 4p16::12p11 → 12pter; 12qter → 12p11::4p16 → 4pter). The two living older brothers have karyotypes identical to that of the father and are also phenotypically normal. The firstborn son died as a neonate. He was born prematurely in breech presentation, weighing 1,590 gm and measuring 40.6 cm in length. His autopsy revealed immature organs, pulmonary congestion with occasional intraalveolar hemorrhage and areas of focal atelectasis, polycystic kidneys of the newborn (group I), a bicuspid aortic valve, and necrosis and hyalinization of the Hassall corpuscles in the thymus. No cytogenetic studies were performed.

At 8 weeks of age, the proband had an incarcerated inguinal hernia. A right herniorrhaphy was performed to remove the infarcted, undescended testicle and epididymis. He suffered a fractured femur without sustaining a fall at 1 year of age. He also had several episodes of sudden jerking of the head and limbs during the first year. These seemed to abate without therapy and the electroencephalogram was normal. He was hospitalized seven times for respiratory distress in the first 15 months of life. He continued to feed poorly and vomited daily for his first 18 months. His private pediatrician prescribed Sedadrops for his almost daily colic.

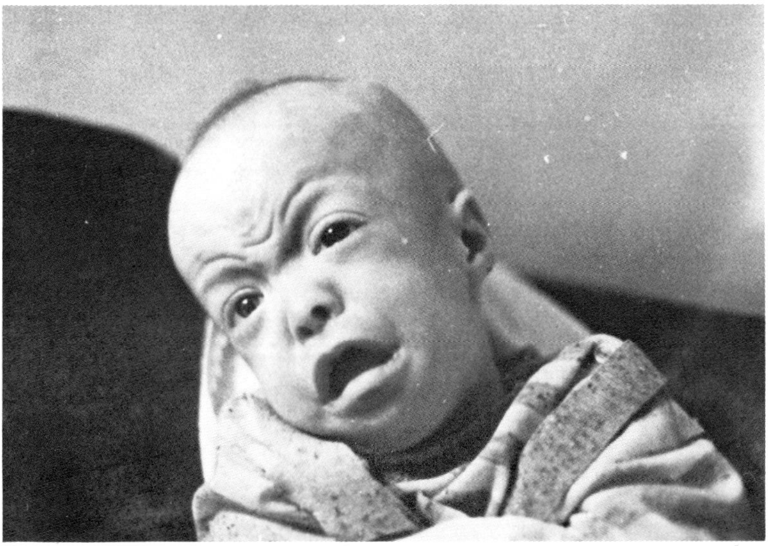

Fig. 4. Proband at 27 months of age.

His growth retardation has been severe. At 6 months he weighed 4,842 gm and was 62.2 cm long, with a head circumference of 39 cm. At 27 months (Fig. 4) he weighs 6.08 kg and is 76 cm long, with a head circumference of 42.7 cm. His inability to feed well has continued, but the vomiting has subsided. His physiognomy has not altered appreciably and he bears little resemblance to other family members. He has been hospitalized for a croup-like illness an additional ten times in the past 12 months.

His developmental delay is likewise severe. He functions at about a 4-month level. He rolls over in both directions but cannot sit alone. He still does not have full head control. He is currently enrolled in a developmental program which includes physical therapy.

COMMENTS

The similar clinical findings noted in this case and the ten other reported cases of 12p trisomy have permitted the delineation of the specific abnormalities seen in this particular malformation syndrome. [1–11] As outlined by Rethoré [12], Tenconi et al [10] and Bijlsma et al [11], these characteristic dysmorphic features include: high, prominent forehead; flat, round face with prominent cheeks; flat nasal bridge; short nose; large philtrum; dysplastic ears; down-turned mouth with everted lower lip; clinodactyly of the 5th finger; abnormally placed thumb and simian crease. Physical growth is normal while psychomotor development is retarded.

In all 11 cases the partial 12p trisomy has resulted from a parental translocation; 6 paternal and 5 maternal, involving chromosome 12 and chromosomes 4, 6, 7, 8 (2 cases), 9, 14 (2 cases), and 21 (3 cases). Bijlsma [11] has summarized the clinical and cytogenetic findings in all of the cases except that of Tenconi et al [10].

However, the proband in this case has many dysmorphic findings other than those of the 12p trisomy syndrome. Many are identifiable as those associated with 4p monosomy (deletion) or Wolf syndrome [12–18]. A deletion involving the chromosomal material from the 4p16 band to the 4p terminus results in this malformation syndrome [18]. In addition to being trisomic for the 12p11 → 4pter segment, the proband is also monosomic for the 4p16→4pter. segment.

Table 1 compares the typical findings seen in 12p trisomy and in 4p deletion with those present in this case. As could be anticipated from his karyotype, the proband emerges as a phenotypic hybrid combining many of the major features of both the 12p trisomy and the 4p deletion syndromes.

TABLE 1. Comparison of Findings in 12p Trisomy and 4p Deletion With Those Present in Proband

Clinical finding	Appearance in 12p trisomy[a]	Appearance in 4p deletion[b]	Appearance in proband
Sex ratio	3:1 male preponderance	1:1 equal distribution	male
Birthweight	normal to corpulent	low	low (2,830 gm)
Birth length	normal	short	short (49.5 cm)
Birth head circumference	normal	microcephalic	microcephalic (30.5 cm)
Postnatal physical growth	normal	severely retarded	severely retarded
Mental development	mild to moderate retardation	severe retardation	severe retardation
Head shape	turricephaly	dolichocephaly	turricephaly
Median scalp defects	–	+	–
High prominent forehead	+	+	+
Deep forehead wrinkles on crying	–	+	+
Flat face	+	±	+
Glabellar hemangioma	–	+	+
Short nose	+	–	±
Wide nasal bridge	+	+	+
"Greek warrior helmet" appearance	–	+	+
Hypertelorism	+	+	+
Epicanthal folds	+	+	+
Ocular malformations	–	+	coloboma
Antimongoloid slant	–	±	+
Nystagmus	±	–	–
Philtrum	broad	deep, narrow with strongly marked pillars	broad with strongly marked pillars
Cleft lip/palate	–	+	–
Down-turned, fish-shaped mouth	+	+	+
Prominent, everted lower lip	+	–	+
Small chin	+	+	+
Retrognathia	–	+	+
Preauricular tag or pit	–	+	+
Low-set ears	±	–	+
Helix	folded	flat	somewhat flattened
Protuberant anthelix	+	–	+
Protruding antitragus	–	±	+
Deep concha	+	–	+
Neck	short	long	short
Trunk	normal	elongated	elongated
Spina bifida	–	+	+ (occulta)

TABLE 1. Comparison of Findings in 12p Trisomy and 4p Deletion With Those Present in Proband (continued)

Clinical finding	Appearance in 12p trisomy[a]	Appearance in 4p deletion[b]	Appearance in proband
Sacral or coccygeal dimple	–	+	+
Limbs	normal to stocky	slender	slender
Dimples at elbows and knees	–	+	+
Proximally placed thumbs	+	+	+
Pes planus	+	±	+
Hypotonia	–	+	+
Ectopia testicles	+	+	+
Hypospadias	–	+	–
Cardiopathy	±	±	–
Seizures	–	+	–
Single palmar flexion crease	~ 75%	~ 25%	+
Hypoplasia of ridges	–	+	+
Elevated atd angle	–	±	+
Increased number of arches	–	+	–

[a]See [10–12, 18]
[b]See [12–18]

REFERENCES

1. Uchida IA, Lin CC: Identification of partial 12 trisomy by quinacrine fluorescence. J Pediatr 82:269-272, 1973.
2. Fryns JP, Van den Berghe H, Van Herck G, Cassiman JJ: Trisomy 12p due to familial t(12p–;6q+) translocation. Humangenetik 24:247–252, 1974.
3. Armendares S, Salamanca F, Nava S, Ramirez S, Cantu JM: The 12p trisomy syndrome. Ann Génét 18:89–94, 1975.
4. Rethoré MO, Kaplan JC, Junien CL, Cruveiller J, Dutrillaux B, Aurias A, Carpentier S, Lafourcade J, Lejeune J: Augmentation de l'activité de la LDH-B chez un garçon trisomique 12p par malségrégation d'une translocation maternelle t(12;14)(q12;p11). Ann Génét 18:81–87, 1975.
5. Tocci P, Hitchcock C, Neadle M, Carlin M: Trisomy 12; In a mentally retarded boy due to paternal t(12–;4p+) translocation. Am J Hum Genet 27:90A, 1975.
6. Hoo JJ: 12p trisomy: A syndrome? Ann Génét 19:261–263, 1976.
7. Alfi OS, Lange M: Trisomy 12p, a clinically recognizable syndrome. Abstract. In Bergsma D, Lowry B (eds): "New Syndromes." New York: Alan R. Liss for The National Foundation–March of Dimes, BD:OAS XIV(3B):231–232, 1977.

8. Nielsen J, Vetner M, Holm V, Askjear SA, Reske-Nielsen E: A newborn child with karyotype 47,XX,+der(12)(12pter→12q12::8q24→8qter), t(8;12)(q24;q12)pat. Hum Genet 35:357–362, 1977.
9. Biederman B, Bowen P, Robertson C, Schiff D: Partial trisomy 12p due to t(12;21)pat translocation. Hum Genet 36:35–41, 1977.
10. Tenconi R, Piovan E, Preto A, Magnabasco R, Baceichetti C: Syndrome +12p, Case Report and Review. Hum Genet 39:97–101, 1977.
11. Bijlsma JB, de France HF, Blecker-Wagemakers LM, Dijkstra PF: Double translocation t(7;12),t(2;6) heterozygosity in one family. Hum Genet 40:135–147, 1978.
12. Rethoré MO: In Yunis JJ (ed): "Chromosomes in Biology and Medicine. Vol. II. New Chromosomal Syndromes." New York: Academic Press, 1977.
13. Wolf U, Reinwein H, Porsch R, Schroter R, Baitsch H: Defizienz an den kurzen armen eines chromosoms nr. 4. Humangenetik 1:397–413, 1965.
14. Wolf U, Porsch R, Baitsch H, Reinwein M: Deletion of short arms of a B chromosome without "cri-du-chat" syndrome. Lancet 1:769, 1965.
15. Wolf U, Reinwein H: Keiresche und cytogenetische differential diagnose den defizienzen anden kurzen armen der B– chromosomen. Z Kinderheilkd 98:235, 1967.
16. Miller OJ, Warburton P, Breg WR: Deletions of group B chromosomes. In Bergsma D (ed): Part V "Phenotypic Aspects of Chromosomal Aberrations." New York: The National Foundation–March of Dimes, BD:OAS V(5):100–105, 1969.
17. Guthrie RD, Aase JM, Asper AC, Smith DW: The 4p– syndrome. A clinically recognizable chromosomal deletion syndrome. Am J Dis Child 122:421–426, 1971.
18. de Grouchy J, Turleau C: "Clinical Atlas of Human Chromosomes." New York: John Wiley, 1977.

Selected Abstracts

COMPLEX ANEUPLOIDY IN VITRO IN A NORMAL FEMALE

Paul J. Benke and Joan Davies

Department of Pediatrics & Mailman Center, University of Miami School of Medicine, Miami, Florida

Thirty percent aneuploidy was demonstrated in phytohemagglutin in (PHA) stimulated lymphocytes from a 35-year-old woman referred for genetic counseling. This patient had a stillborn fetus with multiple congenital anomalies. Chromosome studies of the patient's two children, mother, father, and two brothers were normal. A similar percentage of aneuploid cells was shown in cultured skin fibroblasts. These differed qualitatively from lymphocytes by having an increased variability in aneuploid chromosomes. Cells harvested by 1) gentle shaking 2) colcemid or velban incubation demonstrated aneuploidy. A and B group chromosomes were not aneuploid, and the X chromosome was the most common aneuploid chromosome. All cloned fibroblastic cells grown up from single cells demonstrated 30–60% aneuploidy. Coating tissue culture tubes with basic charged polymers did not prevent aneuploidy in PHA-stimulated lymphocytes.

Aneuploidy in this patient may be secondary to an unknown factor of in vitro tissue culture conditions. However, these findings may also represent a new mutation of mitotic cell division and chromosome distribution which may be related to clinical conditions of aneuploidy.

Birth Defects: Original Article Series, Volume XIV, Number 6C, **pages 407–429**

DE NOVO PARTIAL MONOSOMY 7Q IN A PATIENT WITH A FAMILIAL BALANCED TRANSLOCATION, 46,XX, del (7) (q33), t(8;9)(q11;q31) pat

Brian Biederman and Peter Bowen

Department of Pediatrics, University of Alberta, Edmonton, Alberta, Canada

The proposita was born at term to a gravida III, 27-year-old mother whose pregnancy was uneventful. The birthweight was 3,150 gm and the Apgar rating was 7 and 9 at 1 and 5 minutes respectively. She was investigated at 12 months of age for growth and developmental retardation and microcephaly. Her length was 63 cm (< 3rd percentile), wt 7.5 kg (< 3rd percentile) and head circumference 39.5 cm (> 4 SD below the mean). Significant findings were bilateral epicanthus, myopia, depressed nasal bridge, red bulbous nose, deep philtrum, relatively large ears with folded helices, increased muscle tone and stretch reflexes and positive Babinski responses. Her developmental status on the Denver schedule corresponded to about 6 months of age.

Chromosome studies with G- and R-bands (acridine orange) showed a deletion of the long arm of chromosome 7 (q33) in the proposita. In addition, the proband as well as the father and a sister have a (?) coincidental balanced translocation t(8;9) (q11;q31). The mother and another sister have a normal karyotype.

This case bears a close resemblance to 7 other patients with a partial deletion of the long arm of chromosome 7 described in the literature. Recurrent features are growth deficiency of prenatal onset, microcephaly, moderate mental retardation, skin laxity, long philtrum, and large ears with folded helices.

FAMILIAL PERICENTRIC INVERSION OF CHROMOSOME 1 IN A CHILD WITH SEPTO-OPTIC PITUITARY DWARFISM AND DE NOVO PERICENTRIC INVERSION OF CHROMOSOME 2

Florence Char, MD and Grace K. Chueng, MS

Department of Pediatrics, University of Arkansas for Medical Sciences

The purpose of this communication is to report a case of familial pericentric inversion of chromosome 1 in a 7-year-old white male with septo-optic pituitary dwarfism and de novo pericentric inversion of chromosome 2.

The patient was born to a 32-year-old mother and her 35-year-old husband. A first pregnancy ended in spontaneous abortion at 3 months gestation. A 4-year-old sister is normal. There is no history of consanguinity. The patient's father is from

Iceland and mother is of Italian descent. There is no history of short stature, mental retardation, or birth defects in the family.

The pregnancy and delivery were uncomplicated. Birthweight was 3.5 kg Birth length was 52 cm. Exotropia with left hypertropia was noted at 2–3 months. Pale and atrophic disks were noted at approximately 1 year of age. Retarded growth was evident by 18 months. Diagnosis of septo-optic dysplasia type of pituitary dwarfism was established in 1972. Pneumoencephalogram, EEG, and IQ were normal. Since then the patient has been receiving growth hormone therapy. At age 7 years his height was 108.5 cm.

Chromosome studies were done on the patient, his immediate family and available relatives. Banding studies were done utilizing the heat-Giemsa technique. Except for the patient, all were phenotypically normal. In the individuals with pericentric inversion of chromosome 1, the break points were at distal 2p and 3q.

In addition to having the pericentric inversion of chromosome 1, the patient was found to have pericentric inversion of chromosome 2 with break points at 2p5 and between 1q and 2q, apparently a de novo pericentric inversion.

PARTIAL TRISOMY OF 6q IN TWO SIBS WITH MULTIPLE DEFECTS

C. Clark, H.R. Cowell, and K. Paulson

Clinical Research Department, Alfred I. duPont Institute, Wilmington, Delaware

Two female sibs presented with a similar clinical picture of severe mental retardation and multiple congenital anomalies including scoliosis, flexion contractures of the hips, knees, elbows and fingers, low-set ears, micrognathia, exotropia, and peculiar facies. This constellation of anomalies has not previously been reported in the literature as a syndrome. Dermatoglyphic findings include a low TRC and an increased frequency of digital arches.

They are the products of a nonsonsanguineous marriage which had resulted in a clinically normal female, a stillborn with the same clinical appearance as the probands and two miscarriages. There is no family history of mental retardation, birth defects, or miscarriages.

Routine Giemsa staining from peripheral leukocytes suggests a normal 46,XX karyogram for both probands, mother and sister. Father demonstrates 46,XY. No irregularity could be detected in the females.

G- and Q- banding of both probands revealed 46,XX,11q+. In addition, G- and Q-banding of the mother and sister demonstrated 46,XX,t(6;11)(q24;q25) and 46,XY for the father.

Routine karyotyping failed to delineate the abnormal chromosome 11 in the probands even though a significant terminal segment of chromosome 6 is translocated to chromosome 11.

We urge laboratories to review cases, in which a chromosome abnormality seems likely, with banding procedures. It is imperative to analyze parents and sibs so that definitive genetic counseling can be given.

PARTIAL TRISOMY 9 – REPORT OF TWO FAMILIES AND FURTHER EVIDENCE FOR PHENOTYPIC HETEROGENEITY

John R. Davis, Steven P. Serlin, and Iris C. Veomett

Department of Pathology and Phoenix Indian Medical Center, University of Arizona Health Sciences Center, Tucson, Arizona

Partial trisomy of chromosome 9 has emerged as one of the more common C group abnormalities. Over 50 cases have been reported using banding techniques. Although many dysmorphic features are shared in common, phenotypic heterogeneity occurs due to the variable size of the partial trisomic fragment. It has been suggested that crucial determinants for the syndrome reside in the short arm. Inclusion of additional length (centromere and proximal long arm) does not intensify the impact of the syndrome, but may actually mollify the effects.

Two families with partial trisomy 9 are reported. In family B the index patient at age 4 months had microcephaly, brachycephaly, typical facies, and hand-feet abnormalities. Congenital heart disease and hydrocephalus were present. After ligation of patent ductus and temporary ventriculoperitoneal shunting, the child is severely retarded at age 34 months. Two cousins were similarly affected; one died in the neonatal period. The 9 trisomic fragment (9pter→p 13) was translocated in tandem to chromosome 18 short arm tip. The kindred included nine balanced translocation carriers spanning four generations.

Family C index patient at birth had cleft lip and palate, brachycephaly, abnormal but not classic facies, hand-feet abnormalities, hirsutism, and an inconsequential cardiac murmur. With modal count 47, a marker chromosome consisted of 9pter→q13. The mother was a balanced translocation carrier whose karyogram included the derivative 9 chromosome; the remainder of chromosome 9 was translocated in balanced fashion to chromosome 20. Pedigree testing is in progress.

These two kindred further illustrate the variable phenotypic effects in partial trisomy 9. If the trisomic fragment includes long arm portions in addition to short arm, the impact may be less severe. These additional cases reflect the great frequency of translocation carrier basis (usually maternal), and the prevalence of multiple generation occurrence.

TRISOMY 18 MOSAICISM ASSOCIATED WITH CONGENITAL PURE RED CELL ANEMIA (DIAMOND BLACKFAN SYNDROME) IN A 10-YEAR-OLD GIRL

Peter A. Duncan, Lawrence R. Shapiro, Patrick L. Wilmot, Mary J. Murphy, and Peter B. Farnsworth

Divisions of Pediatrics and Pathology, Westchester County Medical Center and the Departments of Pediatrics and Pathology, New York Medical College

A white female infant was evaluated at 7 months of age for an anemia which was first noted at 4 months. This anemia was unresponsive to iron therapy. Her birthweight was low for gestational age, and she exhibited slow physical and psychomotor development, microcephaly, and a congenital dislocation of the left hip at 7 months of age. The mother was 44 years old at the child's birth, and had taken an antiemetic drug during the first trimester of the pregnancy. The occurrence of anemia, microcephaly, and other malformations in the family were denied except for an unspecified type of mental retardation in a paternal cousin of the proposita. At 10 years of age, the patient's height, weight, and head circumference were below the 3rd percentile, but the head measurement was disproportionately smaller. She functioned at approximately 50% of her anticipated developmental level.

At 2 years of age, a standard chromosome analysis of the patient revealed a two-cell-line mosaicism (46/47) in the blood, skin, and bone marrow. Chromosomal analysis at 7 years of age, utilizing the Giemsa-banding technique, revealed a 46,XX/47,XX,+18 mosaicism in the patient. A study of the mother's chromosomes was normal (46,XX).

Congenital pure red cell anemia was diagnosed on the basis of normal leukocytes and platelets, reticulocytopenia, prolonged postnatal elevation of fetal hemoglobin, persistence of macrocytes in the blood, a markedly increased myeloid:erythroid ratio with normal megakaryocytes in the bone marrow, and responsiveness to prednisone.

A review of the literature has revealed that a chromosome abnormality was found in 8 of 45 previously reported cases of congenital pure red cell anemia; however, none of these abnormalities involved chromosome 18.

ABNORMALITY OF 16TH CHROMOSOME – ITS PHENOTYPIC EXPRESSION

Nancy L. Golden and Robert M. Bilenker

Department of Pediatrics, Case Western Reserve University School of Medicine, Cleveland, Ohio

A genotypic abnormality of the 16th chromosome in which there is extra material on the short arm has been identified. The extra material is light staining and its source is unknown. Both parents have normal chromosomes. The prematurely born black male found to have the abnormal 16th chromosome was investigated because of multiple congenital anomalies. He had midfacial hypoplasia with a flat nose and nasal bridge, midline cleft palate, micrognathia, and small hyperteloric eyes. His ears were asymmetric and low set. Skeletal anomalies included a short neck with cervical hemivertebrae, short clavicles and small acromion processes, and long fingers. He had fixed flexion contractures of multiple joints (arthrogryposis). The infant may have abnormal muscle tissue as he has developed recurrent apneic spells. Developmentally, he has made no gains: at the age of 8 months, he is functioning at a newborn level. There are few reports of living patients with abnormalities of the 16th chromosome. Most of these abnormalities result in spontaneous abortion. This particular abnormality (16p+) has not been previously reported.

ABSENCE OF DERMAL RIDGES AND MOSAICISM FOR SEX CHROMOSOMES

Ranjan Y. Goradia, Brian K. Davis, and Rolando J. Deleon

Department of Biology, Virginia Polytechnic Institute and State University, Blacksburg, VA

Four generations of absence of dermal ridges are shown in the above pedigree. No ridges are present anywhere on the fingers, palms, toes, or soles. Miscarriages and and deaths at birth are common. Among 19 pregnancies in the first generation, 17 ended in miscarriage or death immediately following birth. One child died at 7 years of age and one survived to adulthood. One pregnancy ended in miscarriage in the second generation and two in the third.

No skin abnormalities except the absence of dermal ridges are present. Preliminary analysis of a skin biopsy section from the palm of the propositus indicates normal histology.

Chromosome analysis of leukocytes from the propositus show mosaicism for 46,XX/47,XY, + Xp–. Similarly, his mother's leukocytes were mosaic for 46, XX/47,XX,+ XP–. More than 20% of the cells were aneuploid in both individuals. The extra chromosome was shown to be the long arm of the X chromosome by trypsin-Giemsa banding.

This is the first case of absence of dermal ridges which has been shown to be associated with chromosome aneuploidy.

SEX CHROMOSOME MOSAICISM IN A MOTHER AND TRISOMY 13 IN HER CHILD

Ranjan Y. Goradia*, John Mann†, and Brian K. Davis*

Biology Department*, Virginia Polytechnic Institute and State University, Blacksburg, VA
Department of Pediatrics†, Kaiser Hospital, Santa Clara, CA

A 31-year-old white female with a history of ovarian dysfunction delivered a male infant with trisomy 13. Her chromosome constitution from cultured leukocytes shows mosaicism for 46,XX/47,XXX. Both of these cell lines include cells with chromosome fragments.

Analysis by Q- or G-banding of 200 cells revealed 18 with chromosome aneuploidy. Among those cells with three X's, two were of the same length, while the third was about 10% longer with the short arm twisted and slightly uncoiled in the centromere region. In one cell, the long chromatids of the third X formed a giant ring with many constrictions while still attached to the twisted short arm. Fragments of chromosomes of varying sizes, including rings and linear chromosomes, were recovered in both cell lines.

Dermatoglyphic analysis of the proposita shows a dermal ridge count of 110, which is the mean count for triplo-X hands. This reduction in dermal ridge count from the normal 46,XX was not from the presence of arches, but rather from four pocket whorls and one double loop.

Physical examination of the proposita was normal except for the history of unpredictable ovulation. A buccal smear showed no abnormalities.

This is the first reported case of a trisomy X mosaic mother giving birth to an autosomal trisomic chiod. Parents of trisomy children should be examined carefully for the presence of mosaicism abnormal chromosomes or fragments. More data, especially from low fertility parents, will give some insight into the prevention of aneuploid birth defects.

OBSERVATIONS IN ANOTHER CHILD WITH 9p– SYNDROME

Dorothy A. Hahn, J. Rogers Byrd, and Chen-Kung Ho

Department of Pediatrics, Medical College of Georgia, Augusta, Georgia

At age 8 months, a black female was referred for evaluation of respiratory symptoms including profuse nasal discharge, difficult breathing, and noisy respirations. Peculiar facies, retardation in growth and psychomotor development were also noted. Study on peripheral leukocytes showed 46,XX,del(9)(p21) karyotype. Details of her dysmorphic features will be presented and compared with those of previously reported cases.

MALE PSEUDOHERMAPHRODITISM ASSOCIATED WITH DOUBLE MOSAICISM IN DOWN SYNDROME

Judith G. Hall, Roberta A. Pagon, David Luthy, and Thomas Norwood

Departments of Pediatrics, Medicine, Pathology and Obstetrics & Gynecology, University of Washington School of Medicine and Division of Medical Genetics, Children's Orthopedic Hospital and Medical Center, Seattle, Washington

We have recently seen an apparent female child with Down syndrome referred for genetic counseling at age 15 months, who proved to be a male pseudohermaphrodite. She was the product of an uncomplicated pregnancy to a 31-year-old father and 27-year-old mother. The family history was negative for congenital anomalies and intersex problems, and a 4-year-old sister is apparently normal. Down syndrome was diagnosed at birth and congenital heart disease was recognized in the newborn period. Frequent respiratory infections occurred during her first year of life. At 15 months she was unable to sit up or stand, but did socialize and was able to turn over. Physical examination (including external genitalia) was typical for a female with Down syndrome. Chromosome studies confirmed trisomy 21, but three cell lines were present: 47,XY,+21/46,X,+21/46,XY. The 46,X,+21 line prevailed in both lymphocyte and fibroblast metaphases with 83 and 67%, respectively. The 46,XY cells were present only in the lymphocytes ($< 20\%$).

Many explanations are possible, but we postulate that the Y chromosome was lost during the development of a trisomy 21 conceptus. The loss must have occurred sufficiently early to be apparent in two divergent somatic tissues, and to be predominant or complete in primordial germ cells as reflected in their inability to organize testosterone producing testicular tissue. A second mitotic error (involving chromosome 21), must have occurred later in development, since it was detected only in the lymphocyte population.

ASSOCIATION OF ENDOCRINE DISTURBANCES IN DOWN SYNDROME AND THE MATERNAL SIDE OF THE FAMILY*

S. Hara, D. N. Singh, M. Bayatpour, and C. M. Reddy

Department of Pediatrics, Meharry Medical College, Nashville, Tennessee

Of a total of 44 cases of Down syndrome, 8 (6 females and 2 males) were found to have thyroid problems; 5 patients were diagnosed as Hashimoto thyroiditis, based on thyroid function tests and antithyroglobulin titer determination; 2 patients were found to have goitrous cretinism; and another patient had exophthalmic thyrotoxicosis. Insulin dependent juvenile diabetes was found in 2 male Down syndrome patients.

Three of the 44 mothers of the Down syndrome were found to have thyroiditis and none of them had diabetes. Among grandmothers, 6 had diabetes, 3 were found to have some type of thyroid disturbance associated with goiter, and 4 had a crippling type of rheumatoid arthritis.

More than 2 members of the maternal side of the 5 families of the Down syndrome patients had either leukemia or carcinomas or both. Occurrence of spontaneous abortions was significantly high among the mothers of the Down syndrome. Twenty-six families had a history of miscarriages and abortions.

A detailed description of our observations and a literature review of an association of endocrine disturbances in the maternal side of the family of the Down syndrome will be presented. As such it may be deduced that a young woman with or without endocrine problems whose mother has a positive history may have more than a random risk of having nondisjunction of chromosomes resulting in, for example, trisomy 21 or other chromosomes.

*This study was supported by Maternal and Child Health Training Grants # 440 & 442 of HEW PHS.

DELETION 10p SYNDROME: COMPARISON OF FINDINGS IN 3 CASES

L. Jackson, H. Rosenblum, and M. Barr

Division of Medical Genetics, Thomas Jefferson University, Philadelphia, Pennsylvania

Two cases of deletion of the short arm of chromosome 10 in children have been reported (Frank et al: these Proceedings, p 87, 1974; Shokeir et al: J Med Genet 12:99–113, 1975). We have followed a third case from age 9 months to 4 years. The child, a boy, was born 5/11/73 after an uneventful pregnancy and forceps delivery to a primigravida mother aged 21 years. The birthweight was 3.5 kg and the neonatal course was uneventful. A delay in developmental milestones at age 7 months precipitated the clinical investigation at age 9 months. In common with the other two cases, the child showed developmental retardation, trigonocephaly with prominent forehead, downward lateral canthal slant, epicanthus, ptosis of the eyelids, flat nasal bridge, high-arched or cleft palate, abnormal ears and widely spaced nipples. Cryptorchidism, congenital heart disease, hypotelorism, and abnormal dermatoglyphics were findings in one or the other of the two reported cases which were not seen in the present case. The chromosomal deletion appeared to be at band p13 in all three cases. The syndrome thus far appears reasonably consistent in its clinical presentation and should be recognizable by the characteristic facies.

ON TRISOMY 9PTER→9Q1

E. C. Jenkins, H. R. Salwen, M. A. Krauss, J. G. Davies, and N. B. Kardon

Department of Human Development and Genetics, New York State Institute for Basic Research in Mental Retardation, Department of Pediatrics, North Shore University Hospital, Cornell University Medical College, New York, New York

A case of de novo trisomy 9pter→q13 was observed. Cytogenetic analysis of G-, R-, Q-, and C-banded preparations revealed a karyotypic description of 47,XY,+del(9)(pter→q13). In addition to the principal characteristics of the 9p trisomy syndrome, the child presented with skeletal and urogenital abnormalities. That the 9q1 region exerts little additional effect upon the phenotype when present in a trisomic state (Sutherland's hypothesis) should be revised since it appears that certain clinical abnormalities such as clubfeet, finger contractures and hypospadias are due to trisomy 9q1.

PARTIAL MONOSOMY AND TRISOMY OF CHROMOSOME 18 IN AN OFFSPRING OF A CARRIER OF PERICENTRIC INVERSION OF 18

Mary Kukolich, Joseph Sears, Raymond Lewandowski, and Carl Mankinen

Texas Genetic Screening and Counseling Service, Department of Mental Health Mental Retardation, 404 W. Oak Street, Denton, Texas

A 6-month-old female infant with 46,XX,−18,+inv18(p11.2q21) is presented. She was referred as a possible trisomy 18. Pregnancy was complicated by polyhydramnios. Physical abnormalities included growth failure, depressed nasal bridge, slightly bulbous and upturned nose, low-set and posteriorly rotated ears, and a high-arched palate with a prominent midline ridge. Other abnormalities noted were pectus excavatum and hypoplasia of the nails of fingers and toes. Dermatoglyphic examination showed 10 arches.

There was a positive family history on the maternal side of neonatal deaths and miscarriages over three generations. Chromosome analysis of the mother and maternal grandmother revealed a pericentric inversion of chromosome 18.

TRISOMY 13 MOSAICISM—A DIFFICULT CLINICAL DIAGNOSIS

Raymond Lewandowski, Jr., Mary Kukolich, Joseph Sears, and Carl Mankinen

Texas Genetic Screening and Counseling Service, Department of Mental Health Mental Retardation, 404 W. Oak Street, Denton, Texas

Trisomy 13 is a chromosomal defect with severe sequelae in affected individuals. Only 5% of this patient population results from chromosomal mosaicism, often with less severe clinical effects. This paper presents a client with trisomy 13 mosaicism, reviews the literature on this defect, and demonstrates the need for careful banding studies because of the overlap in phenotype with the other D-group trisomies.

Physical examination revealed normal head circumference, broad forehead, hypertelorism, sparse eyebrows, upward slanting palpebral fissures, broad-based nose, large ears, prominent antihelix, large chest, clinodactyly of 5th digits, proximally implanted thumbs, hyperconvex long fingernails, dimples, hemangiomas, absent flexion creases on thumbs, decreased TFRC, t′ triradius.

47,XX,t(1;21) IN A CHILD WITH DOWN SYNDROME

Patricia L. Monteleone, Sister Leo Rita Volk, Gurbax Sekhon, John Grzegocki, Michael Tietjens, and Hari Sekhon

Department of Pediatrics, St. Louis University School of Medicine, Department of Pediatrics, Washington University School of Medicine, and St. Mary's Health Center, St. Louis, Missouri

A newborn infant with multiple congenital anomalies presented recently in our nursery. The child was the third offspring of a 33½-year-old mother and 35-year-old father. There was no history of henetic disease, consanguinity, or congenital anomalies in the family. Physical examination revealed a 2,610 gm, full-term infant with epicanthal folds, flattened nasal bridge, fat pad at nape of the neck, protruding tongue, hypotonia, and simian creases, suggestive of Down syndrome. Chromosomal analysis revealed 47 chromosomes with a translocation of the extra No. 21 chromosome to the centromere of the long arm of one of the No. 1 chromosomes and permanent breakage of chromosome 1 at the centromere.

Karyotypes of the infant as well as other family members will be presented.

THE 8p- SYNDROME

Jacob A. Reiss*, Phillip Brenes*, R. Ellen Magenis†, and Jan Chamberlin†

Department of Pediatrics*, Permanente Clinic, Kaiser Foundation Health Plan of Oregon, Portland, Oregon, and Crippled Children's Division and Division of Medical Genetics†, University of Oregon Health Sciences Center, Portland, Oregon

An infant boy with minor facial abnormalities, severe congenital heart disease and failure to thrive was found to have a deletion of the small arm of chromosome 8. This infant was born to a gravida 1 para 1 17-year-old woman whose pregnancy was unremarkable. Birthweight was 2,500 gms. Although the immediate postpartum course was uneventful, the infant fed poorly, did not thrive and developed tachypnea, dyspnea, and episodes of cyanosis as a neonate. At approximately one month of age he underwent cardiac catheterization which revealed an endocardial cushion defect, and pulmonary hypertension.

The cardiac defect and other minor physical abnormalities as well as his failure to thrive out of proportion to his cardiac disease prompted karyotype analysis. The karyotype is felt to be 46,XY,del(8)(:p21→qter). This is a de novo deletion; both parent's karyotypes are normal. The patient's clinical course was marked by severe physical and psychomotor retardation, seizures, multiple respiratory illnesses and frequent hospitalizations, terminating in death with bronchopneumonia at 20 months of age. Permission for autopsy was not granted.

The clinical picture included a prominent, oddly folded, right pinna, epicanthal folds, a cupid's bow mouth and an irregular sensory nystagmus. Initially the infant had deep palmar and plantar furrows which became less dramatic with time. He had bilateral inguinal hernias. This infant resembled a young girl presented at these meetings one year ago who had a similar deletion of the short arm of chromosome 8. Linkage data was obtained for this family and the results of the glutathione reductase tests will be presented.

INCONSISTENCIES IN THE TRISOMY CHROMOSOME 13q+ PHENOTYPE

Cheryl Rockman-Greenberg and Dagmar Kalousek

Departments of Medical Genetics and Pathology, Montreal Children's Hospital, Montreal, Quebec

Trisomy for distal 13q and monosomy for proximal 4p is reported in a 1,795 gm female infant born after a 41-week gestation to a gravida 5 para 1 aborta 3 30-year-old Greek mother and 38-year-old Greek father. Clinical features of this small for gestational age baby, now 6 months old, include severe growth and psychomotor retardation, microcephaly with small anterior fontanel, bilateral inferonasal colobomata of the irides, marked periorbital and nuchal hemangiomata, beaked nose with prominent nasal bridge and long philtrum, and bilateral bridged simian creases. The dermatoglyphic features were not unusual except for fibular arch and fibular loop hallucal patterns and a whorl in the left thenar area. IVP revealed left-sided hydronephrosis, probably secondary to ureteropelvic obstruction. The Giemsa-banded karyotype is 46,XX,der4,t(4;13)(p16;q14), ie there was trisomy for the distal two-thirds of the long arm of chromosome 13 and partial monosomy for the short arm of chromosome 4. This arose as a result of abnormal maternal meiotic segregation as the mother is a carrier of a balanced reciprocal translocation involving chromosomes 4 and 13.

The high incidence of chromosomal anomalies, particularly balanced translocations, in habitually aborting couples is discussed. Our patient demonstrates many of the clinical features of patients previously described with the clinical entity of trisomy for the distal portion of chromosome 13. However, the variations among reported cases emphasize the difficulty of assigning particular parts of this malformation syndrome to specific chromosome regions.

MENTAL RETARDATION ASSOCIATED WITH CRANIOFACIAL-SKELETAL-CARDIOVASCULAR AND NEUROLOGIC ABNORMALITIES (46,XX,9qh-)

Carlos F. Salinas*, Albert J. Pfister*, Ashby B. Taylor, Robert L. Lorenzo†, and Ronald J. Jorgenson***

Department of Oral Medicine, Section of Clinical Genetics*, Department of Pediatrics, Section of Pediatric Cardiology**, Department of Radiology, Section of Pediatric Radiology†, Medical University of South Carolina, Charleston, South Carolina

The purpose of this paper is to report an unusual case showing a constellation of congenital abnormalities that at first suggest a chromosomal aberration.

The proposita is a 2-year 7-month-old white female. She is the product of an uncomplicated full-term pregnancy of her G1P1AO 18-year-old mother. Both parents are young, healthy, and nonconsanguineous.

Abnormalities noted at birth included low birthweight (1,814 gm), short length (45 cm), microcephaly, strawberry hemangioma on the vertex of the scalp, low-set ears, large nose, small mandible, poor sucking reflex, rocker bottom feet, tendency of the 2nd fingers to overlap the 3rd finger, transversal palmar crease on the left hand, coarse systolic murmur over the precordium.

The patient has been followed since birth and development has been markedly delayed. She continues below the 3rd percentile for weight, height, and head circumference. The strawberry hemangioma of the scalp cleared spontaneously. She had bilateral inguinal herniorrhaphies at 4 months of age. Peculiar craniofacial features include an inverted triangular prominence of the forehead, a striking bulbous nose with anteverted nostrils, blepharochalasis, pseudoblepharophimosis, slightly low set ears with malformed pinnae, and a long philtrum (2.0 cm). The intraoral examination showed a delayed dental eruption and an intact palate.

The infant has developed spastic diplegia and seizures which currently are under control.

Despite absence of clinical cardiac dysfunction at 2 years 7 months, the child has a variety of cardiovascular anomalies: absence of the hepatic portion of the inferior vena cava (IVC) with azygous drainage of the IVC to the superior vena cava, symmetric tracheal bifurcation with bilateral hyparterial bronchi, a left aortic arch with aberrant origin of the right subclavian artery, and the low thoracic aorta crosses to the right of the spine and descends in the abdomen on the right with the IVC.

Other radiologic findings include 13 pairs of ribs without additional vertebral or other bony abnormalities. The sternum shows partial fusion between manubrium and upper sternal segments.

Chromosomal studies done with Giemsa staining, Q- and T- banding techniques show a normal female karyotype (46,XX). However, preliminary results using G-banding techniques suggest a deletion of the heterochromatic region in the long arm of a single No. 9 chromosome (9qh-).

46,XY, r(4) FOUND IN A 64-YEAR-OLD MALE WITH BORDERLINE NORMAL INTELLIGENCE

Joseph Sears, Ray Lewandowski, Mary Kukolich, and Carl Mankinen

Texas Genetics Screening and Counseling Service, Department of Mental Health Retardation, Denton, Texas

Patient was born December 1912 being the 3rd of 6 sibs. Three month premature home delivery was described with birthweight between 3 and 4 pounds. Early developmental history is not available since both parents are deceased. Patient attended regular classes from 1st through 6th grades and was transferred to special classes for the next 3 years. As a young man, he was able to function well both socially and occupationally in a rural area.

Physically he is small in stature and weight, has microcephaly, male type balding, full mouth dentures, facial asymmetry, ptosis OD, long philtrum, sparse facial and body hair, deep-set eyes, high palate, enlarged tongue with deep grooves. Mild kyphosis, melon-shaped torso, abundance of girdle adipose tissue, soft small testes and penis were found.

Limb anomalies consist of absence of right thumb, hypoplastic left thumb and radial deviation of hands.

Chromosome studies were done on patient and all his sibs, utilizing routine banding procedures. A review of literature for r4, 4p–, and 4q– is accomplished with comparison of clinical features and the difficulties of evaluating an elderly patient for dysmorphic features.

AN UNUSUAL CHROMOSOME TRANSLOCATION IN A CASE WITH CML

Gurbax S. Sekhon, Kutay Taysi, and Sri Moedjono

Department of Pediatrics, Division of Medical Genetics, Washington University School of Medicine, St. Louis, Missouri

The Philadelphia chromosome has been a consistent finding in bone marrow cells of 85% of patients with CML. The Philadelphia chromosome was initially thought to be a simple deletion of part of the long arm of chromosome 22. More recently, deleted chromosome 22 material has been identified on the long arm of chromosome 9. However, examples of translocation of the deleted material from chromosome 22 to chromosomes 2, 6, 13, 17, 19, 21, and 22 have also been reported.

We present here a case of CML with an unusual chromosome translocation that to our knowledge has not been reported before. This translocation involves the short arm of chromosome 4 and the long arm of chromosome 22. The case is a 26-year-old woman who had CML diagnosed at the age of 18 years. Her first blast crisis occurred shortly after the diagnosis was established and remission was successfully induced with the admission of vincristine and prednisone. In January, 1977 she underwent splenectomy in preparation for possible bone marrow transplantation; following which she developed intermittent episodes of jaundice. Early March, 1977 she was admitted with impending blast crisis and possible disseminated intravascular coagulopathy (DIC). She is currently on prednisone, vincristine, and hydroxyurea.

TRISOMY 10p DUE TO FAMILIAL PERICENTRIC INVERSION

Gurbax S. Sekhon and William S. Sly

Deprtment of Pediatrics, Division of Medical Genetics, Washington University School of Medicine, St. Louis, Missouri

Pericentric inversions of human chromosomes are currently being recognized with increasing frequency. Pericentric inversion in a parent may lead to the formation of chromosomally unbalanced gametes, provided crossing-over occurs within the inverted segment.

A pericentric inversion was observed in the father and the paternal grandfather of the proband whose karyotype showed a 10q+. The paternal grandfather of the proband had produced 9 children, none of whom had the severe malformations seen in the proband, though one child (so far unexamined cytogenetically) had congenital hydrocephalus. Analysis of the rearranged segments by various techniques showed a duplication of the short arm of chromosome 10 in the proband, supporting the interpretation that unbalanced gametes formed during paternal meiosis. The clinical features were: microcephaly, sloping forehead, hypertelorism, redundant skin at the nape of the neck, flat nasal bridge, low-set ears, cleft lip and palate, hypoplastic fingernails, clubfoot, cryptorchidism, and absence of left 12th rib. Postmortem examination also revealed multiple congenital cardiac defects, absence of corpus callosum, and hypoplastic kidneys.

The clinical findings of our patient are consistent with previous reports suggesting the existence of a characteristic 10p trisomy syndrome.

47,XY,+21/47,XY,+13 MOSAICISM

Rawatmal B. Surana

Department of Pediatrics, Howard University, College of Medicine, Washington, D.C.

A case of unusual mosaicism is reported in a black male infant with clinical diagnosis of Down syndrome. The propositus was the product of a full-term uncomplicated pregnancy. The mother was 22 years old and father was 24 years old at the time of birth. Both parents are healthy, not married, and have another 20-month-old male child.

Pertinent clinical findings, observed over a period of 7 months, include: mental retardation, failure to thrive, brachycephaly, palpebral fissures turned upwards and outwards, hypertelorism, unilateral simian line, bilateral inguinal hernia, normal external genitalia, and marked hypotonia.

Since birth, chromosome studies have been done on four occasions over a period of 6 months and 500 metaphases have been analyzed both by routine karyotyping as well as by G-staining method. In 86% of the cells chromosome count was 47 with an extra chromosome in G-group identified as number 21. In 12% of the cells, the chromosome count was also 47 but the extra chromosome was number 13. In 2% of the cells chromosome loss was random.

In all the cells of the propositus and his mother, one of the number 9 chromosomes had very prominent secondary constriction.

Chromosomal analysis on both parents and male sib was normal. Genetic marker studies (courtesy, Wilma A. Bias, PhD) were not suggestive of a mixed population of cells.

Our patient does not have clinical features suggestive of trisomy 13 such as eye anomalies, scalp defects, polydactyly, etc. Because no normal cell line and no cell line with 48 chromosomes containing extra 21 and 13 chromosomes have been found in our patient, it is difficult to explain the abnormality in this patient by one single event.

PARTIAL TRISOMY 1q IN TWO BROTHERS DUE TO MATERNAL BALANCED TRANSLOCATION (1q-; 6p+)

Kutay Taysi and Gurbax Singh Sekhon

Department of Pediatrics, Division of Medical Genetics, Washington University School of Medicine, St. Louis, Missouri

Extra chromosome material on the short arm of chromosome 6 (46,XY,6p+) was found in two mentally retarded, 30- and 39-year-old half brothers with dysmorphic features. The phenotypically normal mother had a balanced translocation between the long arm of chromosome 1 and the short arm of chromosome 6: 46,XX,t (1q;6p)(q32)(p25). Another phenotypically normal son of the mother had the same balanced translocation.

Two affected probands had normal No. 1 chromosomes, and one of their No. 6 chromosomes had additional material in the short arm. Apparently they are trisomic for the long arm segment of chromosome 1, distal to q32, due to abnormal segregation of the maternal translocation chromosome. They both have similar clinical features which include mental retardation, short stature, low frontal hairline, synophrys, high-arched palate, simple, prominent ears, micrognathia, cubitus valgus, and dermatoglyphic abnormalities.

Partial trisomy of the long arm of chromosome 1 has been reported in few cases with various severe congenital malformations, and short survival time. The mild dysmorphic features and long survival time in the present cases may suggest that the clinical diversity of the partial trisomy for the long arm of chromosome 1 are due to length differences of the trisomic segments and their position.

PERICENTRIC INVERSION OF CHROMOSOME 7 AND A BALANCED 10:15 TRANSLOCATION IN SKIN AND BLOOD OF AN INDIVIDUAL WITH ACUTE MYELOGENOUS LEUKEMIA (AML)

Kathleen E. Toomey, Jerome I. Rotter, T. Mohandas, and Michael M. Kaback

Department of Pediatrics, UCLA-Harbor General Hospital, Torrance, California

The significance of chromosome aberrations which have been found in association with malignant processes remains obscure. Proposed hypotheses regarding their relationship include: 1) Cells with chromosome abnormalities are more susceptible to carcinogenic agents (virus, chemicals, or radiation). 2) The duplication or deletion of genetic material is the primary etiologic factor in malignancy. 3) Chromosome aberrations are the result of the disease process itself; and 4) Chromosome aberrations and predisposition to malignancy, perhaps through increased chromosome instability, are both the effect of a common genetic factor.

The majority of previous cytogenetic studies in leukemic conditions have demonstrated abnormalities in bone marrow preparations with normal peripheral blood and/or skin karyotypes. We wish to present the occurrence of acute myelogenous leukemia in a 19-year-old male with multiple minor congenital anomalies and a complex chromosome rearrangement in peripheral blood lymphocytes (detected while in remission) and skin fibroblasts.

The patient is a tall, thin, white male with mild mental retardation. He has an unusual facies with a high nasal root, small palpebral fissures, an antimongoloid slant and micrognathia. The ears are small, low-set, and posteriorly rotated. There is kyphoscoliosis, proximally inserted thumbs, and pes cavus bilaterally. He has first degree hypospadias, cryptorchidism, and a shawl scrotum. Acute myelogenous leukemia developed at age 19 and, when seen, he was on maintenance therapy. Chromosome analysis with Q-banding of PHA-stimulated lymphocytes and skin fibroblasts showed a pericentric inversion of chromosome 7 and an apparently balanced translocation between chromosomes 10 and 15 in all cells examined. His karyotype is 46,XY,inv(7)(p15q22);t(10;15)(q24;q26). Karyotypes prepared from blood cultures of his parents and brother were normal. The presence of the same rearrangement in fibroblasts precludes its being an effect either of the disease process or of his chemotherapy. Loss of chromosome 7 has been one of the more frequently reported abnormalities in leukemic cells. To our knowledge, however, this is the second report of a pericentric inversion of human chromosome 7 in blood or somatic lines.

An early embryonic event leading to this chromosomal rearrangement may have concurrently induced a mutation either directly related to the break point involved in the translocation, or at a different site which manifests itself in the malignant transformation. Ascertainment of other such individuals will undoubtedly aid in clarifying the relationship between malignancy and chromosomal aberration.

A CHILD WITH UNUSUAL CONGENITAL ANOMALIES AND 46,XY,16p+

William G. Wilson, Philip D. Buchanan, Kathleen W. Rao, and Arthur S. Aylsworth

Department of Pediatrics, University of North Carolina School of Medicine, Chapel Hill, North Carolina

This black male infant was the 2,269 gm product of an uncomplicated pregnancy, born at 36 weeks gestation by assisted breech delivery. The mother was 27 years old and primiparous. Prenatal and family history were unremarkable.

The child was noted to have multiple anomalies at birth including large anterior and posterior fontanels, short palpebral fissures, posteriorly rotated and pointed ears with bilateral small superior tubercles and preauricular tags, flat nasal bridge with deficient tip, widened and creased alae nasi, short columnella, broad alveolar ridges, small mandible, limitation of extension at the elbows, hips, and knees, camptodactyly of the 3rd and 4th fingers, narrow hyperconvex nails, and shallow dermal ridges. Ophthalmologic examination was otherwise normal. Neurologic examination revealed increased tone in the limbs and poor suck, moro, and cry.

The child had transient tachypnea, followed by intermittent focal seizures in the first week of life. Skull x rays, EEG, and plasma amino acids were all normal.

Chromosome analysis of lymphocyte and skin fibroblast metaphases show the child to have a 46,XY,16p+ chromosome constitution in all cells. Quinacrine mustard banding shows partial deletion of the short arm of chromosome 16: del(16)(p12→pter). Extra chromosome material has been translocated onto the short arm of chromosome 16 producing a partial trisomy involving unidentified chromosome material. This extra chromosome material does not stain with the C-banding technique. The heterochromatic region of both No. 16 chromosomes is normal.

Chromosome analysis of lymphocyte metaphases from both parents was normal and showed no abnormalities or variants in the No. 16 chromosomes. This abnormal chromosome 16 appears to represent an unbalanced reciprocal translocation chromosome rather than a chromosomal variant. This patient, therefore, has a partial deletion of the short arm of chromosome 16 with partial trisomy of unidentified chromosome material. Abnormalities of the short arm of chromosome 16 are very rare and this patient provides a unique opportunity to correlate clinical abnormalities with a No. 16 short-arm deletion.

TERMINAL DELETION (q22→qter) OF CHROMOSOME 18q IN A 15-YEAR-OLD FEMALE

F. S. Yen, Philip Podruch, and Bernard Weisskopf

Child Evaluation Center, Department of Pediatrics, University of Louisville, School of Medicine, Louisville, Kentucky

A 15-year-old girl is reported with a 46,XX,del(18)(q22) karyotype from peripheral leukocyte culture. Chromosome studies on her parents revealed normal karyotypes.

The mother was 36 and the father 35 years old at the time of her birth. Birthweight was 3,485 gm. She is the second of three living children. She is at the 3rd percentile for height (152 cm) and the 5th percentile for weight (43 kg). Her head circumference of 52.5 cm is at the 5th percentile. The clinical features include narrow and tortuous ear canals, narrowing of the nares, and a high and asymemetric palate. An increased carrying angle of the arms, pes cavus, slight scoliosis to the left and minimal joint deformity were noted. In infancy, hypotonia and a decrease in muscle mass were reported. Presently, a decrease in the muscle mass is still apparent, although the tone has apparently been improved. Her present level of intellectual functioning is in the high-mild to low-borderline range of mental retardation (WISC D=65, P=75, FS=69). She was noted to have an emotional reaction characterized by depression and withdrawal.

BALANCED AND UNBALANCED X-AUTOSOME TRANSLOCATIONS

B. Zabel

Department of Pediatrics, University of Mainz, West Germany

Case I: Partial trisomy for the distal part of the long arm of chromosome 15 due to a balanced X-15-translocation in the mother: 46 X,t(X;15)(p22;q15)mat (75% inactivation mainly of the X part of the X-15 translocation chromosome; 25% inactivation of the normal X chromosome.) Clinical features: 6-month-old female with striking facial anomalies including depressed nasal bridge, high-arched palate, low-set ears, mild micrognathia; cardiac defect, diaphragmatic hernia, statomotor retardation.

Case II: 18-month-old female with 46,X,t(Xp–;21p+); breakpoints; Xp11/21p11. There is a cell line (30%) which is disomic for the short arm of the X chromosome due to the X-inactivation pattern (70%:inactivation of the normal X chromosome; 30% inactivation of the Xp–). Clinical features include hydrocephalus, prominent forehead, high-arched palate, dysplastic ears, muscular hypotonia, severe statomotor and psychomotor retardation.

PARTIAL TRISOMY OF CHROMOSOME 5

B. Zabel

Department of Pediatrics, University of Mainz, West Germany

A) Short arm

Mother and sister of patient are carriers of a balanced translocation 46,XX, t(3;5)(9p27;p13).

Patient has partial trisomy for the short arm of chromosome 5 (5q13→pter). Clinical features of the 12-month-old female patient include macrocephaly, flat face, depressed nasal bridge, macroglossia, muscular hypotonia, statomotor and psychomotor retardation, cerebral seizures.

B) Long arm

Father and sister of patient are carriers of a balanced translocation 46,XY or XX,t(5;22)(q33;13).

Patient has partial trisomy for the long arm of chromosome 5 (5q33→qter). Clinical features of the 10-month-old male patient include microcephaly, hypertelorism, depressed nasal bridge, antimongoloid slanting of the palpebral fissures, small mouth, large upper lip, low-set dysplastic ears, limited mobility of the hips, clubfeet, hernia, vitium cordis, generalized hypotrophy, failure to thrive, statomotor and psychomotor retardation.

The patients are compared with each other and with two published patients with partial trisomy 5q.

CRI-DU-CHAT SYNDROME – A CORRELATION BETWEEN ABNORMALITIES AND SIZE OF DELETION

Mary Esther Carlin, MD and Michele M. Neadle, CT

Department of Pediatrics – Mailman Center for Child Development, University of Miami School of Medicine, Miami, Florida

Twelve patients (8 females; 4 males) with cri-du-chat syndrome (5p-) were studied to correlate abnormalities with the amount of chromosomal deletion. All were retarded but the younger patients and those with consistent early educational intervention scored higher on IQ testing. Institutionalized individuals and those with significant perinatal difficulties had low scores. Those whose deletion is in the 5th (terminal) band have a significantly higher IQ and are able to use language to communicate effectively. Some function well enough to learn minimal academic skills and/or work at sheltered employment facilities. Those with a larger deletion scored as shown:

	Younger Patients (< 3.5 years)						Older Patients (> 3.5 years)						
Band	A	Age (months)	IQ	B	C	D	A	Age (years)	IQ	B	C	D	E
5	16	34	35	+	–	±	10	13.5	49	±	–	–	35.7
							10	12	34	±	–	–	
							12	12	24	–	+	+	
4	40	15	50	+	–	–	34	5.5	11	–	+	++	11.3
							35	15	13	–	+	+	
							38	17	10	–	+	–	
3	42	39	15	+	–	–	48	23.5	16	–	+	±	12.5
	49	22	39	–	–	±	53	8.5	9	–	+	+	

A = approximate % deletion; B = early stimulating education; C = institutionalized; D = perinatal difficulties; E = average IQ per band

Most of the physical abnormalities were present in 75% or more and none except height and weight seemed related to the deletion size. The number of distinct dermatoglyphic abnormalities found was increased in those whose deletions were in the 4th and 3rd bands. Normal hexosaminidase B levels in all substantiated that the genetic locus for that enzyme is below the 3rd band. Early determination of the amount of chromosome deletion may help predict future potential.

Index